Dynamic Modeling and Econometrics in Economics and Finance

Volume 27

In recent years there has been a rapidly growing interest in the study of dynamic nonlinear phenomena in economic and financial theory, while at the same time econometricians and statisticians have been developing methods for modeling such phenomena. Despite the common focus of theorists and econometricians, both lines of research have had their own publication outlets. The new book series is designed to further the understanding of dynamic phenomena in economics and finance by bridging the gap between dynamic theory and empirics and to provide cross-fertilization between the two strands. The series will place particular focus on monographs, surveys, edited volumes, conference proceedings and handbooks on:

- Nonlinear dynamic phenomena in economics and finance, including equilibrium, disequilibrium, optimizing and adaptive evolutionary points of view; nonlinear and complex dynamics in microeconomics, finance, macroeconomics and applied fields of economics.
- Econometric and statistical methods for analysis of nonlinear processes in economics and finance, including computational methods, numerical tools and software to study nonlinear dependence, asymmetries, persistence of fluctuations, multiple equilibria, chaotic and bifurcation phenomena.
- Applications linking theory and empirical analysis in areas such as macrodynamics, microdynamics, asset pricing, financial analysis and portfolio analysis, international economics, resource dynamics and environment, industrial organization and dynamics of technical change, labor economics, demographics, population dynamics, and game theory.

The target audience of this series includes researchers at universities and research and policy institutions, students at graduate institutions, and practitioners in economics, finance and international economics in private or government institutions.

More information about this series at http://www.springer.com/series/5859

Gilles Dufrénot · Takashi Matsuki
Editors

Recent Econometric Techniques for Macroeconomic and Financial Data

Springer

Editors
Gilles Dufrénot
Aix-Marseille University
Marseille, France

Takashi Matsuki
Osaka Gakuin University
Osaka, Japan

ISSN 1566-0419 ISSN 2363-8370 (electronic)
Dynamic Modeling and Econometrics in Economics and Finance
ISBN 978-3-030-54251-1 ISBN 978-3-030-54252-8 (eBook)
https://doi.org/10.1007/978-3-030-54252-8

This Springer imprint is published by the registered company Springer Nature Switzerland AG
The registered company address is: Gewerbestrasse 11, 6330 Cham, Switzerland

Introduction

This volume presents a collection of chapters providing a comprehensive overview of some recent econometric methods used to investigate the dynamics of macroeconomic and financial time series.

A first topic studied by the authors is the measurement issues raised by nonlinearities and non-stationarities, two typical properties that have been widely evidenced in economic time series. Alternative approaches to the existing methodologies that have not been yet widely explored in economic and financial research are presented here. This book presents the concepts of quantile spectrum and co-spectrum which allow tackling both the nonlinearities and non-stationary dynamics in the data. They extend the classical spectral analysis based on Fourier transforms to Laplace transforms of copulas which are better suited to behaviors implying extreme events, asymmetric dynamics, non-mixing processes. The contributions also explore the implication of nonlinearity for cointegration analysis in macroeconomics and combine Markov-switching models and Lévy processes to account for switching regimes driven by jump-diffusion processes in the financial markets. To account for non-stationary behaviors, the authors introduce several new concepts that generalize some former techniques. They indeed propose new testing methods of stochastic cycles based on the concept of long memory, introduce state-space models to estimate unobserved non-stationary components in time series and panel data and revisit the methodology of dynamic hierarchical factor models in panel data series.

A second new topic investigated in this book concerns the stochastic volatility models. They analyze the implications of detrending common factors on multivariate GARCH estimates, propose a new model of dynamic beta models in finance and analyze the implication of the inclusion of the theory of storage for the volatility models used to study the dynamics of commodity markets. Further, large events in financial data are investigated using Pareto-type models.

A third aspect discussed relates to some measurement problems in two fields that have gained a great importance over the recent years: the econometrics of commodity markets and of globalization. They propose a comparison of the commodity price datasets and compare the implications of different data transformations (deflating approaches, trend–cycle decomposition, aggregation through averaging). Further, they compare the implications of measuring globalization at a macro-aggregate level or by using data at a micro-level. The implications for applied econometrics can differ significantly, which they illustrate here taking the example of the exchange rate pass-through.

This book is designed for applied econometricians and economists in both the fields of macroeconomics and finance. For this purpose, the emphasis is put on the implication of applying these new econometric methods to topics that matter to the financial economists: price dynamic in equity, commodity, exchange rate markets and portfolio analysis. In the field of macroeconomics, they revisit the debates on the trend–cycle decomposition, growth analysis, monetary policy and international trade.

This book is divided into two parts (first part on Macroeconometrics and second part on Financial Econometrics). Each is composed of seven chapters by leading scholars in the field.

Contributions in First Part: Macroeconometrics

Chapter "Quantile and Copula Spectrum: A New Approach to Investigate Cyclical Dependence in Economic Time Series" contains an extensive presentation of quantile and copula spectrum with illustration to macroeconomic data. The authors explain why this approach allows to tackle the drawbacks of the classical spectrum which is based on too restrictive assumptions, notably weak stationarity and linearity of economic processes. New spectral density functions are introduced based on ranked-based periodogram, copula spectral density and quantile regressions with hidden periodicities. The authors finally provide empirical examples to show how the different techniques can be practically implemented on financial data.

Chapter "On the Seemingly Incompleteness of Exchange Rate Pass-Through to Import Prices: Do Globalization and/or Regional Trade Matter?" discusses the effects of measurement problems of globalization on the exchange rate pass-through (ERPT) in the EU. The chapter contains two important results. The first result casts some doubt on previous results in the literature: The ERPT generally observed in the literature is not at play when intra-EU trade (regional globalization) is correctly controlled for. The second result is that exchange rate changes still exert important pressure on domestic prices.

Chapter "A State-Space Model to Estimate Potential Growth in the Industrialized Countries" proposes a new methodology based on state-space models to discuss a hot topic in the policy arena: What is the natural interest rate of the industrialized countries? Because this variable is unobserved, they are considered as a state variable. The authors choose a multivariate approach by jointly estimating this natural rate, potential growth, structural unemployment in a semi-structural model where the measurement equations capture the influence of financial markets and monetary policy. The authors explain why the issue of trend–cycle decomposition is important and propose an estimation method.

Chapter "Detecting Tranquil and Bubble Periods in Housing Markets: A Review and Application of Statistical Methods" provides a critical review of recent empirical methodologies on the measurement of bubbles and introduces a generalization of existing bound testing approaches in the literature to detect the presence of bubbles in housing markets. The author uses a threshold ADRL models to distinguish between positive and negative bubbles and goes forward beyond the standard interpretation of a bubble that periodically collapses. This also allows him to choose the timing of mild bubbles endogenously. The methodology leads to the conclusion that there have been a greater number of short-term periods of explosive bubbles than usually suggested in the literature.

Chapter "An Analysis of the Time-Varying Behavior of the Equilibrium Velocity of Money in the Euro Area" presents a time-varying parameter state-space model to answer the following question: Can we trust monetary aggregates to stabilize prices and economic activity? The authors investigate the relationship between money supply and inflation by considering some key assumptions for the econometric investigation: potential nonlinearities in the relationships between the money supply and inflation and the asymmetries of the business cycle phases between expansions and recessions. According to the authors, permanent income per-capita explains the trend in M3 velocity in the Eurozone, thereby confirming a finding of previous studies that the decline in potential output is amongst the most relevant explanatory factors of the "inflation puzzle."

In Chapter "Revisiting Wealth Effects in France: A Double-Nonlinearity Approach," a double-nonlinearity approach is proposed to investigate the relationship between wealth and private consumption with an illustration on the French households. The authors propose a time-varying and multiple regime cointegration specification with asymmetric adjustment and complex forms of wealth effects. They find evidence of unstable wealth effects, specifically during the subprime crisis.

Chapter "Productivity Spillovers in the Global Market" proposes a global vector autoregressive model to investigate the extent to which the US productivity has been affected by its own economic growth and by cross-country spillover effects of productivity shocks. To identify the shocks, the authors propose a new methodology based on structural generalized impulse response functions in order to control for the changing relationships between countries at different times (throughout the 1990s, the dot-com bubble in 2001, the financial crisis of 2008 and the years after). They find evidence of heterogeneous and short-lived effects of productivity spillovers.

Contributions in Second Part: Financial Econometrics

Chapter "Commodity Prices in Empirical Research" provides a survey of econometric techniques used to study the commodity markets. There is no existing such overview in the literature yet. The author offers an overview of the empirical strategies and challenges characterizing a set of 4 key economic questions raised by commodities from both financial and macroeconomic perspectives: the secular decline of relative commodity prices, commodity currencies, volatility models designed to incorporate the implications of the theory of storage and the financialization of commodity markets. He discusses and compares standard data transformations, including trend–cycle decompositions, deflating and frequency preferences.

In Chapter "Modeling Time-Varying Conditional Betas. A Comparison of Methods with Application for REITs," the authors apply an innovative method allowing to model the conditional beta of real estate. They analyze the evolution over time of different betas and their reactions to observed shocks in the equity markets. They discuss the drawbacks of the classical econometric methods based on rolling window techniques and propose new models to estimate time-varying betas: indirect dynamic conditional betas, state-space models, autoregressive conditional betas. They propose some empirical illustrations on real estate investment trusts in the USA and Europe over the post-2008 financial crisis, from 2009 to 2019.

Chapter "Revisiting the Glick-Rogoff Current Account Model: An Application to the Current Accounts of BRICS Countries" revisits Glick-Rogoff's model, in which productivity shocks act as a key driver of current account changes and apply the model to the BRICS countries. The authors extended the Glick-Rogoff model with five macroeconomic variables, namely financial deepening, old dependency ratio, young dependency ratio, net foreign assets and trade openness. They find two important results. First, productivity is only important in non-turbulent environments. The Glick-Rogoff model performs well in the period prior to the global financial crisis but loses all explanatory power in the sample period, which includes the global financial crisis. Second, other macroeconomic variables are important determinants regardless of the inclusion of the crisis in the sample period.

In Chapter "Cycles and Long-Range Behaviour in the European Stock Markets," the authors develop a methodology to study stochastic cycles in stock prices. They use the framework of long-memory models by exploiting the theory of fractional integration models. They propose a strategy to test the assumption of persistent cycle than the usual unit root test considered in the literature. Monte Carlo simulations are shown, and applications to stock markets are provided using recursive estimation methods.

In Chapter "A Non-linear Approach to Measure the Dependencies Between Bitcoin and Other Commodity Markets," the authors combine Markov-switching models with Levy jump-diffusion. They search to capture the different sub-periods of crises over the business cycle and to test the relevance of regime switching measures with respect to independent pure-jump process. This parametric model is used to evaluate spillovers and jumps transmission between commodities and cryptocurrencies. They consider various crashes and over the business cycle that are captured by jumps.

Chapter "Typology of Nonlinear Time Series Models" proposes a survey of nonlinear time series models and addresses the issue of nonlinear cointegration in a multivariate framework. For the conditional mean, the author proposes a typology that distinguishes three families of models: nonlinear autoregressive models, nonlinear moving average models and bilinear models. Details about methods of estimation and tests are surveyed, and an empirical illustration is proposed to money multiplier and its components in India.

Finally, Chapter "Pareto Models for Risk Management" proposes a new contribution on Pareto models for risk management. The authors generalize the standard Pareto models in the literature by introducing a new framework based on second-order approximation. The authors introduce extended Pareto distributions and investigate several econometric estimators (Hill's estimator and maximum likelihood). These models are used to model large events (probable maximum losses, high quantiles and expected shortfalls). They show how their concepts can be applied to evaluate insurance and reinsurance pricing, and downside risk measures. Several illustrations to real data are provided.

Gilles Dufrénot
Takashi Matsuki

Contents

Contributors

Marcel Aloy Aix-Marseille University (Aix-Marseille School of Economics), CNRS & EHESS, Marseille, France

Thomas Brand CEPREMAP, Paris, France

Mariam Camarero INTECO, University Jaume I, Valencia, Spain

Guglielmo Maria Caporale Brunel University London, London, UK

Jean-François Carpantier Aix-Marseille School of Economics, Marseille, France

Arthur Charpentier Université du Québec à Montréal (UQAM), Montréal (Québec), Canada

Aditi Chaubal Indian Institute of Technology Bombay, Mumbai, India

Olivier Damette University of Lorraine, Nancy, France

Gilles Dufrénot Aix-Marseille School of Economics and CEPII, Marseille, France

Emmanuel Flachaire Aix-Marseille Université AMSE, CNRS and EHESS, Marseille, France

Luis A. Gil-Alana University of Navarra, Pamplona, Spain;
Universidad Francisco de Vitoria, Madrid, Spain

Stéphane Goutte CEMOTEV, UVSQ, Paris-Saclay, France

Fredj Jawadi University of Lille, Lille, France

Benjamin Keddad PSB—Paris School of Business, Paris, France

Nazmus Sadat Khan The World Bank (Macro, Trade and Investment Global Practise) and University of Muenster, Dhaka, Bangladesh

Floris Laly UCLouvain (Louvain School of Management), LFIN-LIDAM, Louvain-la-Neuve, Belgium

Sébastien Laurent Aix-Marseille University (Aix-Marseille School of Economics), CNRS & EHESS, Marseille, France;
Aix-Marseille Graduate School of Management – IAE, Aix-en-Provence, France

Christelle Lecourt Aix-Marseille University (Aix-Marseille School of Economics), CNRS & EHESS, Marseille, France

Antonia López-Villavicencio EconomiX-CNRS, University of Paris Nanterre, Nanterre, France

Takashi Matsuki Osaka Gakuin University, Suita, Japan

Antoine Mayerowitz Paris School of Economics, Paris, France

Valérie Mignon EconomiX-CNRS, University of Paris Nanterre, Nanterre, France;
CEPII, Paris, France

Jun Nagayasu Graduate School of Economics and Management, Tohoku University, Sendai-city, Miyagi, Japan

Carlos Poza Universidad Francisco de Vitoria, Madrid, Spain

Juan Sapena Catholic University of Valencia, Valencia, Spain

Kimiko Sugimoto Konan University, Nishinomiya, Japan

Cecilio Tamarit INTECO, University of Valencia, Valencia, Spain

Yushi Yoshida Faculty of Economics, Shiga University, Shiga, Japan

Weiyang Zhai Faculty of Economics, Shiga University, Shiga, Japan

Macroeconometrics

Quantile and Copula Spectrum: A New Approach to Investigate Cyclical Dependence in Economic Time Series

Gilles Dufrénot, Takashi Matsuki, and Kimiko Sugimoto

1 Introduction

Why using quantile spectrum. The purpose of traditional spectrum analysis is to analyze periodic behavior of a stationary time series. Let $X_t, t \in \mathbb{Z}$ be a single time series and $\gamma_X(j)$ be its autocovariance function at lag j, where $\gamma_X(j) = E(X_0 X_j) - \{E(X_0)\}^2$. To this end, the spectrum or spectral density at frequency λ is estimated as

$$f_X(\lambda) = \frac{1}{2\pi} \sum_{j \in \mathbb{Z}} \gamma_X(j) \cos(j\lambda), \quad \lambda \in (-\pi, \pi]. \tag{1}$$

To ensure this function continuous and symmetric about 0, the assumption that the autocovariance function $\gamma_X(j)$ is absolutely summable is imposed on this process. In other words, X_t is assumed to have a finite second moment.

When X_t is covariance stationary, its first and second moments exist and depend not on time t but on its difference. The Wold decomposition theorem shows that a certain covariance stationary series, e.g., X_t can be decomposed into deterministic and stochastic components, e.g., $X_t = \mu + \sum_{j=1}^{\infty} \phi_j \varepsilon_{t-j}$, where μ is a constant mean and ε_t is an i.i.d. random variable and $\sum_{j=1}^{\infty} |\phi_j| < \infty$. This also implies the linearity of the time series. Therefore, even if we transform a time series in the time domain

G. Dufrénot
Aix-Marseille School of Economics, Marseille, France

T. Matsuki (✉)
Osaka Gakuin University, Suita, Japan
e-mail: matsuki10@gmail.com

K. Sugimoto
Konan University, Nishinomiya, Japan

G. Dufrénot and T. Matsuki (eds.), *Recent Econometric Techniques for Macroeconomic and Financial Data*, Dynamic Modeling and Econometrics in Economics and Finance 27, https://doi.org/10.1007/978-3-030-54252-8_1

to the frequency domain, the existence of the second moment and the linearity of the series are implicitly assumed.

In practice, we often encounter the situation where financial time series and macroeconomic time series are not properly shown by ordinary spectrums. These time series are often considered to have heavy tail distributions, infinite moments, nonlinearity, or non-stationarity.

For example, Fig. 1 shows the density functions of three stock returns of the Dow Jones industrial average index, CAC (Cotation Assistée en Continu) 40 index and Nikkei 225 average stock price index, which are solid lines. As a reference, the standard normal density function is also shown as a dotted line. Although the return of the Dow Jones average index more densely distributes around its mean value compared to the standard normal distribution, those of the CAC and Nikkei stock price indices apparently have thicker or heavier tails of their distributions. Like this case, when we estimate single spectrums or cross spectrums of several time series that have peculiarities shown in Fig. 1, ordinary procedures of a spectrum density estimation may not be suited to precisely capture cyclical data variation or data dependent structures due to the restriction of normality. To deal with such cases, we will introduce some novel techniques in the frequency domain in the subsequent sections.

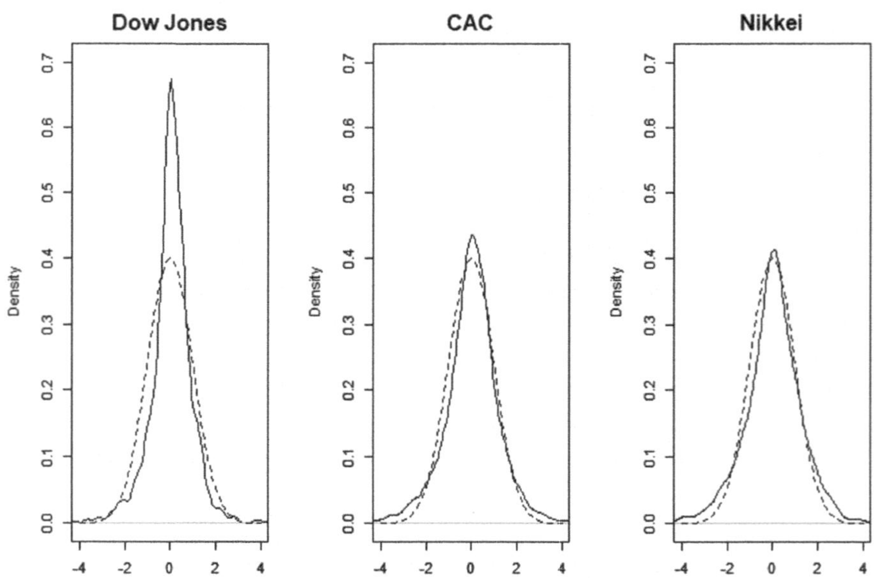

Note: The number of observations is 2,516 and the Epachenikov kernel is used for estimation.

Fig. 1 Distribution of the returns

Periodogram analyses for understanding and interpreting the properties of macroeconomic and financial series have received a renewed interest in recent years. This interest can be explained by the development of new tools allowing to capture a variety of phenomena: regime changes, structural breaks, volatility giving rise to clusters, non-stationary and nonlinear dynamics and time-irreversibility. The classical spectral analysis, based on Fourier transforms of autocovariance and autocorrelation functions, cannot capture such phenomena, because it is based on some assumptions that are too restrictive to provide a full characterization of the data. For a significant number of economic series, the Gaussianity hypothesis does not fully explain the dependencies in the series when their distributions exhibit extreme events, strong asymmetries and regime-switching dynamics.

To remedy this drawback, several alternative approaches to the classical spectral analysis have been proposed in the literature. The literature is vast (periodograms for locally stationary processes, polyspectrum analysis, evolutionary spectral methods, time-varying second-order spectra). In this chapter, we focus on a thread of the literature which has proposed new types of frequency domain concepts to analyze joint distribution functions at different lags. New concepts such as *Laplace*, *quantile*, *copula* spectrum are increasingly used to detect periodicities in time series by considering their entire distribution rather than focusing on the first two moments.

The survey proposed in this chapter uses some materials provided by researchers during the last decade. Just to mention a few, Hong (1999, 2000), Li (2008, 2012), Hagemann (2013) propose a concept of Laplace spectrum and Laplace periodogram, which are median-related spectral analysis tools. This was extended to cross-spectrum by Hagemann (2013) to detect cycles at different points of a distribution without imposing assumptions on the existence of the moments. Lim and Oh (2015) generalize this approach by introducing a concept of *composite* quantile spectrum (weighted linear combinations of quantile spectra). Lee and Subba Rao (2012), Dette et al. (2015), Kley et al. (2016) have proposed the concept of copula-based spectrum which does not require any distributional assumption such as the existence of finite moments. Local copula spectral density functions are defined as the Fourier transform of copula cross-covariance kernels. Baruník and Kley (2019) propose an extension to the multivariate case. They introduce quantile cross-spectral quantities (quantile coherency) to capture different dependence structures between time series across quantile.

2 Harmonic Regression Models and Laplace Periodograms

For a given time series $\{Y_t\}$ of length n and its frequency $\omega \in (0, \pi)$, the ordinary periodogram is defined as

$$G_n(\omega) := \frac{1}{n}\left|\sum_{t=1}^{n} Y_t \exp(-it\omega)\right|^2. \tag{2}$$

In the above equation, if $\omega = 2\pi k/n$, where k is a certain integer, it can also be expressed as

$$G_n(\omega) = \frac{1}{4}n\left\|\widetilde{\boldsymbol{\beta}}_n(\omega)\right\|^2 = \frac{1}{4}n\widetilde{\boldsymbol{\beta}}_n'(\omega)\widetilde{\boldsymbol{\beta}}_n(\omega), \tag{3}$$

where $\|\cdot\|$ denotes the Euclidian norm, and $\widetilde{\boldsymbol{\beta}}_n(\omega)$ denotes the least squares estimator in the linear model with regressors $\boldsymbol{x}_t(\omega) = [\cos(\omega t), \sin(\omega t)]'$, corresponding to an L_2-projection of the observed series onto the harmonic basis, which are obtained as the solution of the following equation.

$$\left\{\widetilde{\lambda}_n(\omega), \widetilde{\boldsymbol{\beta}}_n(\omega)\right\} := \operatorname{argmin}_{\lambda\in\mathbb{R},\boldsymbol{\beta}\in\mathbb{R}^2} \sum_{t=1}^{n} \left(Y_t - \lambda - \boldsymbol{x}_t'(\omega)\boldsymbol{\beta}\right)^2. \tag{4}$$

When the OLS criterion is replaced by the least absolute deviation (LAD) criterion in the harmonic regression, the LAD coefficient $\ddot{\boldsymbol{\beta}}_n(\omega)$ is obtained as follows:

$$\left\{\ddot{\lambda}_n(\omega), \ddot{\boldsymbol{\beta}}_n(\omega)\right\} := \operatorname{argmin}_{\lambda\in\mathbb{R},\boldsymbol{\beta}\in\mathbb{R}^2} \sum_{t=1}^{n} \left|Y_t - \lambda - \boldsymbol{x}_t'(\omega)\boldsymbol{\beta}\right|. \tag{5}$$

By using $\ddot{\boldsymbol{\beta}}_n(\omega)$, Li (2008) has defined the Laplace periodogram as

$$L_n(\omega) := \frac{1}{4}n\left\|\ddot{\boldsymbol{\beta}}_n(\omega)\right\|^2. \tag{6}$$

Therefore, both $G_n(\omega)$ and $L_n(\omega)$ are obtained by the squared norm (or sum of squares) of harmonic regression coefficients multiplied by some constant terms. In particular, the Laplace periodogram inherits the robustness properties of linear LAD regression. Just as the OLS estimator is used to characterize the sample mean, the LAD estimator applied captures the behavior of the observation around the median (0.5 quantile).

Li (2008) has derived the asymptotic normality and useful related theorems of the Laplace periodogram, which are very useful to consider asymptotic behaviors of several periodograms. His results are based on the concept of zero-crossings.

Definition (*Stationarity in zero-crossings*) The lagged zero-crossing rate of a random process $\{\varepsilon_t\}$ between t and s is defined as $\gamma_{ts} := P(\varepsilon_t\varepsilon_s < 0)$, and $\{\varepsilon_t\}$ is called to be stationary in zero-crossings if and only if γ_{ts} depends only on $t - s$, that is, $\gamma_{ts} = \gamma_{t-s}$ for all t and s. γ_τ is called as the lag-zero-crossing rate of $\{\varepsilon_t\}$ and $S(\omega) := \sum_{\tau=-\infty}^{\infty}(1 - 2\gamma_\tau)\cos(\omega\tau)$ is called as the zero-crossing spectrum of $\{\varepsilon_t\}$.

Using the definition described above, a strictly stationary process is also stationary in zero-crossings.

Theorem (Asymptotic normality of the Laplace periodogram) *Let* $\ddot{\boldsymbol{\beta}}_n(\omega)$ *and* $L_n(\omega)$ *be defined by Eqs.* (5) *and* (6) *with* $Y_t = \varepsilon_t (t = 1, \ldots, n)$, *where* $\{\varepsilon_t\}$ *is a* *random process with a common marginal distribution function* $F(x)$ *and density* $f(x)$ *such that* $F(0) = 1/2$ *and* $f(0) > 0$ *and such that the assumption of continuous differentiability in a neighborhood of* $x = 0$ *is satisfied. We also assume that (i)* $\{\varepsilon_t\}$ *is either an m-dependent process stationary in zero-crossings or a linear process of the form* $\varepsilon_t = \sum_{j=-\infty}^{\infty} \phi_j e_{t-j}$, *where* $\{e_t\}$ *is an i.i.d. random sequence with* $E(|e_t|) < \infty$ *and* $\{\phi_j\}$ *is an absolutely summable deterministic sequence satisfying* $\sum_{|j|>m} |\phi_j| = O(n^{-1})$ *for some* $m = O(n^{\delta})$ *and* $\delta \in [0, 1/4)$, *and (ii) in either case its zero-crossing rates* $\{\gamma_\tau\}$ *satisfy* $\sum_{\tau=0}^{\infty} |1 - 2\gamma_\tau| < \infty$. *Let* $S(\omega)$ *be the zero-crossing spectrum of* $\{\varepsilon_t\}$, *and let* $\{\omega_1, \ldots, \omega_q\}$ *be a set of distinct values in* $(0, \pi)$ *that may depend on n but that satisfy the condition (iii)* $D_{jkn} := n^{-1} \sum_{t=1}^{n} \mathbf{x}_t(\omega_j)\mathbf{x}_t'(\omega_k) = \frac{1}{2}\delta_{j-k}\mathbf{I} + O(1)$. *Assume further that* $S(\omega_j) > 0$ *for all j. Then,* as $n \to \infty$,

$$n^{\frac{1}{2}} \operatorname{vec}\{\ddot{\boldsymbol{\beta}}_n(\omega_j)\}_{j=1}^{q} \xrightarrow{d} N(\mathbf{0}, 2\eta^2\mathbf{S}) \text{ and } \{L_n(\omega_j)\} \xrightarrow{d} \left\{\frac{1}{2}\eta^2 S(\omega_j) Z_j\right\},$$

where $\xrightarrow{d}$ *denotes convergence in distribution,* $\eta^2 := 1/(4f^2(0))$.
$\mathbf{S} := \operatorname{diag}\{S(\omega_1), S(\omega_1), \ldots, S(\omega_q), S(\omega_q)\}$, *and* $Z_j \sim$ i.i.d.$\chi^2(2)$ $(j = 1, \ldots, q)$.

Proof See Li (2008).

This theorem implies that the asymptotic mean of $L_n(\omega)$ is equal to $L(\omega) = \eta^2 S(\omega)$, which is the Laplace spectrum of $\{\varepsilon_t\}$. When $\{\varepsilon_t\}$ is stationary in second moments, $G_n(\omega) \xrightarrow{d} \frac{1}{2}\sigma^2 R(\omega)\chi^2(2)$ (Brockwell and Davis 1991), where the autocorrelation spectrum $R(\omega) := \sum_{\tau=-\infty}^{\infty} \rho_\tau \cos(\omega\tau) = 2\sum_{\tau=0}^{\infty} \rho_\tau \cos(\omega\tau) - 1$. In addition, the asymptotic mean of $G_n(\omega)$, which is called the power spectrum, is $G(\omega) = \sigma^2 R(\omega)$. The Laplace spectrum $L(\omega)$ is the counterpart of the power spectrum $G(\omega)$. The former is proportional to the zero-crossing spectrum $S(\omega)$, while the latter is proportional to the autocorrelation spectrum $R(\omega)$. The zero-crossing spectrum is obtained as a Fourier transform of the zero-crossing rates, and the autocorrelation spectrum is obtained as a Fourier transform of the autocorrelation coefficients.

More importantly, η^2 is distribution-dependent, but in general a finite η^2 does not require a finite variance of the asymptotic distribution. This is a theoretical advantage of the Laplace periodogram because it does not require the existence of any moments to have a well-defined asymptotic distribution.[1] This implies that in practice the Laplace periodogram is expected to be more robust to high volatilities of financial data.

[1]When $\{\varepsilon_t\}$ is a white noise process with a finite variance σ^2, $L(\omega) = \eta^2$ $(= 1/(4f^2(0))$ and $G(\omega) = \sigma^2$. Obviously, to obtain the spectrum as the mean of the asymptotic distribution, the ordinary periodogram needs the existence of a finite variance, while the Laplace periodogram needs only the condition of $f(0) > 0$.

The median-based approach can be generalized to any quantile regression. Li (2012) has extended the approach to arbitrary quantiles with $0 < \tau_1 = \tau_2 < 1$. A quantile regression estimator $\widehat{\boldsymbol{\beta}}_{n,\tau}(\omega)$ is obtained as the following solution:

$$\left\{\widehat{\lambda}_{n,\tau}(\omega), \widehat{\boldsymbol{\beta}}_{n,\tau}(\omega)\right\} := \operatorname{argmin}_{\lambda\in\mathbb{R},\boldsymbol{\beta}\in\mathbb{R}^2} \sum_{t=1}^{n} \rho_\tau\left(Y_t - \lambda - \boldsymbol{x}_t'(\omega)\boldsymbol{\beta}\right), \tag{7}$$

where the check function $\rho_\tau(u) := u(\tau - I(u \leq 0)) = (1-\tau)|u|I(u \leq 0) + \tau|u|I(u > 0)$ for $\tau \in (0, 1)$ (see Koenker 2005). Li (2012) has defined the quantile periodogram (of the first kind) at quantile level τ as

$$Q^{I}_{n,\tau}(\omega) := \frac{1}{4}n\left\|\widehat{\boldsymbol{\beta}}_{n,\tau}(\omega)\right\|^2. \tag{8}$$

This is a scaled version of the squared norm of the quantile regression coefficients corresponding to the trigonometric regressors. With the special choice of $\tau = 1/2$ (median or LAD regression), the quantile periodograms are reduced to the Laplace periodograms.

An alternative quantile periodogram (of the second kind) at quantile level τ is defined as

$$Q^{II}_{n,\tau}(\omega) := \sum_{i=1}^{n} \rho_\tau\left(Y_t - \widehat{\lambda}_{n,\tau}\right) - \sum_{i=1}^{n} \rho_\tau\left(Y_t - \widehat{\lambda}_{n,\tau}(\omega) - \boldsymbol{x}_t'(\omega)\widehat{\boldsymbol{\beta}}_{n,\tau}\right) \tag{9}$$

where $\widehat{\lambda}_{n,\tau}$ is the sample τth quantile given by $\widehat{\lambda}_{n,\tau} := \operatorname{argmin}_{\lambda\in\mathbb{R}} \sum_{t=1}^{n} \rho_\tau(Y_t - \lambda)$.

The main difference with Eq. (8) is that, in Eq. (9), we consider the net effect of the trigonometric regressors. It can be shown that

$$\left\{Q^{I}_{n,\tau}(\omega)\right\} \xrightarrow{d} \left\{\frac{1}{2}\eta_I^2 S(\omega_j) Z_j^1\right\}, Z_j^1 \sim \text{i.i.d.}\chi^2(2),\ j = 1, \ldots, q, \tag{10a}$$

$$\left\{Q^{II}_{n,\tau}(\omega)\right\} \xrightarrow{d} \left\{\frac{1}{2}\eta_{II}^2 S(\omega_j) Z_j^2\right\}, Z_j^2 \sim \text{i.i.d.}\chi^2(2), j = 1, \ldots, q, \tag{10b}$$

where $\eta_I^2 = \tau(1-\tau)/\kappa^2$, $\eta_{II}^2 = \tau(1-\tau)/\kappa$, are scaling constants.

Proof See Li (2012).

$Q^{I}_{n,\tau}(\omega)$ and $Q^{II}_{n,\tau}(\omega)$ provide information about the cyclical behavior of the time series $\{Y_t\}$ around its τth quantile level. They both have asymptotic exponential distributions with mean $\eta_I^2 S(\omega_j)$ and $\eta_{II}^2 S(\omega_j)$ which are called *quantile spectra*. For fixed values of τ, $S(\omega_j)$ is the Fourier transform of the autocorrelation function of $\left\{\operatorname{sgn}(Y_t - \lambda_{n,\tau})\right\}$, or the spectral cumulative representation of the serial dependence of $\{Y_t\}$ in terms of the bivariate cumulative probabilities

$$F_\tau(\lambda, \lambda) = P(Y_t \le \lambda, Y_{t+\tau} \le \lambda). \tag{11}$$

Dette et al. (2015) have generalized the quantile periodogram for arbitrary $(\tau_1, \tau_2) \in (0, 1)^2$ (not necessarily equal), which is defined as

$$L_{n,\tau_1,\tau_2}(\omega) := \frac{1}{4} n \widehat{\boldsymbol{\beta}}'_{n,\tau_1}(\omega) \begin{pmatrix} 1 & i \\ -i & 1 \end{pmatrix} \widehat{\boldsymbol{\beta}}_{n,\tau_2}(\omega). \tag{12}$$

This can be regarded as an extended version of the Laplace periodogram. The asymptotic properties of $L_{n,\tau_1,\tau_2}(\omega)$ can be established under two assumptions.

Notation Consider the sequence $\{Y_t\}_{t=1}^n$ where each Y_t is a measurable function defined by the probability space $(\Omega, \mathcal{F}, P)$. We define $Y_i^j := \{Y_t\}_{t=i}^j$, where i and j are integers. The σ-field generated by X_i^j is σ_i^j and P_i^j the joint distribution of Y_i^j. Moreover, define $q_\tau = P^{-1}(\tau)$, $\tau \epsilon [0, 1]$ the τth quantile of the distribution P.

Assumption 1 $\{Y_t\}$ is strictly stationary and β-mixing. The stationarity hypothesis implies that, $\forall t$ and $\forall i, j$, (non-negative integers), the vectors Y_t^{t+i} and Y_{t+j}^{t+i+j} have the same distribution. β-mixing can be defined as follows (there are several equivalent definitions in the literature). For each positive integer n, we have:

$$\lim_{k\to\infty} \beta(k) := \sup \left\| P_1^t \otimes P_{t+k}^n - P_{t,k} \right\|_{TV} = O\left(k^{-\delta}\right), \quad \delta > 1 \tag{13}$$

where $\|\cdot\|_{TV}$ is the total variation mean, $P_{t,a}$ is the joint distribution of (X_1^t, P_{t+a}^n). $\beta(k)$ is the distance between the joint distribution of random variables separated by k time units and a distribution under which these random variables are independent. Many time series satisfy Assumption 1 (for instance ARMA processes and stochastic volatility models like GARCH).

Assumption 2 The distribution function P_t of Y_t and the joint distribution function P_t^{t+K} of (Y_t, Y_{t+k}) are twice differentiable with uniformly bounded derivatives.

Theorem *Asymptotic normality of* $\widehat{\boldsymbol{\beta}}_{n,\tau}(\omega_j)$
Under Assumptions 1 and 2, we have

$$n^{\frac{1}{2}} \operatorname{vec}\left\{\widehat{\boldsymbol{\beta}}_{n,\tau}(\omega_j)\right\}_{j=1}^q \xrightarrow{d} N(\boldsymbol{O}, M) \tag{14}$$

where $M := M_{\tau_1,\tau_2}(\omega_1, \omega_2)$ *is the covariance matrix defined by*

$$M_{\tau_1,\tau_2}(\omega_1, \omega_2) := \begin{cases} \pi \begin{pmatrix} \Re \dot{f}_{\tau_1,\tau_2}(\omega) & \Im \dot{f}_{\tau_1,\tau_2}(\omega) \\ -\Im \dot{f}_{\tau_1,\tau_2}(\omega) & \Re \dot{f}_{\tau_1,\tau_2}(\omega) \end{pmatrix}, & \text{if } \omega_1 = \omega_2 = \omega \\ \begin{pmatrix} 0 & 0 \\ 0 & 0 \end{pmatrix}, & \text{if } \omega_1 \ne \omega_2. \end{cases}$$

$\dot{f}_{\tau_1,\tau_2}(\omega)$ is the scaled version of the spectral density function:

$$\dot{f}_{\tau_1,\tau_2}(\omega) = \frac{f_{\lambda_1,\lambda_2}(\omega)}{\left(f_{\lambda_1}(\omega) f_{\lambda_2}(\omega)\right)}, \lambda_1 = P^{-1}(\tau_1), \lambda_2 = P^{-1}(\tau_2),$$

and $f_{\lambda_1,\lambda_2}(\omega) = \frac{1}{2\pi}\sum_{k=-\infty}^{+\infty}\gamma_k(\tau_1, \tau_2)\mathrm{e}^{-ik\omega}, \quad (\tau_1, \tau_2)\epsilon(0, 1)^2.$

$\gamma_k(\tau_1, \tau_2)$ is the Laplace cross-covariance kernel of lag k defined by

$$\gamma_k(\tau_1, \tau_2) := \mathrm{COV}\left[I(Y_t \le \tau_1), I(Y_{t+k} \le \tau_2)\right].$$

$\Re\dot{f}_{\tau_1,\tau_2}(\omega)$ and $\Im\dot{f}_{\tau_1,\tau_2}(\omega)$ are the real and imaginary parts of $\dot{f}_{\tau_1,\tau_2}(\omega)$.

Proof See Dette et al. (2015).

Theorem *Asymptotic normality of $L_{n,\tau_1,\tau_2}(\omega)$*
Under Assumptions 1 and 2, we have

$$L_{n,\tau_1,\tau_2}(\omega_1, \omega_2)\begin{cases} \sim \pi\dot{f}_{\tau_1,\tau_2}(\omega)Z, \quad Z \sim \text{i.i.d.}\chi^2(2), \; if \; \tau_1 = \tau_2 \\ \xrightarrow{d} \frac{1}{4}(Z_{11}, Z_{12})\begin{pmatrix} 1 & i \\ -i & 1 \end{pmatrix}\begin{pmatrix} Z_{21} \\ Z_{22} \end{pmatrix}, \quad if \; \tau_1 \ne \tau_2. \end{cases} \tag{15}$$

where $(Z_{11}, Z_{12}, Z_{21}, Z_{22})$ is a Gaussian vector with mean **0** *and covariance matrix*

$$\Sigma(\omega) = \begin{pmatrix} \dot{f}_{\tau_1,\tau_2}(\omega) & 0 & \Re\dot{f}_{\tau_1,\tau_2}(\omega) & \Im\dot{f}_{\tau_1,\tau_2}(\omega) \\ 0 & \dot{f}_{\tau_1,\tau_2}(\omega) & -\Im\dot{f}_{\tau_1,\tau_2}(\omega) & \Re\dot{f}_{\tau_1,\tau_2}(\omega) \\ \Re\dot{f}_{\tau_1,\tau_2}(\omega) & -\Im\dot{f}_{\tau_1,\tau_2}(\omega) & \dot{f}_{\tau_1,\tau_2}(\omega) & 0 \\ \Im\dot{f}_{\tau_1,\tau_2}(\omega) & \Re\dot{f}_{\tau_1,\tau_2}(\omega) & 0 & \dot{f}_{\tau_1,\tau_2}(\omega) \end{pmatrix}$$

Proof See Dette et al. (2015).

3 Sample and Smoothed Laplace Periodogram

Define the following new variable of interest called a *quantile crossing indicator*:

$$V_t(\tau, q(\tau)) = \tau - I\{Y_t < q(\tau)\}. \tag{16}$$

If the distribution function of Y_t is continuous and increasing at

$$q(\tau) := \inf\{y : P(Y \le y)\}, \tag{17}$$

$V_t(\tau)$ is bounded, stationary and mean zero random variable. Using Koenker and Basset's approach, we define an estimate of $V_t(\tau)$ as follows:

$$\widehat{V}_t(\tau) = V_t\left(\tau, \hat{q}_n(\tau)\right) \tag{18}$$

where $\hat{q}_n(\tau) = \operatorname{argmin}_{q\in\mathbb{R}} \sum_{t=1}^{n} \rho_\tau(Y_t - y)$, $\rho_\tau(x) = x\{\tau - I(x < 0)\}$. $\hat{q}_n(\tau)$ is the estimate of the τth quantile. The τth quantile periodogram is given by

$$Q_{n,\tau}(\omega) := \frac{1}{2\pi}\left|\frac{1}{\sqrt{n}}\sum_{t=1}^{n}\widehat{V}_t(\tau)e^{-it\omega}\right| = \frac{1}{2\pi}\sum_{|j|<n}\hat{r}_{n,\tau}(j)\cos(\omega j), \qquad (19)$$

where $i^2 = -1$ and $\hat{r}_{n,\tau}(j) = \frac{1}{n}\sum_{t=|j|+1}^{n}\widehat{V}_t(\tau)\widehat{V}_{t-|j|}(\tau)$, $|j| < n$. $Q_{n,\tau}(\omega)$ is an unbiased estimate of the τth spectral density, but is not consistent. A consistent estimator is obtained by smoothing the periodogram using kernel functions (all the results below are taken from Hagemann 2013).

We obtained a smoothed τth quantile periodogram as

$$\widehat{Q}_{n,\tau}(\omega) = \frac{1}{2\pi}\sum_{|j|<n}\lambda(j/b_n)\hat{r}_{n,\tau}(j)\cos(\omega j). \qquad (20)$$

$\lambda(j/b_n)$ is a lag window and b_n is a bandwidth parameter. It is known from the literature on spectral analysis that an optimal lag window leading a non-negative periodogram is the so-called quadratic spectral window defined as

$$\lambda_{QS}(x) = \frac{25}{12\pi^2x^2}\left\{\frac{\sin\left(\frac{6\pi x}{5}\right)}{\frac{6\pi x}{5}} - \cos\left(\frac{6\pi x}{5}\right)\right\}. \qquad (21)$$

The following results hold (for the proofs, see Hagemann 2013).

Result 1: Confidence interval

Define

$$\overline{Q}_{n,\tau}(\omega,k) = \frac{1}{2k+1}\sum_{|j|<k}\widehat{Q}_{n,\tau}\left(\omega + \frac{2\pi j}{n}\right),\ k \in \mathbb{Z}. \qquad (22)$$

Then, a confidence interval of the smoothed periodogram with a probability of $(1-\alpha)\%$ is

$$\left[\frac{(4k+2)\bar{Q}_{n,\tau}(\omega,k)}{Z_1}, \frac{(4k+2)\bar{Q}_{n,\tau}(\omega,k)}{Z_2}\right],$$
$$Z_1 \sim \text{i.i.d.}\,\chi^2_{1-\frac{\alpha}{2}}(4k+2),\ Z_2 \sim \text{i.i.d.}\,\chi^2_{\frac{\alpha}{2}}(4k+2) \qquad (23)$$

Result 2: Asymptotic normality

Assume that

(i) $F_Y(y) := P(Y \leq y)$ is Lipschitz continuous in a neighborhood of $q(\tau) := \inf\{y : P(Y \leq y) \geq \tau\}$ and has a continuous density at $q(\tau)$.
(ii) There is some n^* such that for $n > n^*$, $F_{\widetilde{Y}}(y) := P\big(\widetilde{Y} \leq y\big)$ is Lipschitz continuous in a neighborhood of $q(\tau)$ and $E\big|Y_n - Y_n'\big| = O(\rho^n)$, for some $\rho \epsilon (0, 1)$ (geometric moment contracting, a property that is satisfied for many linear and nonlinear stochastic models).
(iii) λ is even and Lipschitz continuous with support $[-1, 1]$, $\lambda(0) = 1$,

$$\lim_{x \to 0}[1 - \lambda(x)]|x|^3 < \infty, b_n \to \infty, b_n = O\left(n^{\frac{1}{4}}\right), n = o\big(b_n^7\big).$$

Then,

$$\sqrt{\frac{\text{int}\left(\frac{n}{2}\right)}{b_n}}\left[\ddot{Q}_{n,\tau}(\omega) - Q_{n,\tau}(\omega)\right] \sim N(0, \sigma^2(\eta)), \tag{24}$$

where $\sigma^2(\eta) = [1 + h(2\eta)]Q_{n,\tau}(\omega) \int_{-1}^{1} \lambda(x)dx$, $h(\eta) = \begin{cases} 1, \text{ if } \omega = 2\pi k, k \in Z \\ 0, \text{ Otherwise.} \end{cases}$

$$\ddot{Q}_{n,\tau}(\omega) = \frac{1}{2\pi \text{ int}\left(\frac{n}{2}\right)} \sum_{|j|<\text{int}\left(\frac{n}{2}\right)} \lambda(j/b_n)\hat{r}_{n,\tau}(j)\cos(\omega j) \tag{25}$$

$\hat{r}_{n,\tau}(j)$ can be used to test whether the spectrum is informative at a given τth quantile. We present here two approaches proposed by Hagemann (2013) to test the flatness of the quantile spectrum based on two Cramer-von Mises tests.

The null hypothesis to test is

$$H_0\colon r_{n,\tau}(j) = 0, \forall j > 0 \,\text{against}\; H_1 : r_{n,\tau}(j) \neq 0, \text{ for some } j > 0. \tag{26}$$

Under H_0 the flatness of the quantile spectrum implies that $Q_{n,\tau}(\omega) \equiv \frac{\tau(1-\tau)}{2\pi}$. The Cramer-von Mises statistic is defined by:

$$\text{CM}_{n,\tau}(j) = \frac{n}{j}\sum_{j=1}^{n-1}\left(\frac{\hat{r}_{n,\tau}(j)}{j}\right)^2. \tag{27}$$

The following quantity can be used to realize the test:

$$\widehat{\text{CM}}_{n,\tau} = \frac{1}{2\pi n}\sum_{j=1}^{n-1} j^{-2}\left(\sum_{t=j+1}^{n-1}(\tau - J_t)\big(\tau - J_{t-j}\big)\right)^2, \tag{28a}$$

where $J_1, J_2, \ldots, J_n$ are independent Bernoulli (τ). Under the assumption that Y_t is i.i.d., the statistics in (28a) has the same distribution as

$$\widetilde{\text{CM}}_{n,\tau} = \frac{1}{2\pi n} \sum_{j=1}^{n-1} j^{-2} \left(\sum_{t=j+1}^{n-1} \widehat{V}_t(\tau)\widehat{V}_{t-j}(\tau) \right)^2 . \tag{28b}$$

In large samples, $V_t(\tau, q(\tau))$ is close to $\tau - I\{Y_t < q(\tau)\}$ in probability. However, in small samples, $\{Y_t < q(\tau)\}$ is a Bernoulli random variable with a success probability τ.

The test is realized through a Monte Carlo procedure which consists of the following steps:

- *Step 1.* Draw n i.i.d. copies of $J_1, J_2, \ldots, J_n$ of Bernoulli (τ) random variables.
- *Step 2.* Compute (28a) with the variables drawn at step 1.
- *Step 3.* Repeat steps 1 and 2, R times with R large.
- *Step 4.* Define $c_{n,\tau}(1-\alpha)$ as the $(1-\alpha)$ empirical quantile of the R realizations of (28a).

$$c_{n,\tau}(1-\alpha) := \inf\{z \in \mathcal{R} : P(C\widetilde{M}_{n,\tau} > z) \leq \alpha\}.$$

The null hypothesis is rejected if $\widehat{CM}_{n,\tau}$ is larger than $c_{n,\tau}(1-\alpha)$.

4 Copula-Based Periodogram and Rank-Based Laplace Periodogram

Laplace periodograms can be used to estimate copula spectra density kernels. We briefly present the methodology here since copula models have become widely used in economics and finance (see Patton 2012 for a review of theory and empirical estimation). One important advantage of copulas is that they do not require any distributional assumption, such as for instance the existence of finite moments.

Let us consider again a strictly stationary time series $\{Y_t\}_{t\in\mathbb{Z}}$ and its marginal distribution function F. In the traditional approach, the spectral density kernels are defined associated with autocovariance kernels of the series. To capture more general features of pairs of Y_t and Y_{t-k}, the clipped processes $(I\{Y_t \leq q\})_{t\in\mathbb{Z}}$ and $(I\{U_t \leq \tau\})_{t\in\mathbb{Z}}$, where $U_t := F(Y_t)$ are introduced; then, the spectral density kernels are defined associated with covariance kernels of these clipped processes, which are shown below.

$$\gamma_k(q_1, q_2) := \text{Cov}(I\{Y_t \leq q_1\}, I\{Y_{t-k} \leq q_2\}), \quad q_1, q_2 \in \overline{\mathbb{R}}, k \in \mathbb{Z}, \tag{29}$$

where $I\{\cdot\}$ denotes the indicator function and $\overline{\mathbb{R}} := \mathbb{R} \bigcup \{-\infty, \infty\}$ the extended real line. The definition described above is the Laplace cross-covariance. The copula cross-covariance is

$$\gamma_k^U(\tau_1, \tau_2) := \mathrm{Cov}(I\{U_t \leq \tau_1\}, I\{U_{t-k} \leq \tau_2\}), \quad \tau_1, \tau_2 \in [0, 1], k \in \mathbb{Z}. \tag{30}$$

By using the Laplace cross-covariance and the copula cross-covariance, researchers can consider more general dependence structures of Y_t and Y_{t-k} that traditional covariance-based methods unable to deal with, such as time-irreversibility, tail dependence, varying conditional skewness or kurtosis, and so on, though various extensions and revisions have been proposed in the L_2-periodograms (Kleiner et al. 1979; Klüppelberg and Mikosch 1994; Mikosch 1998; Katkovnik 1998; Hong 1999; Hill and McCloskey 2014).

Under summability conditions on γ_k and γ_k^U, the population spectral densities are defined as follows.

$$\mathfrak{f}_{q_1, q_2}(\omega) := \frac{1}{2\pi} \sum_{k=-\infty}^{\infty} \gamma_k(q_1, q_2) \mathrm{e}^{-ik\omega}, \quad q_1, q_2 \in \overline{\mathbb{R}}, \omega \in \mathbb{R}, \tag{31}$$

$$\mathfrak{f}_{q_{\tau_1}, q_{\tau_2}}(\omega) := \frac{1}{2\pi} \sum_{k=-\infty}^{\infty} \gamma_k^U(\tau_1, \tau_2) \mathrm{e}^{-ik\omega}, \quad \tau_1, \tau_2 \in [0, 1], \omega \in \mathbb{R}, \tag{32}$$

where $q_{\tau_i} := F^{-1}(\tau_i)$ $(i = 1, 2)$. The former equation is called the Laplace spectral density, and the latter one is called the copula spectral density. These formulas are Fourier transforms of $\{\gamma_k\}$ and $\{\gamma_k^U\}$; therefore, based on the Fourier theorem, the inverse Fourier transform provides

$$\gamma_k(q_1, q_2) = \int_{-\infty}^{\infty} \mathrm{e}^{ik\omega} \mathfrak{f}_{q_1, q_2}(\omega) \mathrm{d}\omega, \tag{33}$$

$$\gamma_k^U(\tau_1, \tau_2) = \int_{-\infty}^{\infty} \mathrm{e}^{ik\omega} \mathfrak{f}_{q_{\tau_1}, q_{\tau_2}}(\omega) \mathrm{d}\omega. \tag{34}$$

Since the clipped processes which take binary values satisfy the strict stationarity assumption, the approach based on the Laplace- and Copula spectral densities is still applicable to detect various statistical relationships in more appropriate manners, which are beyond the scope of the traditional spectral approach. In this aspect, the vast amount of theoretical work has been conducted by Li (2008, 2012, 2013, 2014), Hagemann (2013), Lee and Subba Rao (2012), Dette et al. (2015), and Kley et al. (2016).

An estimate of copula spectrum can be obtained by computing the copula periodogram $L_{n,\tau_1,\tau_2}^U(\omega)$, associated with the series $U_1, \ldots, U_n$, where $U_t := F(Y_t)$, is defined. $L^{U}_{n,\tau_1,\tau_2}(\omega)$ is obtained in Eq. (12) by replacing $\widehat{\boldsymbol{\beta}}_{n,\tau}$ by the $\widehat{\boldsymbol{\beta}}^{U}_{n,\tau}$, which is obtained below.

$$\left\{\widehat{\lambda}_{n,\tau}(\omega), \widehat{\boldsymbol{\beta}}^{U}_{n,\tau}(\omega)\right\} := \operatorname{argmin}_{\lambda\in\mathbb{R},\boldsymbol{\beta}\in\mathbb{R}^2} \sum_{t=1}^{n} \rho_\tau\big(U_t - \lambda - \boldsymbol{x}'_t(\omega)\boldsymbol{\beta}\big). \tag{35}$$

Since the distribution function $F(Y_t)$ is unknown, it is replaced by the rank of Y_t. In this case, the periodogram is called the empirical or rank-based Laplace periodogram, which is defined associated with the normalized rank series $n^{-1}R_1^{(n)}, \ldots, n^{-1}R_n^{(n)}$, where $R_t^{(n)}$ is the rank of Y_t among $Y_1, \ldots, Y_n$. The rank-based Laplace periodogram $L^{R}_{n,\tau_1,\tau_2}(\omega)$ is obtained in Eq. (12) by replacing $\widehat{\boldsymbol{\beta}}_{n,\tau}$ by $\widehat{\boldsymbol{\beta}}^{R}_{n,\tau}$. We have:

$$\left\{\widehat{\lambda}_{n,\tau}(\omega), \widehat{\boldsymbol{\beta}}^{R}_{n,\tau}(\omega)\right\} := \operatorname{argmin}_{\lambda\in\mathbb{R},\boldsymbol{\beta}\in\mathbb{R}^2} \sum_{t=1}^{n} \rho_\tau\left(n^{-1}R_t^{(n)} - \lambda - \boldsymbol{x}'_t(\omega)\boldsymbol{\beta}\right). \tag{36}$$

5 The Multivariate Case

Now we consider the multivariate case of quantile spectral densities and periodograms based on the copula- and Laplace-related concepts, which have been already introduced in the univariate case in the previous sections. Let $\{\boldsymbol{X}_t\}_{t\in\mathbb{Z}}$ be a d-variate strictly stationary process, with components $X_{t,l}$, $l = 1, \ldots, d$; i.e., $\boldsymbol{X}_t = \big(X_{t,1}, \ldots, X_{t,d}\big)'$. $X_{t,l}$ has its marginal distribution function $F_l(q)$ and inverse function $q_l(\tau) := F_l^{-1}(\tau) := \inf\{q \in \mathbb{R} \colon \tau \le F_l(q)\}$, where $\tau \in [0, 1]$. The matrix of quantile cross-covariance, $\Gamma_k(\tau_1, \tau_2) := \left(\gamma_k^{l_1 l_2}(\tau_1, \tau_2)\right)_{l_1,l_2=1,\ldots,d}$, where $\gamma_k^{l_1 l_2}(\tau_1, \tau_2)$ is the cross-covariance of a pair of $\big(X_{t,l_1}, X_{t,l_2}\big)$, which is as follows.

$$\gamma_k^{l_1 l_2}(\tau_1, \tau_2) := \operatorname{Cov}\big(I\{X_{t,l_1} \le q_{l_1}(\tau_1)\}, I\{X_{t-k,l_2} \le q_{l_2}(\tau_2)\}\big), \tag{37}$$

where $l_1, l_2 \in \{1, \ldots, d\}$, $k \in \mathbb{Z}$, and $\tau_1, \tau_2 \in [0, 1]$. The quantile-based quantities are functions of τ_1 and τ_2, which are quantiles of a quantile regression. In the frequency domain, under approximate mixing conditions, the quantile cross-spectral densities are

$$\mathfrak{f}_{q_{\tau_1}, q_{\tau_2}}(\omega) := \left(\mathfrak{f}^{l_1 l_2}_{q_{\tau_1}, q_{\tau_2}}(\omega)\right)_{l_1,l_2=1,\ldots,d}, \tag{38}$$

where

$$\mathfrak{f}^{l_1 l_2}_{q_{\tau_1}, q_{\tau_2}}(\omega) := \frac{1}{2\pi} \sum_{k=-\infty}^{\infty} \gamma_k^{l_1 l_2}(\tau_1, \tau_2) \mathrm{e}^{-ik\omega}, \tag{39}$$

$l_1, l_2 \in \{1, \ldots, d\}$, $\omega \in \mathbb{R}$, and $\tau_1, \tau_2 \in [0, 1]$. Each quantile cross-spectral density, i.e., $\mathfrak{f}^{l_1 l_2}_{q_{\tau_1}, q_{\tau_2}}(\omega)$, is a complex-valued function. As considered in traditional spectral analysis, its real and imaginary parts are referred to as the quantile cospectrum and quantile quadrature spectrum.

To measure dynamic dependence structure of the two processes $\{X_{t,l_1}\}_{t\in\mathbb{Z}}$ and $\{X_{t,l_2}\}_{t\in\mathbb{Z}}$, the quantile coherency is defined as follows.

$$\mathcal{R}^{l_1 l_2}_{q_{\tau_1}, q_{\tau_2}}(\omega) := \frac{\mathfrak{f}^{l_1 l_2}_{q_{\tau_1}, q_{\tau_2}}(\omega)}{\left(\mathfrak{f}^{l_1 l_1}_{q_{\tau_1}, q_{\tau_1}}(\omega) \mathfrak{f}^{l_2 l_2}_{q_{\tau_2}, q_{\tau_2}}(\omega)\right)^{1/2}}, \tag{40}$$

where $(\tau_1, \tau_2) \in [0, 1]^2$. The modulus squared of this quantile coherency, e.g., $\left|\mathcal{R}^{l_1 l_2}_{q_{\tau_1}, q_{\tau_2}}(\omega)\right|^2$, is referred to as the quantile coherence.

When we use the empirical distribution function of $X_{t,l}$, i.e., $\widehat{F}_{n,l}(x) := n^{-1} \sum_{t=0}^{n-1} I\{X_{t,l} \le x\}$, the rank-based copula cross-periodograms (CCR periodograms) are defined as

$$I^{l_1 l_2}_{n,R}(\omega; \tau_1, \tau_2) := \frac{1}{2\pi n} d^{l_1}_{n,R}(\omega; \tau_1) d^{l_2}_{n,R}(-\omega; \tau_2), \tag{41}$$

where $l_1, l_2 \in \{1, \ldots, d\}$, $\omega \in \mathbb{R}$, $(\tau_1, \tau_2) \in [0, 1]^2$, and

$$d^{l}_{n,R}(\omega; \tau) := \sum_{t=0}^{n-1} I\left\{\widehat{F}_{n,l}(X_{t,l}) \le \tau\right\} \mathrm{e}^{-i\omega t} = \sum_{t=0}^{n-1} I\left\{R^{(n)}_{t,l} \le n\tau\right\} \mathrm{e}^{-i\omega t}, \tag{42}$$

where $l = 1, \ldots, d$, $\omega \in \mathbb{R}$, $\tau \in [0, 1]$, and $R^{(n)}_{t,l}$ is also the rank of $X_{t,l}$ among $X_{0,l}, \ldots, X_{t-1,l}$.

Kley et al. (2016) have shown that the CCR periodogram does not have the consistency when it is used to estimate the quantile cross-spectral density $\mathfrak{f}^{l_1 l_2}_{q_{\tau_1}, q_{\tau_2}}(\omega)$. On the other hand, its smoothed versions, i.e., smoothed CCR periodogram shown below gains the consistency by correcting biases (see Kley et al. 2016, Theorem 3.5).

$$\widehat{G}^{l_1 l_2}_{n,R}(\omega; \tau_1, \tau_2) := \frac{2\pi}{n} \sum_{s=1}^{n-1} W_n(\omega - 2\pi s/n) I^{l_1 l_2}_{n,R}\left(\frac{2\pi s}{n}; \tau_1, \tau_2\right), \tag{43}$$

where W_n denotes a sequence of weight functions. The smoothed CCR periodogram also maintains the asymptotic normality. In addition, note that fixing l_1, l_2 and

τ_1, τ_2, the smoothed CCR periodogram is asymptotically equivalent to traditional smoothed periodogram determined from the unobservable bivariate time series $\left(I\left\{F_{l_1}\left(X_{t,l_1}\right) \leq \tau_1\right\}, I\left\{F_{l_2}\left(X_{t,l_2}\right) \leq \tau_2\right\}\right)$ $(t = 0, \ldots, n-1)$. By using this asymptotic normality property, the pointwise asymptotic confidence intervals for the real and imaginary parts of the spectrum can be computed for a pair of (τ_1, τ_2).

The consistent estimators of quantile coherency are also defined as

$$\widehat{\mathcal{R}}^{l_1 l_2}_{n,R,q_{\tau_1},q_{\tau_2}}(\omega) := \frac{\widehat{G}^{l_1 l_2}_{n,R}(\omega; \tau_1, \tau_2)}{\left(\widehat{G}^{l_1 l_1}_{n,R}(\omega; \tau_1, \tau_1)\widehat{G}^{l_2 l_2}_{n,R}(\omega; \tau_2, \tau_2)\right)^{1/2}}. \tag{44}$$

The difference between this coherency and $\mathcal{R}^{l_1 l_2}_{q_{\tau_1},q_{\tau_2}}(\omega)$, with bias correction terms, asymptotically converges to a normal distribution, which implies the asymptotic consistency (see Baruník and Kley 2019, Theorem 4.1).

6 Empirical Example

This section shows an example of the quantile-based spectral analysis for the stock returns by using the R package "QUANTSPEC version 1.2.1". The following returns of daily stock average indexes (Dow Jones Industrial Average, CAC 40, and Nikkei 225) were taken from "Factiva.com" during the post-period of global financial crisis from July 27th 2009 to March 27th 2020 (2516 observations).

We first plot the following three types of data for each stock index: (1) Y_t: returns (2) $\text{Cov}(Y_{t+k}, Y_t)$: autocovariances of the returns, and (3) $\text{Cov}\left(Y^2_{t+k}, Y^2_t\right)$: autocovariances of the squared returns. Figure 2 shows the stock prices and their returns of three stock average indexes, DJ (Dow Jones Industrial Average in the United States), CAC (CAC 40 in France), and Nikkei (Nikkei 225 in Japan). Each return seems to have zero-mean with some outliers.

Their highly volatile periods correspond to "Flash crash" in May 2010, "Black Monday" in August 2011, "China shock" in August 2015, "Brexit" in June 2016, and "VIX shock" in February 2018, and "Coronavirus shock" in March 2020. Additionally, the highly volatile period, especially limited to Nikkei (Japanese market), corresponds to the "East Japan great earthquake" in March 2011.

Figure 3 shows their autocovariances with lag k. DJ has significantly negative serial correlations (Lag $= 1, 3, 5, 8$, or 19) and positive correlations (Lag $= 2, 9$, or 11). CAC has a significantly negative serial correlation (Lag $= 5$) and a positive correlation (Lag $= 6$). Nikkei seems to have no serial correlation. Thus, only Japanese stock market appears to be uncorrelated. This is a typical characteristic of many financial returns, as long as we use a linear measure of dependence.

Figure 4 shows the autocovariances of the squared returns, i.e., their volatilities. In the series of all volatilities, we can find significant and persistent autocovariances. These squared returns are clearly correlated. However, all autocovariances persist

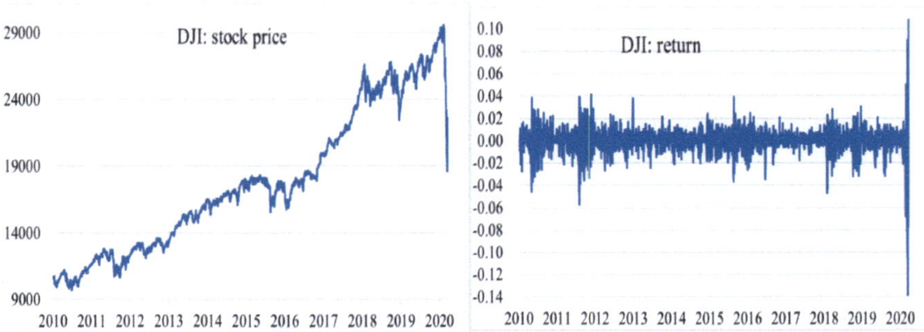

Min: -0.1384 Max: 0.1076 Median: 0.0007 Mean: 0.0003 SD: 0.0109

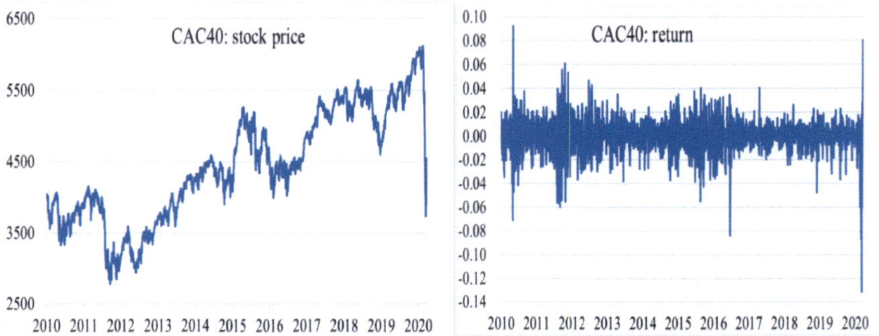

Min: -0.1310 Max: 0.0922 Median: 0.0006 Mean: 0.0001 SD: 0.0133

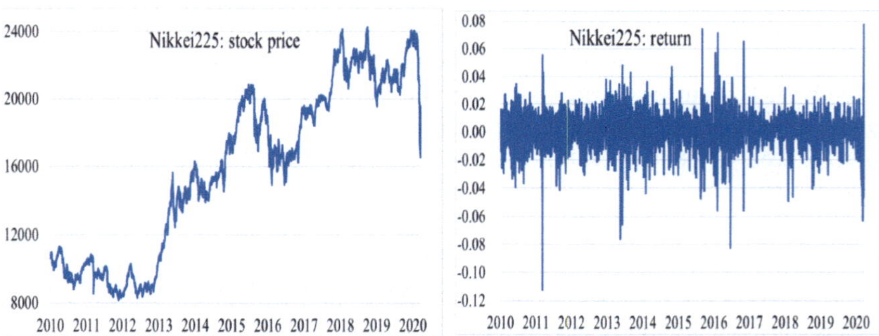

Min: -0.1115 Max: 0.0773 Median: 0.0006 Mean: 0.0003 SD: 0.0137

Fig. 2 Stock prices and returns: Dow Jones Industrial Average, CAC 40 and Nikkei 225 from 09/27/2009 to 03/27/2020

until at least lag 14 (more than 2 weeks). The persistency of their volatilities suggests that an ARCH or GARCH model will be required if we focus on the traditional approach of financial analyses. In this section, we focus on another approach, i.e., quantile-based spectral analysis.

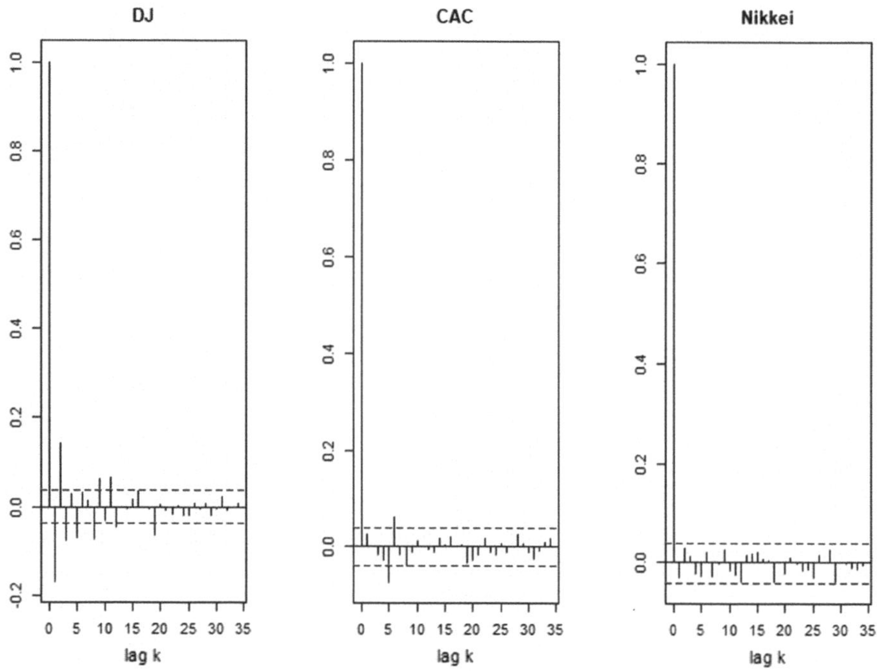

Fig. 3 Autocovariances of stock returns

Before going on quantile-based spectral analysis, we estimate the traditional smoothed periodograms of the returns in Figure 5 (Non-parametric model modified by Daniell smoothing). All the periodograms show some peaks, but do not near zero frequency. CAC and Nikkei have peaks around 0.15 and 0.35 frequencies, which means highly volatile stock returns in 42 and 18 periods ($2\pi/0.15$ and $2\pi/0.35$). DJ also fluctuates but its amplitude is much smaller than the other two. The feature of less volatile US market has already been observed in Fig. 1.

Figure 6a–c show the copula rank periodograms $I_{n,R}^{\tau_1,\tau_2}(\omega)$ for $\tau_1, \tau_2 \in \{0.1, 0.5, 0.9\}$, all Fourier frequencies $\omega \in (0, \pi)$, 200 moving blocks bootstrap replications with block length 40 (about 2 months). In Figure 7a–c, we draw the smoothed copula rank periodograms $\widehat{G}_{n,R}(\omega; \tau_1, \tau_2)$ by using the computed quantile periodogram with the Epanechnikov kernel and bandwidth 0.07.[2] In addition, pointwise confidence intervals are obtained by a normal approximation to the distribution of the estimator. The periodograms suggest serial dependent structure of each stock return below the 10, 50, and 90% quantiles.

The figures show that the copula spectra of three stock returns are not flat, showing clear evidence of serial dependency. This result turns out to be different from their

[2] $I_{n,R}^{\tau_1,\tau_2}(\omega)$ is one-variate case of $I_{n,R}^{l_1 l_2}(\omega; \tau_1, \tau_2)$ for $\boldsymbol{X}_t = X_{t,1}$ in Eq. (41), which is called the CR periodogram. Its smoothed version $\widehat{G}_{n,R}(\omega; \tau_1, \tau_2)$ is also a special case of $\widehat{G}_{n,R}^{l_1 l_2}(\omega; \tau_1, \tau_2)$ in Eq. (43).

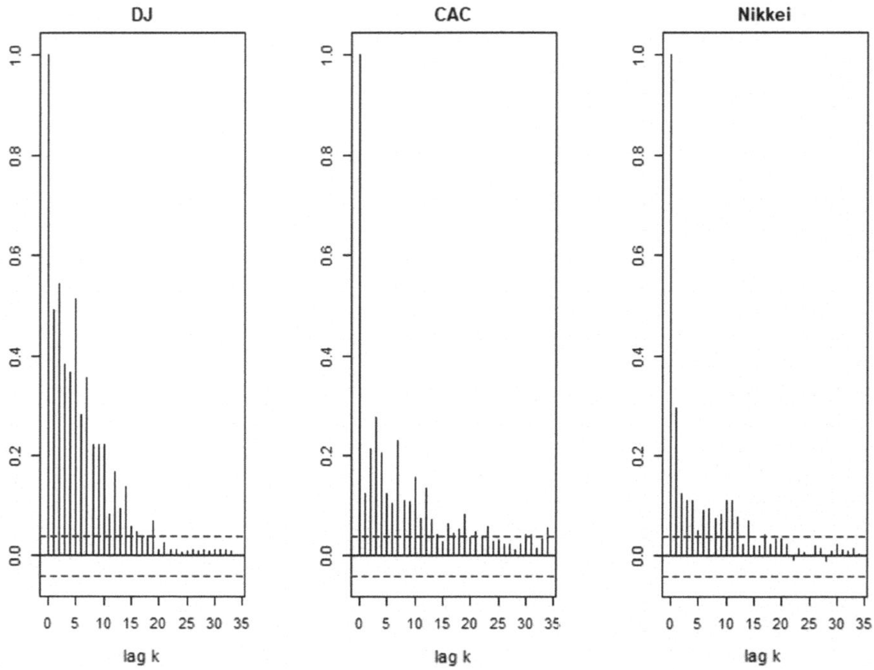

Fig. 4 Autocovariances of squared returns

slight evidence of small autocovariances in Fig. 3. The top left and bottom right figures show the serial dependency structure in negative and positive extreme events. The copula rank periodograms in Fig. 6a–c show that high serial dependency is observed near 0.0 for some cases. The smoothed copula rank periodograms in Fig. 7a–c also suggest that both negative and positive extreme events have some power to change the serial dependency structures in all stock markets. In addition, the copula rank periodograms (or its smoothed versions) indeed have apparent peaks in some frequencies and fluctuate over frequencies as well.

We also show the rank-based Laplace periodograms and its smoothed versions in Fig. 8. Figure 8a–c correspond to the rank-based Laplace periodograms $L^R_{n,\tau_1,\tau_2}(\omega)$. Figure 8d–f correspond to the smoothed rank-based Laplace periodograms $\widehat{f}_{n,R}(\omega;\tau_1,\tau_2)$.[3] Obviously, compared to the (smoothed) traditional L_2-periodograms shown in Fig. 5, the rank-based Laplace periodograms or its smoothed rank-based Laplace periodograms seem to better capture the correlation structures in the stock return series and provide much richer views.

As is evident in these figures, one important aspect is that the copula rank periodograms and the rank-based Laplace periodograms, including their smoothed

[3]The smoothed rank-based Laplace periodogram is defined as $\hat{f}_{n,R}(\omega;\tau_1,\tau_2) := \frac{2\pi}{n}\sum_{s=1}^{n-1} W_n(\omega - 2\pi s/n) L^R_{n,\tau_1,\tau_2}\left(\frac{2\pi s}{n}\right)$, where W_n denotes a sequence of weight functions.

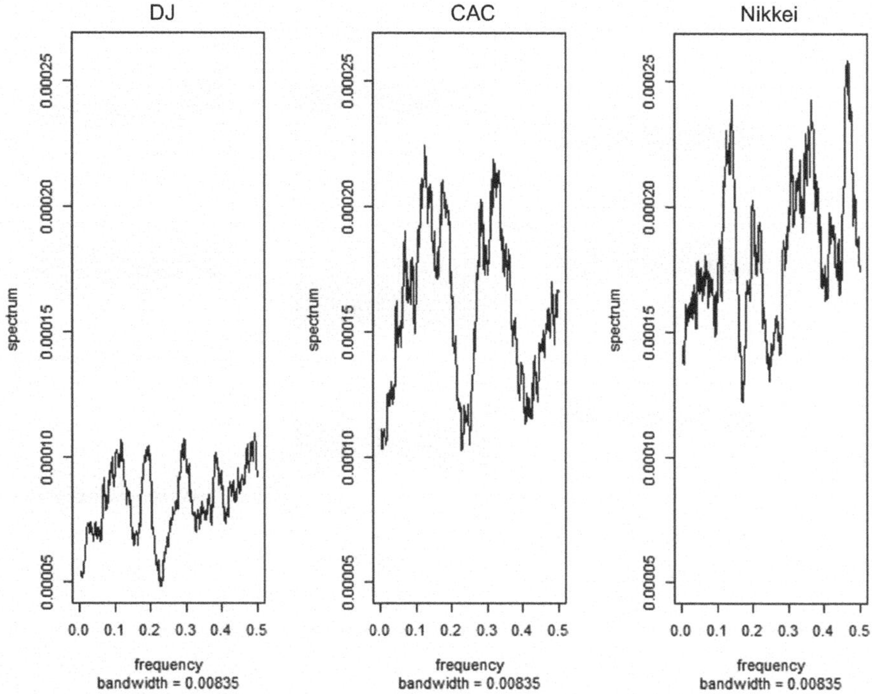

Fig. 5 Smoothed periodograms

versions, are quantile dependent, i.e., their values change across quantile τ. The periodograms estimated at the tails of the distribution ($\tau_1 = \tau_2 = 0.1$ and $\tau_1 = \tau_2 = 0.9$) suggest strong correlations of the returns. We indeed have a significant frequency near zero with a spectrum which decreases across frequencies. This is a typical pattern of a long-memory process.

Next, to grasp the sources of systemic risk, we refer to the behavior of joint quantiles in stock return distributions. The volatility of stock market returns is regarded as time-varying. Thus, common volatility can be obtained as a result of the dependence of concerned two stock markets. Figure 9 shows the rank-based copula cross-periodograms $I_{n,R}^{l_1 l_2}(\omega; \tau_1, \tau_2)$ [Eq. (41)], and the smoothed rank-based copula cross-periodograms $\widehat{G}_{n,R}^{l_1 l_2}(\omega; \tau_1, \tau_2)$ [Eq. (43)].

Figure 9 shows the frequency dynamics in quantiles of the joint distribution of the returns. The copula rank cross-periodograms (or its smoothed version) estimated at the tails of the distribution ($\tau_1 = \tau_2 = 0.1$ and $\tau_1 = \tau_2 = 0.9$) suggest a persistent and strong dependences of the returns for the concerned two markets. Moreover, during the normal period (i.e., the 0.5|0.5 combinations of quantile levels of the joint distribution), US-France, US-Japan, and France-Japan markets have two similar peaks around 0.15 and 0.35 (cycles per day) (0.9 and 2.2 frequencies).

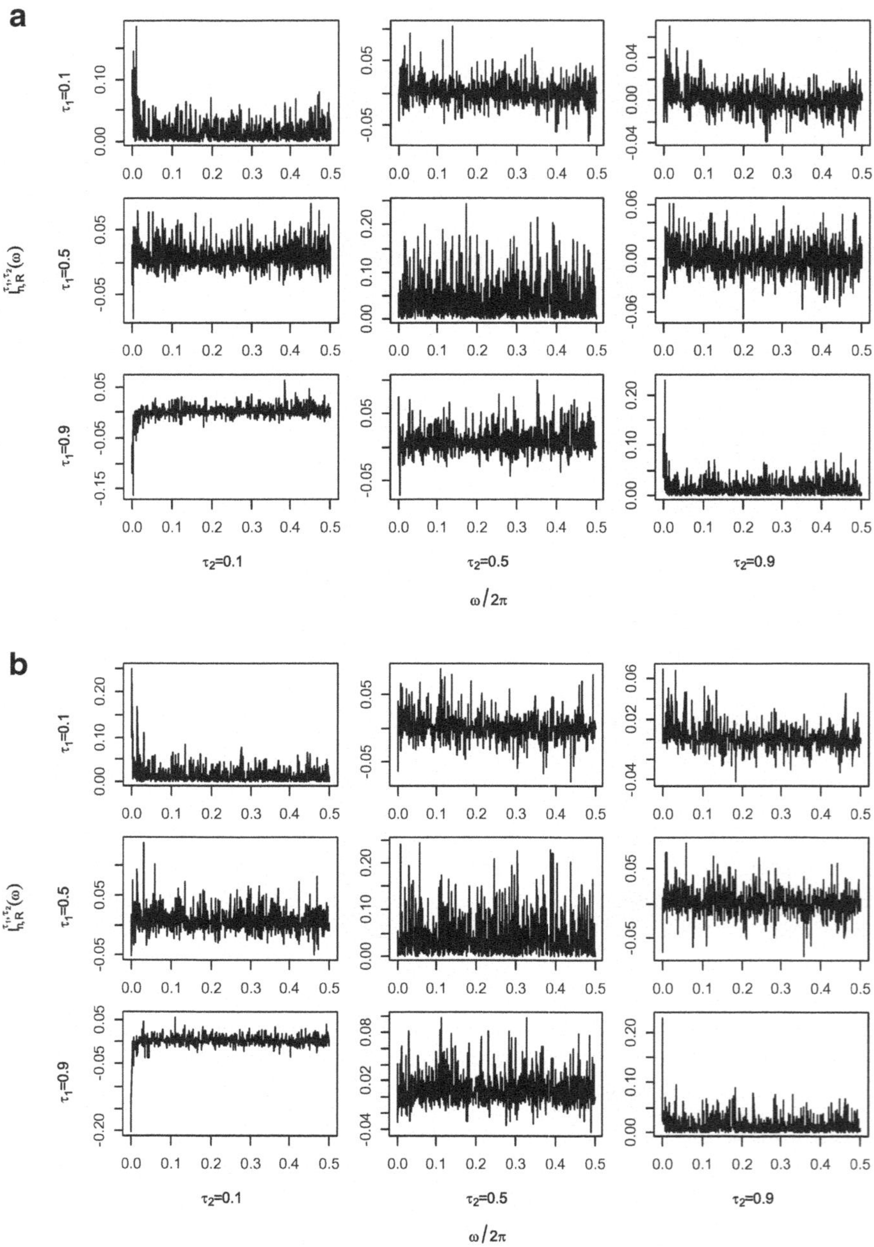

Fig. 6 **a** Copula rank periodogram: Dow Jones. **b** Copula rank periodogram: CAC40. **c** Copula rank periodogram: Nikkei 225

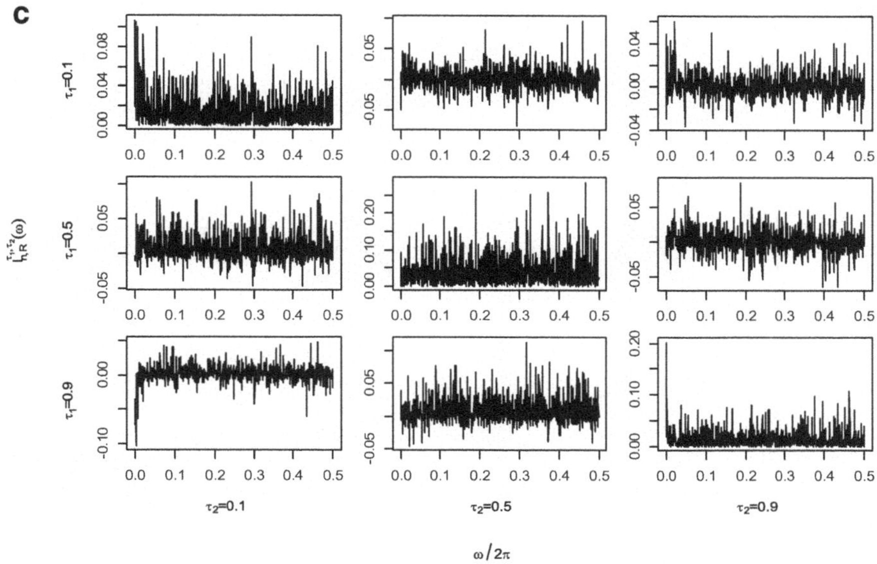

Fig. 6 (continued)

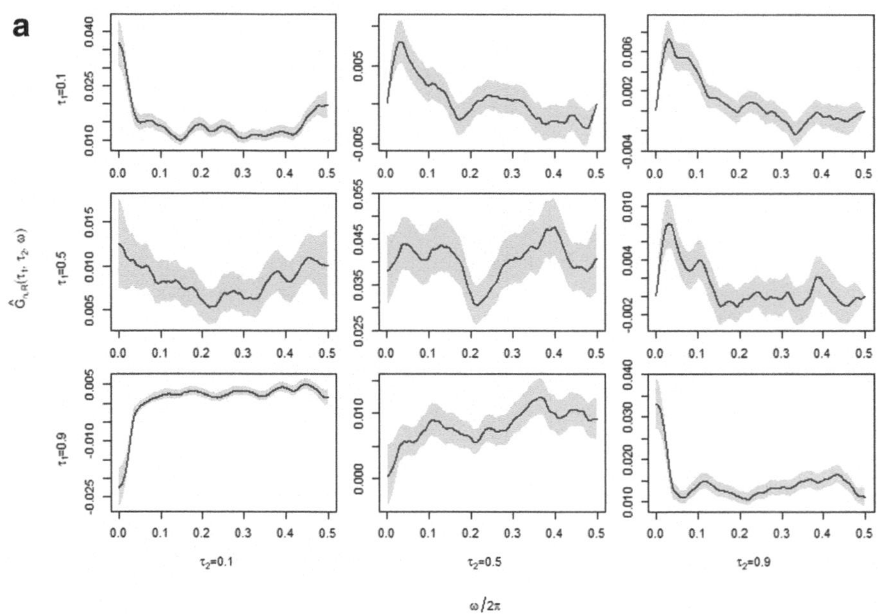

Fig. 7 **a** Smoothed Copula rank periodogram: Dow Jones. **b** Smoothed Copula rank periodogram: CAC 40. **c** Smoothed Copula rank periodogram: Nikkei 225

b

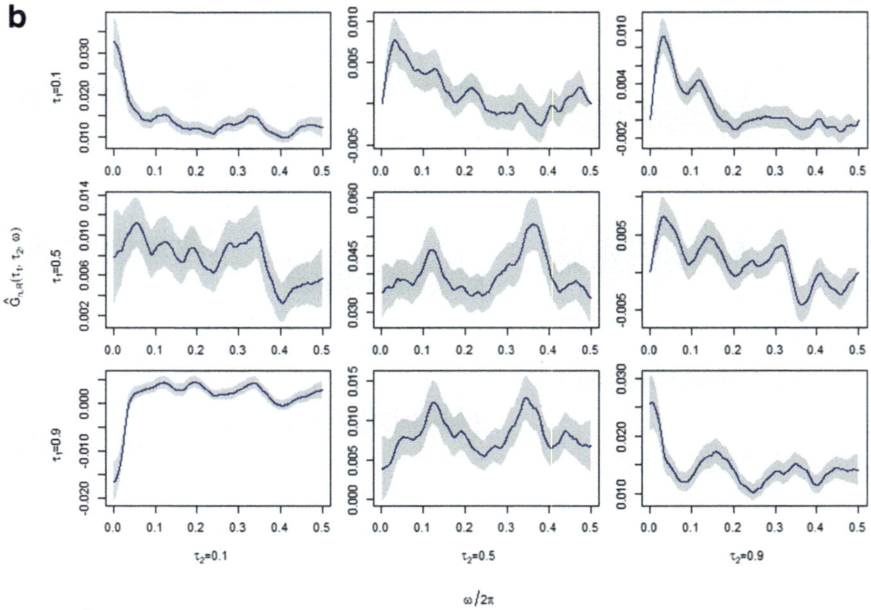

c

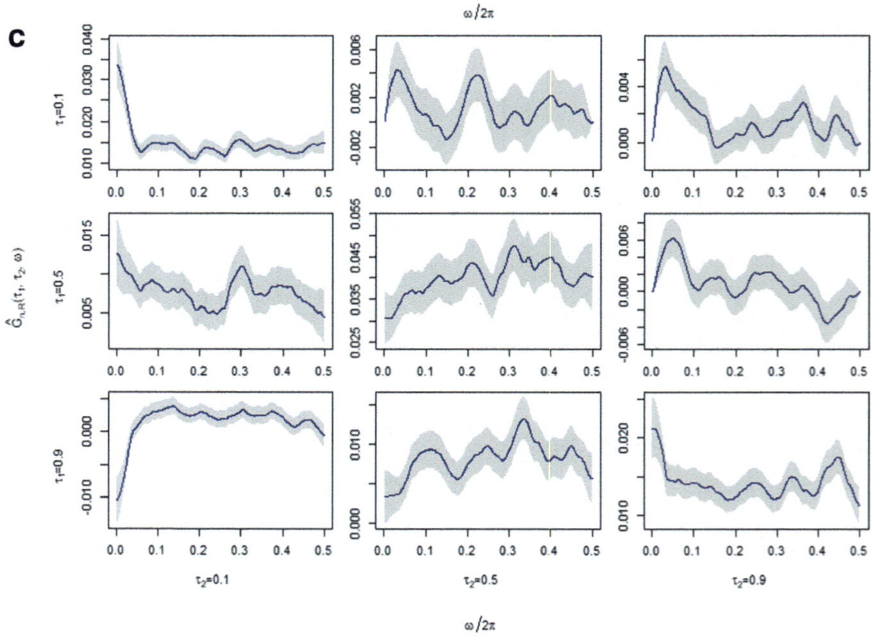

Fig. 7 (continued)

a Dow Jones: $L^R_{n,\tau_1,\tau_2}(\omega)$

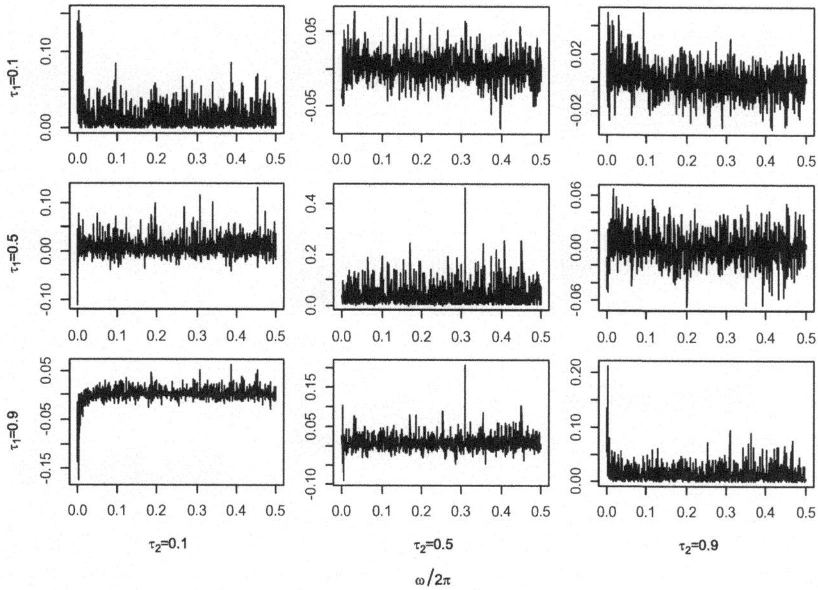

b CAC 40: $L^R_{n,\tau_1,\tau_2}(\omega)$

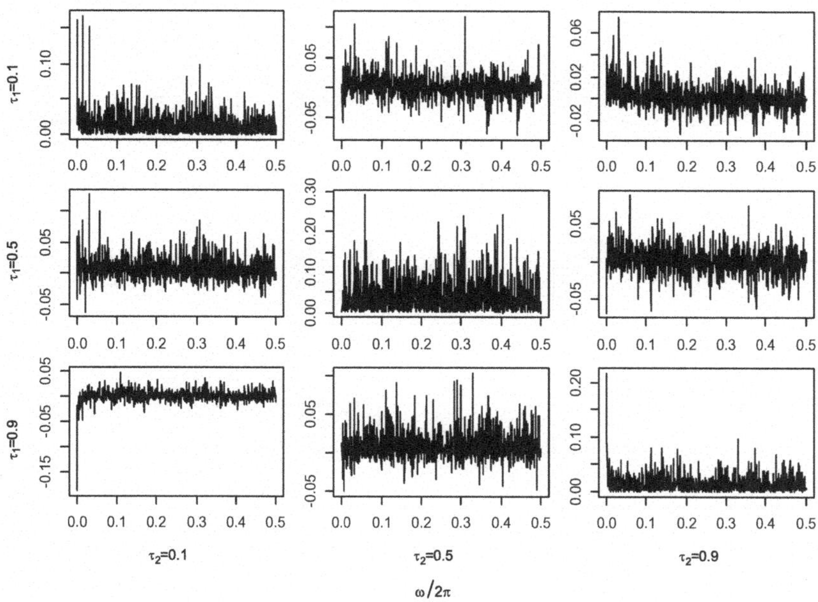

Fig. 8 Rank-based Laplace periodograms $L^R_{n,\tau_1,\tau_2}(\omega)$ and smoothed rank-based Laplace periodograms $\hat{f}_{n,R}(\omega;\tau_1,\tau_2)$

c Nikkei 225: $L^{R}_{n,\tau_1,\tau_2}(\omega)$

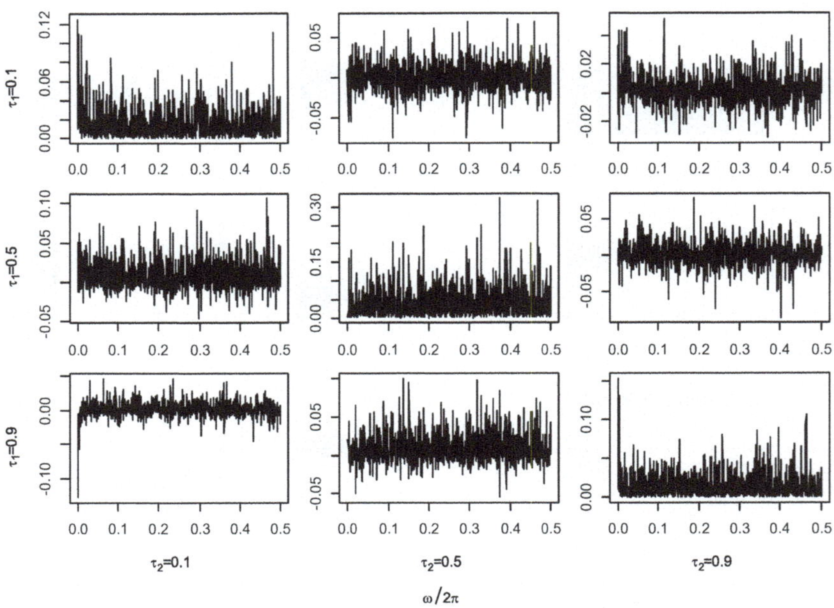

d Dow Jones: $\hat{f}_{n,R}(\omega;\tau_1,\tau_2)$

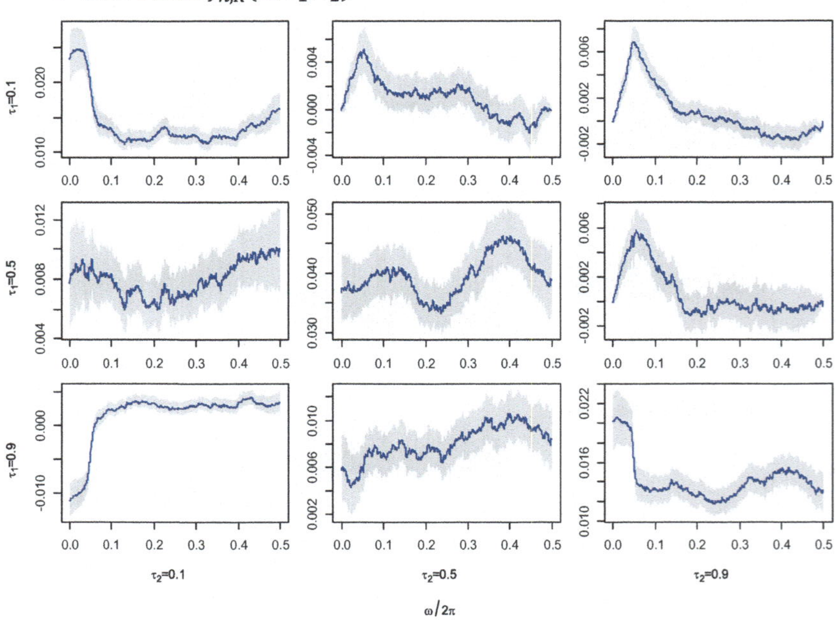

Fig. 8 (continued)

e CAC 40: $\hat{f}_{n,R}(\omega;\tau_1,\tau_2)$

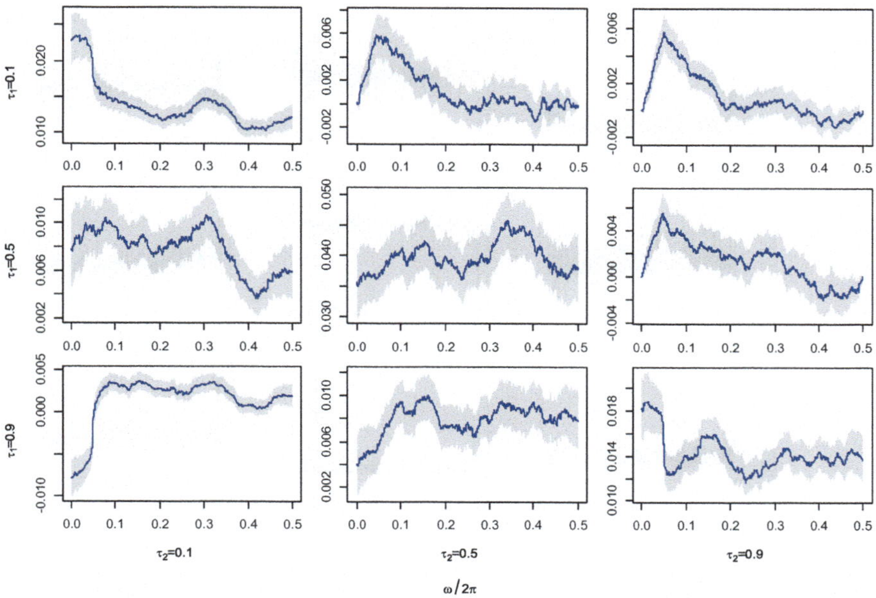

f Nikkei 225: $\hat{f}_{n,R}(\omega;\tau_1,\tau_2)$

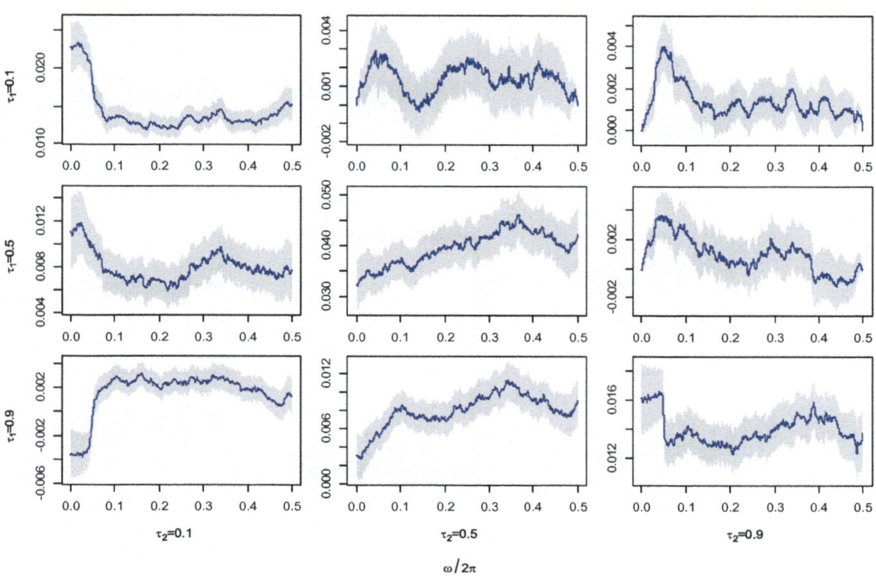

Fig. 8 (continued)

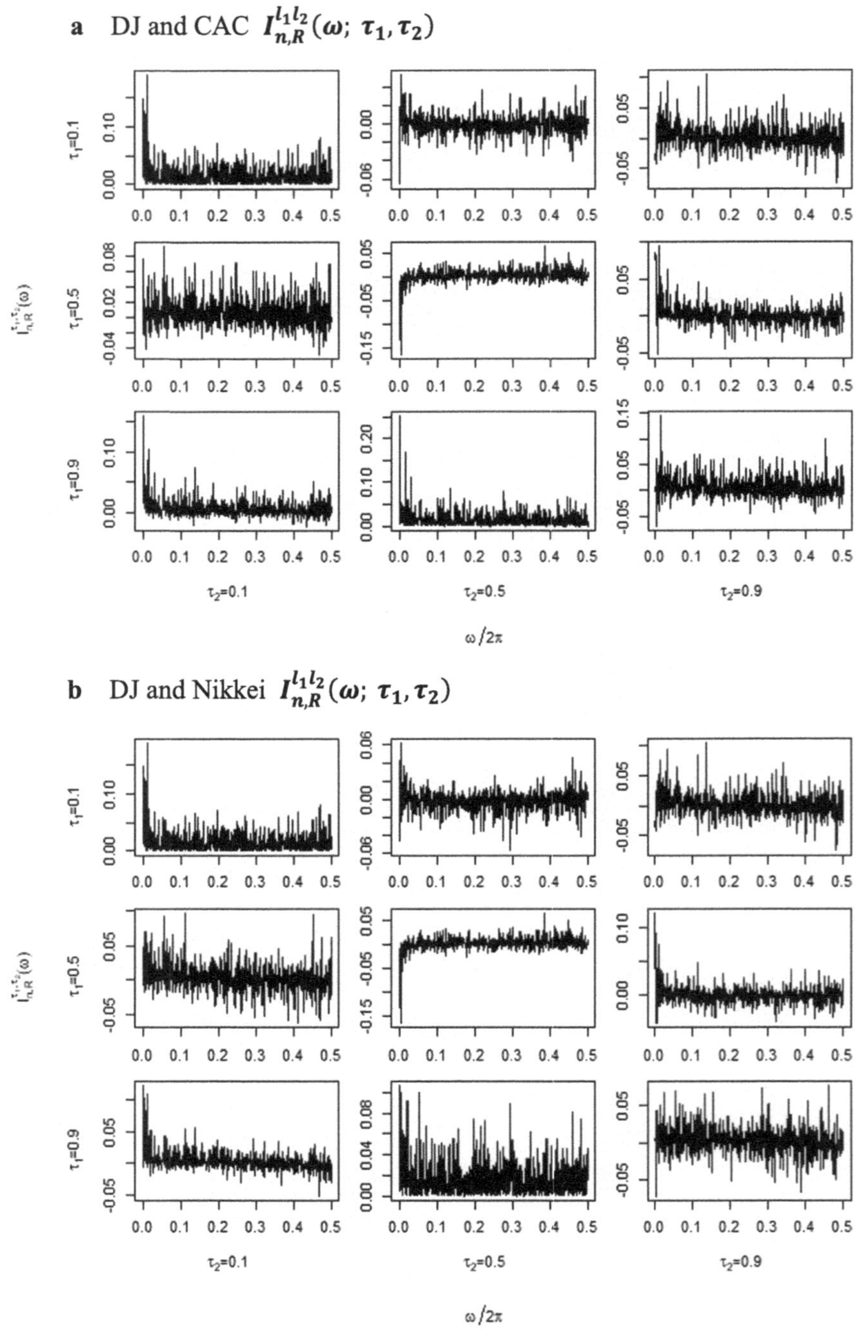

Fig. 9 Rank-based copula cross-periodograms $I^{l_1 l_2}_{n,R}(\omega;\tau_1,\tau_2)$ [Eq. (41)], and the smoothed rank-based copula cross-periodograms $\widehat{G}^{l_1 l_2}_{n,R}(\omega;\tau_1,\tau_2)$ [Eq. (43)]

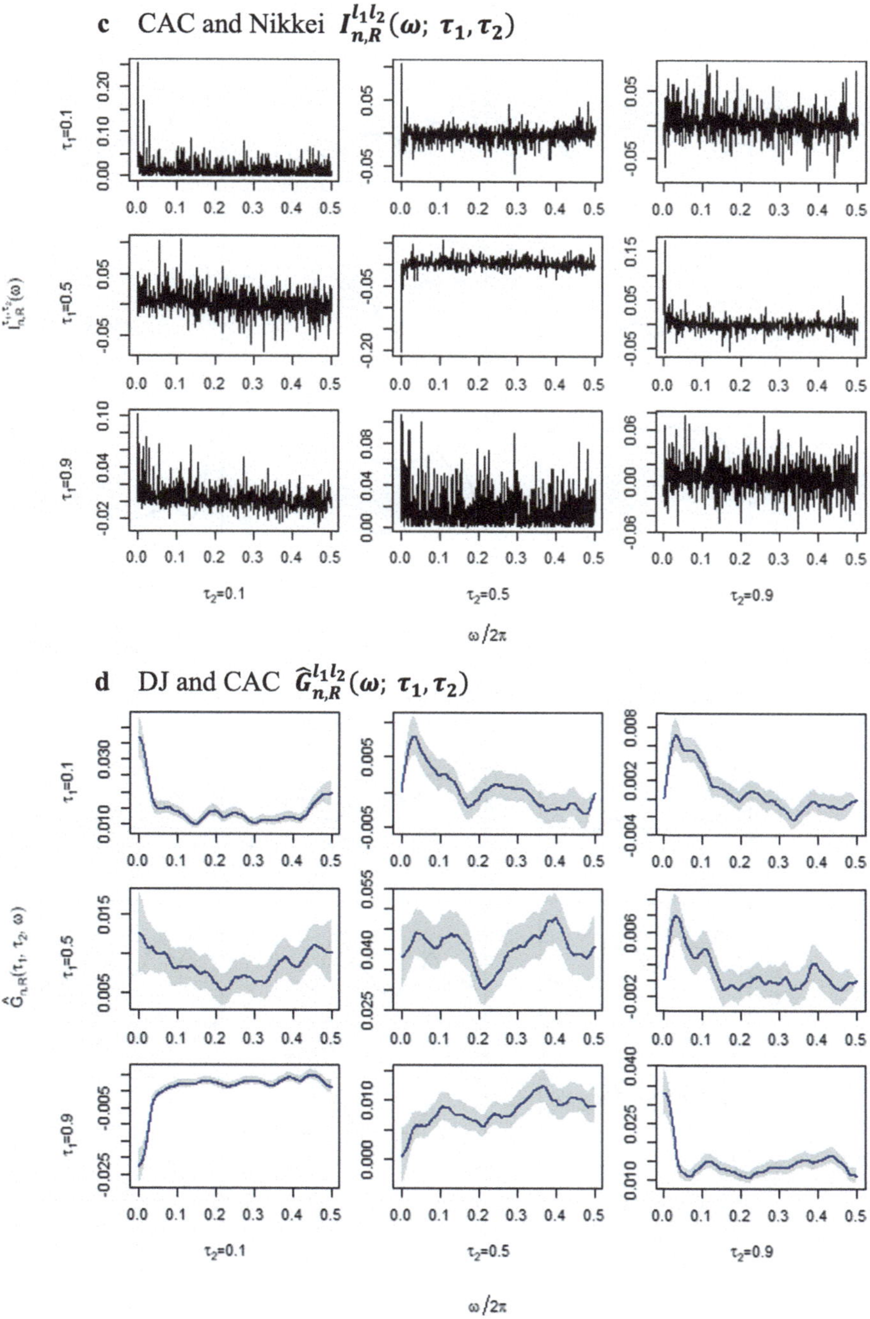

Fig. 9 (continued)

e DJ and Nikkei $\widehat{G}^{l_1 l_2}_{n,R}(\omega;\ \tau_1, \tau_2)$

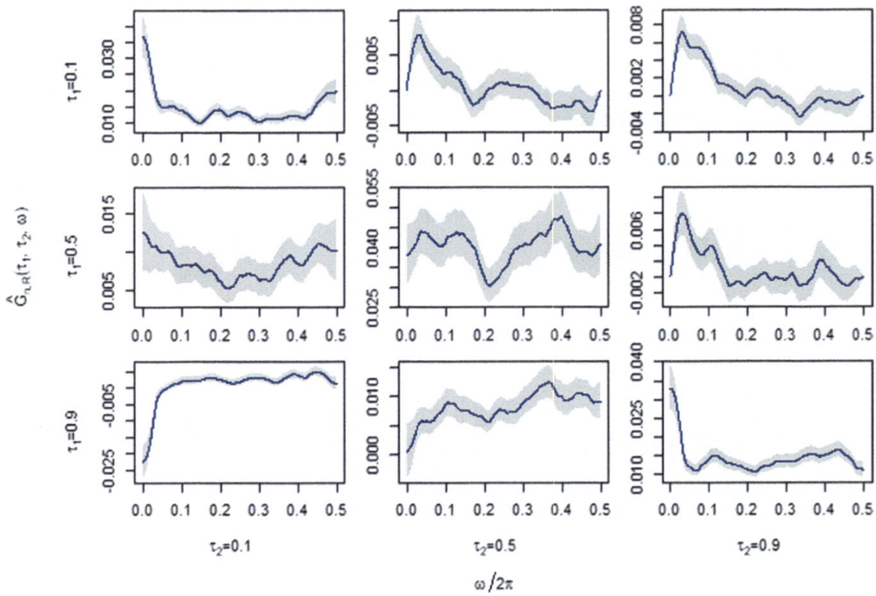

f CAC and Nikkei $\widehat{G}^{l_1 l_2}_{n,R}(\omega;\ \tau_1, \tau_2)$

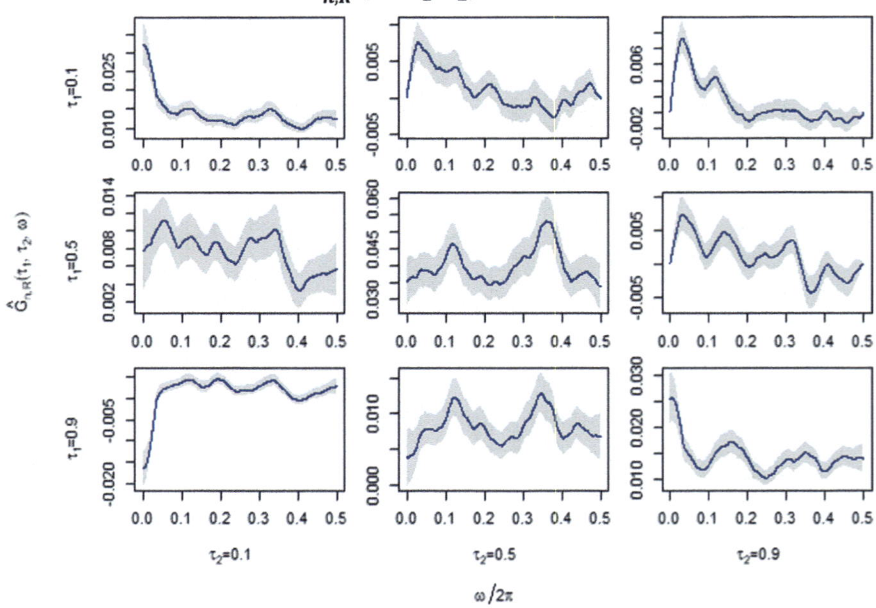

Fig. 9 (continued)

Finally, we refer to the quantile coherency. The left panels of Fig. 10a–c show the quantile coherency estimates for the 0.1|0.1, 0.5|0.5, 0.9|0.9 combinations of quantile levels of the joint distribution for (a) Dow Jones and CAC 40, (b) Dow Jones and Nikkei 225, and (c) CAC 40 and Nikkei 225. The right panels of Fig. 10a–c focus only on the 0.1|0.9 combination of quantile levels. Figure 10 also plots

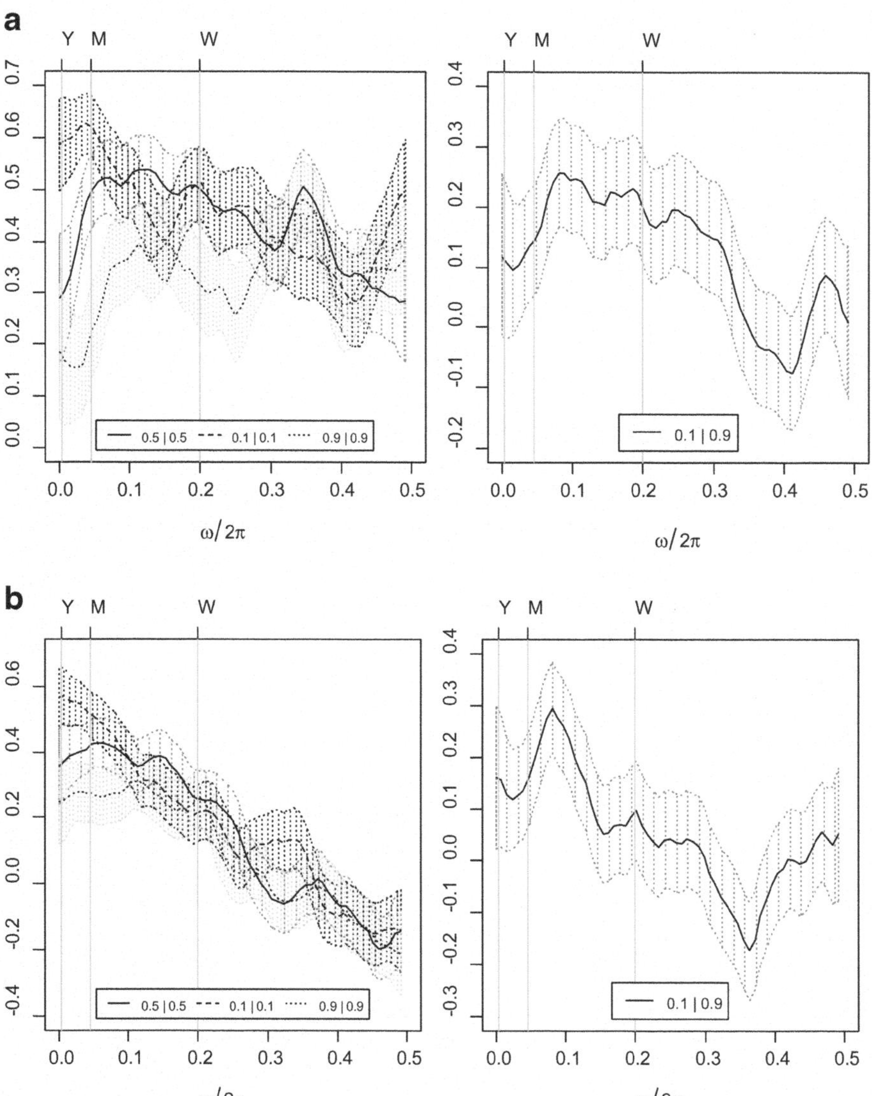

Fig. 10 **a** Quantile coherency for Dow Jones and CAC 40. **b** Quantile coherency for Dow Jones and Nikkei 225. **c** Quantile coherency for CAC 40 and Nikkei 225

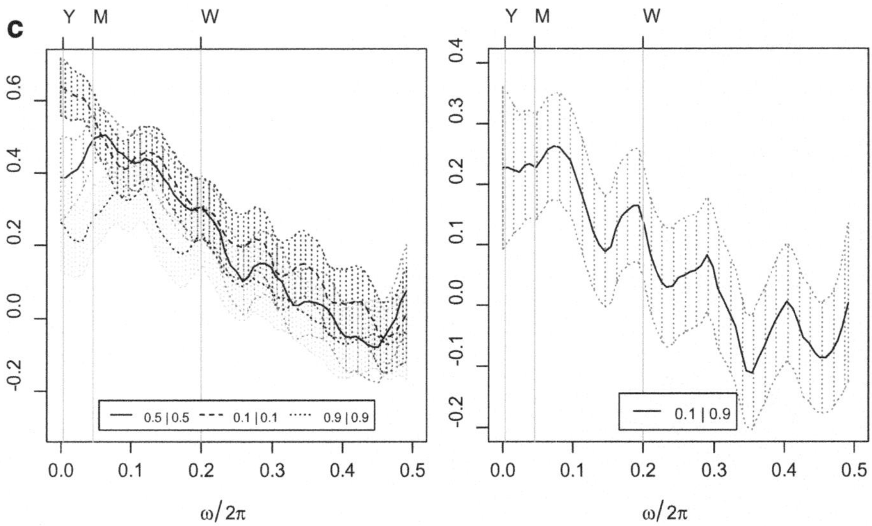

Fig. 10 (continued)

the x-axis in daily cycles and shows the frequencies that correspond to yearly (Y), monthly (M), and weekly (W) periods for the purpose of confirming how weekly, monthly, or yearly cycles are connected across quantiles of the joint distribution. Figure 11 shows the quantile coherency estimates for three fixed yearly, monthly,

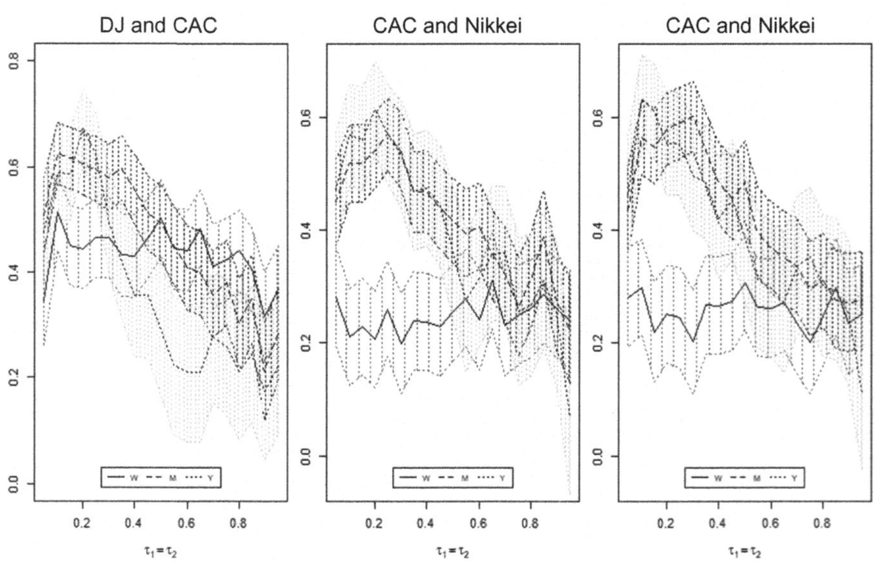

Fig. 11 Quantile coherency for three fixed yearly, monthly, and weekly periods

and weekly periods which correspond to $\omega \in 2\pi\left\{\frac{1}{250}, \frac{1}{22}, \frac{1}{5}\right\}$ at all quantile levels $\tau_1 = \tau_2 \in \{0.05, 0.1, \ldots, 0.95\}$.

Figure 10 shows that the cycles at the lower quantiles relatively have more strong dependence than those at the upper quantiles in all combinations of the markets. The lower quantiles are more strongly related in periods longer than at least one month. The same result can be obtained in Fig. 11. In other words, large negative returns brought by extreme events seem to explain mainly the monthly or yearly common cycles of the concerned markets.

7 Conclusion

This chapter has presented some recent techniques which are becoming more and more influential in the analysis of time series. Quantile spectrum is a concept that can be used when one suspects high nonlinear and non-stationary interactions between financial and economic time series. There are several advantages compared to the standard spectrum. First, one does not need to make any distributional assumptions, especially by imposing the existence of finite moments. Secondly, we can grasp the correlations at different locations of joint distribution of series, by considering their cyclical dynamics. Thirdly, we avoid biased projections of the correlations in time series.

We voluntarily focus on techniques and methodologies that are easy to apply. One approach is based on harmonic regressions and leads to a new family of periodograms and spectra, i.e., Laplace periodograms and spectra. A second approach is based on the Fourier transform of copulas which are widely used in the analysis of the dependence of series. Empirical applications are made easy by the existence of software proposing such analyses (see Kley 2016).

Quantile spectrum analysis opens new areas of research. First, these approaches could be generalized to polyspectrum, when the frequency analysis applies to moments higher than 2. Secondly, quantile-based periodograms can also be used to display correlations in time series with non-Gaussian distributions (Student-t, Weibull, Gumbel, Fréchet, etc.). Recent applications of quantile spectrum for nonlinear and GARCH-type models are studied in Li (2019).

References

Baruník, J., & Kley, T. (2019). Quantile coherency: A general measure for dependence between cyclical economic variables. *Econometrics Journal, 22*(2), 131–152.

Brockwell, P. J., & Davis, R. A. (1991). *Time series: Theory and methods* (2nd ed.). New York: Springer.

Dette, H., Hallin, M., Kley, T., & Volgushev, S. (2015). Of copulas, quantiles, ranks and spectra: An L1-approach to spectral analysis. *Bernoulli, 21*(2), 781–831.

Hagemann, A. (2013). Robust spectral analysis. arXiv e-prints. https://arxiv.org/abs/1111.1965.

Hill, J. B., & McCloskey, A. (2014). Heavy tail robust frequency domain estimation, mimeo.
Hong, Y. (1999). Hypothesis testing in time series via the empirical characteristic function: A general spectral density approach. *Journal of the American Statistical Association, 94*(448), 1201–1220.
Hong, Y. (2000). Generalized spectral tests for serial dependence. *Journal of the Royal Statistical Association. Series B, 62,* 557–574.
Katkovnik, V. (1998). Robust M-periodogram. *IEEE Transactions on Signal Processing, 46*(11), 3104–3109.
Kleiner, B., Martin, R. D., & Thomson, D. J. (1979). Robust estimation of power spectra. *Journal of the Royal Statistical Society B, 41*(3), 313–351.
Kley, T. (2016). Quantile-based spectral analysis in an object-oriented framework and a reference implementation in R: The Quantspect package. *Journal of Statistical Software, 70*(3), 1–27.
Kley, T., Volgushev, S., Dette, H., & Hallin, M. (2016). Quantile spectral processes: Asymptotic analysis and inference. *Bernoulli, 22*(3), 1770–1807.
Klüppelberg, C., & Mikosch, T. (1994). Some limit theory for the self-normalized periodogram of stable processes. *Scandinavian Journal of Statistics, 21*(4), 485–491.
Koenker, R. (2005). *Quantile regression.* Cambridge: Cambridge University Press.
Lee, J., & Subba Rao, S. S. (2012). The quantile spectral density and comparison-based tests for nonlinear time series. arXiv e-prints. https://arxiv.org/abs/1112.2759.
Li, T.-H. (2008). Laplace periodogram for time series analysis. *Journal of the American Statistical Association, 103*(482), 757–768.
Li, T.-H. (2012). Quantile periodograms. *Journal of the American Statistical Association, 107*(498), 765–776.
Li, T.-H. (2013). *Time series with mixed spectra: Theory and methods.* Boca Raton: CRC Press.
Li, T.-H. (2014). Quantile periodogram and time-dependent variance. *Journal of Time Series Analysis, 35*(4), 322–340.
Li, T. H. (2019). Quantile-frequency analysis and spectral divergence metrics for diagnostic checks of time series with nonlinear dynamics. arXiv:1908.02545.
Lim, Y., & Oh, H.-S. (2015). Composite quantile periodogram for spectral analysis. *Journal of Time Series Analysis, 37,* 195–211.
Mikosch, T. (1998). Periodogram estimates from heavy-tailed data. In R. A. Adler, R. Feldman, & M. S. Taqqu (Eds.), *A practical guide to heavy tails: Statistical techniques for analyzing heavy tailed distributions* (pp. 241–258). Boston: Birkhäuser.
Patton, A. J. (2012). A review of copula models for economic time series. *Journal of Multivariate Analysis, 110*(C), 4–18.

On the Seemingly Incompleteness of Exchange Rate Pass-Through to Import Prices: Do Globalization and/or Regional Trade Matter?

Antonia López-Villavicencio and Valérie Mignon

JEL Classification E31 · F31 · F4 · C22

1 Introduction

The exchange rate pass-through (ERPT from now on), understood as the extent to which an exchange rate change is reflected in import and consumer prices, is a central concept in international trade and macroeconomics, both from theoretical and empirical viewpoints [see Knetter (1989), Campa and Goldberg (2005), and Burstein and Gopinath (2013)]. A large body of this literature puts forward that ERPT is incomplete and has been steadily declining over the past few decades.

The partial character of ERPT has received two main explanations: a macroeconomic justification (Monacelli 2005) in which the incompleteness comes from nominal rigidities leading to unresponsiveness in prices in the short run, and a microeconomic explanation linking the incomplete ERPT to an increasing degree of pricing-to-market behavior of firms (Betts and Devereux 2000). A common explanation for the decreasing ERPT is that expectations of inflation have become much more solidly anchored in recent years. Indeed, in the context of a stable and predictable monetary policy environment, nominal shocks play a vastly reduced role in driving fluctua-

A. López-Villavicencio · V. Mignon (✉)
EconomiX-CNRS, University of Paris Nanterre,
200 avenue de la République, 920001 Nanterre, France
e-mail: valerie.mignon@parisnanterre.fr

A. López-Villavicencio
e-mail: alopezvi@parisnanterre.fr

V. Mignon
CEPII, Paris, France

G. Dufrénot and T. Matsuki (eds.), *Recent Econometric Techniques for Macroeconomic and Financial Data*, Dynamic Modeling and Econometrics in Economics and Finance 27, https://doi.org/10.1007/978-3-030-54252-8_2

tions in prices and the exchange rate (Taylor 2000). Thus, a stable monetary policy environment—supported by an institutional framework that allows the central bank to pursue a credible and independent policy—contributes to explaining why even sizeable depreciations of the nominal exchange rate have exerted small effects on prices: when the inflation environment is more stable, firms resist passing exchange rate changes on to prices. Similar arguments are developed in Gagnon and Ihrig (2004), Bailliu and Fujii (2004), Devereux et al. (2004), Ihrig et al. (2006), Marazzi and Sheets (2007), Bouakez and Rebei (2008), Devereux and Yetman (2010) and Dong (2012) where the size of pass-through is a function of the stance of monetary policy.

Recently, López-Villavicencio and Mignon (2017) show that both the level and volatility of inflation, as well as the adoption of an inflation target or the transparency of monetary policy decisions clearly reduce ERPT to consumer prices. However, they find that uncertainty about domestic monetary policy is less relevant for the pass-through to import prices. Other factors than monetary ones may thus be at play in explaining the dynamics of ERPT to import prices.

Based on this evidence, we investigate in this paper if trade integration affects the pass-through to import prices by either increasing the sensitivity to external conditions or by affecting the pricing power of firms. Indeed, many authors have suggested that the process of globalization brought about important changes in the behavior of some major macroeconomic variables such as inflation, output and interest rates (Milani 2012). In particular, the globalization hypothesis, in contrast to the traditional explanation centered on monetary policy credibility, is believed to affect the pass-through of foreign marginal costs and the exchange rate into import prices.

More precisely, two theoretical effects with opposite consequences are at play. According to the first effect, globalization impacts inflation dynamics through its influence on the degree of competition. Specifically, globalization—which refers here to a rising share of goods sold by foreign firms in the domestic market or factors leading to higher trade integration—impacts imported inflation dynamics through its effect on ERPT into import prices. As imported goods represent a large fraction of consumption and intermediate goods, the overall price index becomes more sensitive to external conditions, namely the combined dynamics of the nominal exchange rate and foreign marginal costs. On the whole, the impact of openness on ERPT is positive. The second channel through which globalization influences the dynamics of inflation is, indirectly, *via* its effect on the pricing strategies of domestic firms selling in the internal market. Contrary to the first mechanism, this pricing-power effect negatively impacts ERPT.

While these theoretical effects are clearly established, their outcome is controversial. Specifically, both higher and lower ERPT may result from greater competition. Following Dornbusch (1987) and Benigno and Faia (2016), globalization reflected by greater competition implies higher ERPT. The intensity of exchange rate pass-through depends on the degree of concentration in the market and, in particular, on the share of foreign products in the domestic market. Indeed, greater competition, due to the rise in the share of foreign products sold in a specific industry raises the degree of exchange rate pass-through. Following this mechanism, globalization accentuates the dependence of imported inflation on external conditions, and Benigno and Faia (2016) show that there is evidence of an increase in ERPT degree exactly at the time at which the globalization process took place. Their theoretical results are confirmed by an empirical analysis on US sectoral data providing evidence that ERPT has increased in at least half of the sectors considered, especially after 1999, i.e., after the pick up of trade liberalization.

On the opposite, Gust et al. (2010) argue that increased foreign competition, i.e. greater trade integration, implies lower ERPT. They propose an open economy dynamic general equilibrium model in which strategic complementarity in price setting plays a key role. Indeed, the firm's pricing decisions do not only depend on its marginal costs but also on the prices set by its competitors. This feature implies that it is optimal for a foreign exporter to vary its markup in response to shocks that change the exchange rate, insulating import prices from exchange rate variations. Increased trade integration makes exporters more responsive to the prices of their competitors, leading to a change in pricing behavior that may contribute to the observed decline in the sensitivity of import prices to the exchange rate, lowering the pass-through to prices. Specifically, in their model, Gust et al. (2010) show that an exporter is encouraged to set a quite high and variable markup when its costs are lower than the other—foreign—firms, and a low and inelastic markup when its costs are high. To complement their theoretical model, Gust et al. (2010) provide empirical evidence linking the fall in pass-through to lower trade costs. Using industry-specific measures of pass-through and trade costs, they show that industries in which the decrease in trade costs has been relatively large have also experienced quite important declines in pass-through.

First, instead of considering indiAs shown, the debate related to the theoretical impact of globalization on ERPT is far from being closed. This absence of clear-cut conclusions is accentuated by the fact that globalization encompasses various forms and meanings, such as openness, competition, regionalization, localization, etc., as we will discuss further. Turning to an empirical viewpoint, the literature that explores the link between globalization and ERPT is very scarce, especially in the non-U.S. case. Our aim in this paper is to fill this gap by running an empirical analysis focusing on countries belonging to the eurozone. As import prices constitute a major transmission channel of changes in the euro on domestic prices and, in turn, inflation and output, analyzing ERPT is of crucial importance in the context of a monetary union. The same exchange rate change may affect eurozone countries differently, depending on their openness to trade degree. Accounting for such different responses of import prices to euro exchange rate changes is important for the conduct of the single monetary policy. It is also worthy of interest with regard to the impact of entering into the union and the success of protocols and processes calling for structural reforms in the EMU (European Monetary Union). Some previous studies have been done in the euro area context among which Schroder and Hufner (2002), Anderton (2003), Hahn (2003), Campa et al. (2005), Campa and Gonzàlez-Mínguez (2006), Faruqee (2006), Ben Cheikh and Rault (2016). In this paper, in addition to overcoming the drawback linked to the short time sample used in these studies (with the exception of Ben Cheikh and Rault (2016), we go further than the existing literature in various ways.

First, instead of considering indifferently all countries of the euro area, we focus on three core economies, depending on their external exposure.[1] Specifically, we consider Belgium, which presents the highest degree of trade openness among the core countries, France, which is characterized by the lowest one, and Germany, which is at an intermediate level corresponding to EMU aggregate openness degree.[2] Second, to provide a complete and robust picture, we consider various indicators of globalization: (i) an increase in the degree of trade openness, (ii) a higher intra-industry trade, (iii) a higher presence of Chinese imports over total imports,[3] (iv) lower import tariffs, and (v) higher intra-EU trade as an indicator for regional trade—which can be interpreted as a cost for outside countries but represents a main driver of globalization. Third, as the relevance of globalization in explaining the pass-through dynamics is difficult to assert when using aggregated prices—at it is the case in the bulk of the literature—we also rely on disaggregated data. Using good-level data based on the one-digit Standard International Trade Classification (SITC) enables us to compare ERPT coefficients across goods and should allow us to identify different strategies in the industries. Fourth, an important novelty of the paper is that it specifically accounts for intra-EU trade, i.e. for regional integration. In particular, a large volume of trade being carried inside the euro area, the import prices affecting domestic prices are only those which are not denominated in euro. We thus go further than the previous literature by controlling for such characteristic, allowing us to assess ERPT dynamics for extra-EU trade.

Relying on quarterly data over the 1992Q1–2016Q2 period—2000Q1–2016Q2 for disaggregated data—, our main findings can be summarized as follows. First, while incomplete ERPT is a well-known result in the existing literature, this is not the case when intra-EU trade is excluded from the analysis. Indeed, we show that exchange rate changes tend to be mostly reflected in import prices at both aggregated and disaggregated levels, meaning that ERPT is not far from being complete. This finding calls into question previous results in studies on European economies that do not account for the large volume of intra-EU trade, and highlights the necessity of controlling for regional trade to derive reliable conclusions. Second, interacting exchange rate changes with globalization indicators shows the absence of clear link between openness and ERPT, except for Germany—a country which has experienced an important rise in its trade exposure leading to greater competition. Third, reasoning at the good level, ERPT is found to be high and complete in various sectors, but

---

[1]In addition, it is worth mentioning that a relatively low dispersion of ERPT levels is expected in the euro area, due to the convergence process implied by the monetary union [see, e.g., Ben Cheikh and Rault (2016)].

[2]To provide some figures, the degree of openness to trade in 2015 amounts at 61.4% for France, 86% for Germany, and 164.2% for Belgium; the mean value for the European monetary union being equal to 85% (source: World Bank; trade is the sum of exports and imports of goods and services measured as a share of gross domestic product).

[3]This indicator based on China is used by Marazzi et al. (2005) who show that Chinese booming exports to the United States play a role in explaining the low ERPT value in the U.S.

its degrees differ across countries. Fourth, while there is sparse evidence that globalization impacts the degree of ERPT at the sectoral level, higher trade openness and lower trade tariffs seem to increase ERPT in some sectors. Lastly, our findings show that, when significant, the effect of a rise in intra-EU trade is negative on ERPT in some sectors, particularly those producing manufactured goods and machinery and transport equipment. This result reflects the wish of domestic firms to maintain their market power and protect against foreign competition in these particular sectors.

This paper is organized as follows. Section 2 briefly describes our methodology. Section 3 presents the data and some stylized facts. Section 4 displays our estimation results, and Sect. 5 concludes the paper.

2 Methodology

The existing literature usually models exchange rate pass-through by considering variations of the following equation:

$$\Delta mp_t = \alpha + \sum_{j=1}^{n} \gamma_j \Delta mp_{t-j} + \rho \Delta y_t + \lambda \Delta mc_t^* + \theta \Delta e_t + \epsilon_t \tag{1}$$

where mp represents import prices, y is a local demand factor, mc^* stands for the exporter marginal cost (i.e., the foreign production costs), e is the nominal effective exchange rate, i denotes the industry and t refers to the period. Our primary concern in this equation is the pass-through elasticity, which corresponds to the coefficient on the exchange rate change, namely θ. The case $\theta = 1$ refers to a complete ERPT, corresponding to a one-for-one pass-through changes in import prices. Incomplete ERPT occurs when $\theta < 1$, i.e., when exporters adjust their markup. Equation (1) is estimated at the aggregated (i.e., country) and product levels using, in the latter case, individual fixed effects. All the variables are expressed in logarithms.

To explore the global factors' dimension of pass-through, our empirical strategy consists in extending the benchmark ERPT equation as follows:

$$\begin{aligned} \Delta mp_t = \alpha + \beta_t + \sum_{j=1}^{n} \gamma_j \Delta mp_{t-j} + \rho \Delta y_t + \lambda \Delta mc_t^* \\ + \theta \Delta e_t + \theta^C (\Delta e_t \times \mathrm{C}_t) + C_t + \epsilon_t \end{aligned} \tag{2}$$

where C is an indicator of trade integration: changes in trade openness, changes in intra-industry trade, changes in tariffs for a country's imports, changes in the weight of China in a country i's exports, and changes in intra-EU imports share. In Eq. (2), we interpret a significant coefficient θ^C as evidence that ERPT is affected by global factors.

3 Data

3.1 Time Sample

The period covered in the present study depends on both the availability and the level of disaggregation of data. Indeed, exchange rate pass-through estimates in the literature are usually confronted with a trade-off between sectoral disaggregation level of data and period coverage (Gaulier et al. 2008). Basically, estimates based on aggregated price data allow for a larger time coverage. However, the use of aggregated price series limits the possibility of identifying the structural determinants of the pass-through (to detect differences regarding price discrimination or product differentiation, for instance). Working on disaggregated price data offers more information at the product or good level, but has a cost in terms of data period availability. In this paper, we rely on both aggregated and disaggregated data for three core eurozone countries, namely Belgium, Germany, and France over the period 1992Q1–2016Q2 in the case of aggregated data, and 2000Q1–2016Q2 for a lower level of disaggregation.

3.2 Variables

Regarding the measure of import prices at the aggregated level, we consider extra-EU[4] import unit value indexes taken from the Eurostat Comext database, defined as the ratio between the value of imported goods in monetary terms and the respective quantity of the goods for extra-EU countries. Turning to the disaggregated (i.e., good) level, import unit value indexes of seven sectors (panels) from the Standard International Trade Classification (SITC) industrial good-level data were obtained from the same source. Sub-sections (i.e., panel members) correspond to two-digit sectors or aggregations of them (see Table 11 in Appendix). One industry has been excluded from the analysis, namely mineral fuels, lubricants and related materials (SITC 3), due to the peculiar nature of the sector.[5]

In Eq. (1), both marginal costs and importer's demand characteristics are highly difficult to evaluate since they are not directly observable, so the use of proxies is common in the literature. In our specification, in the spirit of Marazzi et al. (2005) and Marazzi and Sheets (2007), we take the aggregated OECD foreign Producer Price Index (PPI) as the proxy for production costs. For the local demand factor, we use the GDP as it is usually done in the literature [see, e.g., Campa et al. (2005)].

[4]Extra-EU refers to transactions with all the countries outside the EU, namely the rest of the world except the 28 EU member states.

[5]In particular, the industry evolution is more likely to be related to legal changes and natural factors rather than trade.

The exchange rate corresponds to the nominal effective exchange rate provided by the Bank of International Settlements (BIS), an increase in the index indicating a depreciation.

3.3 Indicators of Globalization

Let us now describe our five indicators of globalization. First, at the aggregated and disaggregated levels, trade openness is defined as the sum of extra-EU exports and imports over GDP. Overall and disaggregated trade values are collected from the Comext database. Note that, according to our previous discussion (see Sect. 1), the sign of the ERPT coefficient can be positive or negative. Indeed, following Benigno and Faia (2016), θ^C should be positive and significant if we expect that higher trade openness implies that there are more foreign firms competing in the destination market. In this case, globalization affects the dynamics of imported inflation through its effect on ERPT into import prices, the rise in the share of foreign products in the domestic market increasing the pass-through degree. However, in case of strategic complementarity in setting prices, a foreign exporter does not want its price to deviate too far from its competitors. Thus, the foreign exporter's price becomes more responsive to the prices of its competitors as its markup increases. As a consequence, it is optimal for a firm to vary its markup more and its price less in response to an exchange rate change. Accordingly, we should observe a reduction of the pass-through of exchange-rate changes to import prices with higher trade integration (θ^C should be negative and significant).

Second, we also evaluate how China's presence in total imports may have affected the pricing decisions of exporters from other countries. We, therefore, consider China's imports share over total imports as well as China's share in each SITC sector [see Marazzi et al. (2005)]. Our third global indicator is a measure of intra-industry trade. Here, the underlying hypothesis is that increasing levels of intra-industry trade reflect higher product differentiation with respect to foreign competitors. Indeed, as shown by Caves (1998), product differentiation leads to increasing levels of intra-industry trade among countries, providing opportunities to develop new market-niches. To test for this hypothesis, we employ the Grubel-Lloyd index of intra-industry trade *IIT* [see Lipsey (1976)], which is computed as follows:

$$\mathrm{IIT}_t = 2 \times \frac{\min(M_t; X_t)}{(M_t + X_t)} \tag{3}$$

where M denotes extra-EU imports, and X stands for extra EU-exports (of each SITC sector in the disaggregated analysis) in the considered country. The index ranges between zero (no intra-industry trade) and one (perfect intra-industry trade), and captures the level of product heterogeneity and trade complementarity between each sector-country pair and the trading partners. We interpret an increase in intra-industry trade as an adjustment to trade liberalization. Indeed, as suggested by Colantone et al.

(2015), the *IIT* index is likely to grow following firms' strategic reactions to global integration, in terms of product differentiation and production off-shoring.[6]

Our fourth measure of globalization is based on trade tariffs at the aggregated and sectoral levels. Although tariffs represent only a fraction of overall trade costs, they remain an important underlying factor towards greater trade integration. With this respect, Gust et al. (2010) argue that, with lower costs, foreign exporters should reduce their prices, and the home country's import share should rise. Due to the decrease of foreign exporters' prices relative to their competitors (i.e., the domestic firms), the formers can increase their markups and still gain market share. On the contrary, the prices for domestic goods augment relative to their competitors, and domestic firms have to cut their markups in reaction to stronger competition from abroad. Higher markups on foreign goods reinforce strategic complementarity, and foreign exporters become more willing to vary their markups in response to cost shocks. Thus, according to Gust et al. (2010), a decline in trade costs should cause a fall in the pass-through (i.e., θ^C should be positive). Alternatively, we could argue that exporters who face high tariff rates will face a higher degree of local competition in the markets to which they export and hence will be more limited in passing exchange rate changes onto the prices that they charge. Reasoning this way, we could expect industries protected with higher tariff rates to have lower ERPT (i.e., θ^C should be negative). At the country level, data on import tariff rates for the European Union are retrieved from the UNCTAD Trade Analysis Information System (TRAINS). Data are annual and correspond to the mean effective applied tariff rate of the following non-agricultural and non-fuel products: manufactured goods, ores and metals, chemical products, machinery and transport equipment, and other manufactured goods. Next, we match each SITC sector and sub-sector, with data obtained from the Integrated Database of the World Trade Organization (WTO) at the Harmonized System (HS) 6-digit product. In particular, our measure of tariff corresponds to the average of Add Valorem Duties.[7]

To provide a complete picture, we consider a last measure of globalization reflecting intra-EU trade, which may be interpreted as a regional—i.e., EU-based—integration measure, representing a cost for the outside countries. We define this measure as the ratio of intra-EU imports over total imports. The underlying idea is as follows. Trade barriers having been removed and euro being adopted as a common currency in the euro area, there is a higher proportion of trade in the same currency and, in turn, a smaller share of "output" exposed to exchange-rate fluctuations. These characteristics should affect the way foreign firms pass exchange-rate shocks onto prices as they reduce the market power of exporters outside the eurozone.

[6] For instance, Bernard et al. (2006) present evidence that companies adjust to increasing import pressure by changing their product-mix towards higher value-added goods, characterized by higher export potential and lower intensity of cost-based foreign competition. Moreover, low value-added goods are increasingly imported, in particular from low-wage countries.

[7] We use the applied tariff, which corresponds to the tariff that is actually charged on an import. The corresponding matchings are available upon request from the authors.

3.4 Descriptive Statistics

Tables 1 and 2 provide some descriptive statistics referring to our five different globalization indicators and their growth over the period under study. At the country level (Table 1), Belgium is the country displaying the highest degree of trade openness. However, the level of trade exposure is increasing everywhere, particularly in Germany in addition to Belgium. France is the less opened economy and is the country that exhibits the lowest trade exposure growth. At the industry level (Table 2), openness is relatively higher for chemicals and related products (SITC 5), machinery and transport equipment (SITC 7) and miscellaneous manufactured articles (SITC 8), and it has especially increased for animal and vegetable oils (SITC 4) in France and Germany and miscellaneous manufactured articles (SITC 8) in all countries.

Looking at the figures in Table 2 another trend seems to emerge: Chinese imports account for about 35% in miscellaneous manufactured articles and, while being much lower in other sectors such as chemicals, they have been increasing over time in all panels. This is particularly the case in machinery, even though the average tariff rate increased slightly over the period. Intra-industry trade, in turn, is very heterogeneous among the different sectors in the three countries but, as it is known in the literature, it tends to be higher for manufactured goods than for raw materials or primary goods [see, e.g., Deese (2016)]. Finally, it is interesting to note that intra-EU imports represent more than 70% in several sectors. However, regional trade has decreased in sectors such as manufacturing or chemicals (except in Germany).

4 Results

Let us first consider the estimation of our baseline Eq. (1). The corresponding results are presented in Table 3. As shown, the pass-through estimates present the expected positive and significant sign: an increase in the nominal exchange rate translates into a depreciation of the currency and should normally be followed by a rise in prices. Moreover, ERPT is higher than 0.9 in Belgium and France, meaning that it is not far from being complete. In other words, there are almost a one-to-one

Table 1 Descriptive statistics on globalization and regional indicators at the country level

Country/Panel	Extra-EU trade								Total trade	
	Trade openness		China's share		IIT		Tariffs		Intra-EU imports	
	Level	Growth	Level	Growth	Level	Growth	Level	Growth	Level	Growth
Belgium	0.480	3.310	0.087	4.905	0.953	0.266	2.183	−1.017	0.682	−0.634
France	0.115	2.071	0.088	7.047	0.663	1.800	2.183	−1.017	0.650	−0.111
Germany	0.267	2.846	0.100	5.585	0.897	−0.615	2.183	−1.017	0.539	−0.259

Note This table reports the average values of the globalization indicators over the considered period. Statistics are based on extra-UE trade (i.e., after having removed intra-EU trade) except in the case of intra-EU imports' share

Table 2 Descriptive statistics on globalization and regional indicators at the industry level

	Extra-EU trade								Total trade	
	Trade openness		China's share		IIT		Trade tariffs		Intra-EU imports	
	Level	Growth	Level	Growth	Level	Growth	Level	Growth	Level	Growth
Belgium										
SITC 0–1	0.004	2.244	0.028	2.400	0.519	1.331	9.780	−0.900	0.791	−0.189
SITC 2	0.002	−0.683	0.045	5.133	0.491	1.749	2.483	−0.166	0.585	−0.324
SITC 4	0.000	−3.104	n.a	n.a	0.742	1.980	5.446	−0.328	0.848	1.540
SITC 5	0.014	3.120	0.064	6.990	0.795	−0.474	4.638	−0.214	0.734	−1.039
SITC 6	0.012	−0.284	0.202	8.214	0.771	−0.176	2.697	−1.717	0.673	−1.267
SITC 7	0.011	−0.666	0.155	9.920	0.747	−0.638	2.622	0.483	0.679	−0.735
SITC 8	0.014	2.882	0.343	3.157	0.483	1.671	3.787	0.295	0.481	−1.582
France										
SITC 0–1	0.001	1.171	0.005	0.889	0.432	0.225	9.780	−0.900	0.810	0.228
SITC 2	0.000	−1.159	0.046	4.000	0.629	0.870	2.483	−0.166	0.594	0.197
SITC 4	0.000	3.895	n.a	n.a	0.587	−0.874	5.446	−0.328	0.742	−0.772
SITC 5	0.003	1.151	0.062	5.927	0.648	−1.421	4.638	−0.214	0.735	−0.172
SITC 6	0.001	−0.848	0.193	8.522	0.723	0.692	2.697	−1.717	0.813	−0.116
SITC 7	0.006	−2.177	0.169	10.805	0.730	−0.246	2.622	0.483	0.723	0.222
SITC 8	0.007	3.342	0.386	3.201	0.383	2.035	3.787	0.295	0.678	−0.033
Germany										
SITC 0–1	0.001	2.719	0.055	3.797	0.563	1.119	9.780	−0.900	0.763	0.273
SITC 2	0.000	−0.064	0.106	0.351	0.646	2.013	2.483	−0.166	0.663	0.921
SITC 4	0.000	6.708	n.a	n.a	0.366	0.260	5.446	−0.328	0.740	−0.132
SITC 5	0.003	3.129	0.087	5.628	0.614	−0.274	4.029	−0.290	0.757	0.261
SITC 6	0.003	0.746	0.185	7.165	0.779	−0.275	2.697	−1.717	0.695	−0.099
SITC 7	0.003	3.129	0.087	5.629	0.614	−0.274	2.622	0.483	0.757	0.261
SITC 8	0.006	2.550	0.350	3.125	0.604	0.061	3.787	0.295	0.487	0.593

Note This table reports the average values of the globalization indicators at the industry level over the considered period. Statistics are based on extra-UE trade (i.e., after having removed intra-EU trade) except in the case of intra-EU imports' share. SITC 0 and 1: Food, beverages and tobacco, SITC 2: Crude materials, inedible, except fuels, SITC 4: Animal and vegetable oils, fats and waxes, SITC 5: Chemicals and related products, SITC 6: Manufactured goods, SITC 7: Machinery and transport equipment, SITC 8: Miscellaneous manufactured articles

Table 3 ERPT coefficients at the country level (extra-EU trade)

	1992Q1–2016Q4		After 2002Q1
	Coeff. (*t*-stat)	Coeff. (*t*-stat)	Coeff (*t*-stat)
Belgium			
θ	0.931 (2.76)	0.883 (2.61)	1.397 (2.76)
Dummy Euro		−1.608 (−1.03)	
France			
θ	0.932 (2.81)	0.961 (2.90)	1.287 (2.90)
Dummy Euro		1.054 (1.13)	
Germany			
θ	0.698 (3.85)	0.763 (3.97)	1.192 (5.16)
Dummy Euro		2.01 (1.14)	

Notes This table reports the estimated ERPT coefficients from Eq. (1). Corresponding t-statistics are given between parentheses. Estimation based on extra-UE trade (i.e., after having removed intra-EU trade). Dummy euro is a dummy variable that takes the value of 1 for 2002Q1 and all subsequent periods, zero otherwise

pass-through changes in import prices. For the sake of completeness, Table 3 also displays the results regarding the estimation of Eq. (1) (i) when a dummy variable for the introduction of the euro is included,[8] and (ii) if we restrict the sample size to the 2002Q1-2016Q4 period. As shown, while the dummy variable is non-significant except for Germany, ERPT degree has strongly increased in all countries after the adoption of the common currency, in line with the results of Benigno and Faia (2016). With regard to the statistics displayed in Tables 1 and 2, this rise in ERPT degree may be partly explained by the reduction in intra-EU trade.

These findings highlight the importance, and even the necessity, of controlling for intra-EU trade to correctly assess the degree of ERPT. To further illustrate and for comparative purposes, Table 12 in Appendix shows the estimation results when intra-EU trade is not controlled for (i.e., when total trade is considered). These results confirm previous findings of the literature regarding the incompleteness of ERPT [see Engel (2002), Campa and Goldberg (2005), Marazzi and Sheets (2007), Bouakez and Rebei (2008) and Gust et al. (2010)]. ERPT is even found to be non significant in France. Clearly, excluding intra-EU trade is thus an imperative to correctly evaluate

[8]Specifically, Eq. (1) is written as follows:

$$\Delta mp_t = \alpha + \sum_{j=1}^{n} \gamma_j \Delta mp_{t-j} + \rho \Delta y_t + \lambda \Delta mc_t^* + \theta \Delta e_t + \delta DUM_t + \epsilon_t \tag{4}$$

where DUMT_t is the dummy variable that takes the value of 1 for 2002Q1 and all subsequent periods, zero otherwise.

the exchange-rate pass-through in European countries: not discarding intra-EU trade when assessing the degree of ERPT to import prices strongly biases the results in favor of incomplete pass-through.

4.1 Accounting for Globalization

To assess the role of globalization at the country level, Table 4 reports the estimation results of Eq. (2). We consider the five aforementioned indicators in favor of globalization, namely: (i) an increase in the degree of trade openness, (ii) a higher intra-industry trade, (iii) a higher presence of Chinese imports over total imports, (iv) lower import tariffs, and (v) higher intra-EU trade as a regional globalization measure.

As shown, the interactive effect between exchange rate changes and globalization is mostly non-significant. This means that an increase in product differentiation with respect to foreign competitors, in the share of products from China in total imports and in the share of intra-UE imports, as well as a decline in import tariffs do not contribute to explaining the ERPT to import prices, at least at the aggregate level. The sole significant interactive terms are obtained when considering growth in trade openness and intra-industry trade for Germany and a higher presence of Chinese imports over total imports in France, suggesting that globalization tends to slightly increase the ERPT degree. Regarding Germany, these findings can be related to the fact that the German economy has known an important rise in trade openness over the period under study, leading to greater competition. In line with the argument developed by Benigno and Faia (2016), this growing competition due to the increase in the share of foreign products pushes up the degree of ERPT. In other words, there is slight evidence that the impact of firms' entry on pass-through outweighs the effect of markup adjustments at the intensive margin.

While overall evidence regarding the pass-through effect of global factors is quite weak, it is worth noticing that their impact can operate through other channels. For instance, as recalled by Marazzi and Sheets (2007), pricing decisions of exporters from other countries may have been affected by the efforts made to remain competitive against China. The Chinese economy has also proven its high capacity to win market share, making credible the threat of its potential competition and constraining other exporters from passing through exchange rate shocks. Besides and at a more general level, if there is heterogeneity prevailing at the industry level, results based on aggregated import prices present aggregation bias, suggesting the importance of assessing ERPT degree at the sectoral level with disaggregated data.

Table 4 Global factors and ERPT coefficients at the country level

	Growth in trade openness		Growth in China's imports share		Growth in intra-industry trade		Growth in tariffs		Growth in intra-EU imports share	
	θ	θ^C	θ	θ^C	θ	θ^C	θ	θ^C	θ	θ^C
	Coeff. (t-stat)	Coeff. (t-stat)	Coeff. (t-stat)	Coeff. (t-stat)	Coeff. (t-stat)	Coeff. (t-stat)	Coeff. (t-stat)	Coeff. (t-stat)	Coeff. (t-stat)	Coeff.
Belgium	1.683 (3.43)	−0.006 (−0.19)	0.869 (2.32)	0.011 (0.36)	1.556 (3.73)	−0.036 (−0.33)	0.890 (2.52)	−0.025 (−1.76)	0.858 (2.12)	0.005 (0.05)
France	0.200 (0.47)	0.030 (0.97)	0.871 (1.87)	0.072 (2.05)	1.385 (3.64)	−0.094 (−1.70)	1.672 (3.82)	0.011 (0.62)	0.566 (1.67)	−0.087 (−0.74)
Germany	0.154 (0.79)	0.035 (2.01)	0.750 (3.29)	−0.004 (−0.21)	0.756 (4.25)	0.130 (2.47)	0.879 (4.24)	0.016 (1.78)	0.529 (2.74)	−0.43 (−0.02)

Notes This table reports the estimated ERPT coefficients from Eq. (2). Corresponding t-statistics are given between parentheses. Estimation based on extra-UE trade (i.e., after having removed intra-EU trade)

4.2 Using Disaggregated Data: Accounting for the Good Level

To complement our previous country-level results, let us now estimate ERPT into import prices at a disaggregated level, using the two-digit level of disaggregation in the SITC classification. Analyzing ERPT at the good level allows us to account for the fact that the shift in the composition of imports towards goods whose prices are less sensitive to exchange rate changes has contributed to the "seemingly" pass-through decline. The corresponding results are reported in Table 5.[9]

As shown, ERPT is found to be quite high or even complete in most sectors. These findings again illustrate the importance of controlling for intra-EU trade in assessing the effect of exchange rate changes to import prices.[10] However, the estimates strongly vary depending on the type of goods. The highest ERPT coefficients are generally obtained for goods belonging to SITC 8, SITC 7 and SITC 2 which are the sectors the most commodity-intensive. On the whole, the exchange rate effect on the prices of imported goods comes principally through its indirect effect on commodity prices: in commodity-intensive sectors, foreign producers have strong market power and face very weak domestic competition, and, consequently, the world price passed on when the exchange rate fluctuates. The declining share of commodity-intensive goods for which ERPT is higher than for other goods, may thus explain the declining pass-through reported in several studies (see, e.g. López-Villavicencio and Mignon (2017) and the references therein). For some industries, such as those concerned with manufactured goods (SITC 6), the pass-through strongly differs between countries—the value of ERPT degree for Belgium being about two times that of France.[11] This can be explained by the fact that these industries are more oriented towards product differentiation, leading to distinct ERPT degrees in different countries.

4.3 Accounting for Globalization at the Good Level

Tables 6, 7, 8, 9 and 10 display the estimation results at the disaggregated level of Eq. (2). As shown, there is no clear-cut evidence regarding how global factors affect the way foreign exporters pass-through increasing costs to their prices. Indeed, when significant, the interactive term mostly indicates that higher trade openness or lower

[9]Note that at the disaggregated level we use panel data techniques. The equations are then estimated by the GMM one-step estimator for each SITC sector and the panel members are the divisions in each Section.

[10]Indeed, working with overall unit value indices at a disaggregated level for the euro area countries, Campa et al. (2005) find an ERPT rarely higher than 60–70%.

[11]It is worth mentioning that the share of manufactured goods ranks at the first place in the import structure of Belgium (in 2015, source: Eurostat), highlighting the fact that a link may exist between ERPT degree and the structure of imports of the considered countries. This hypothesis is also supported by the fact that for Germany and France, high ERPT degrees are observed for industries belonging to SITC 8 and 7, which also play a key role in the import structure of these two countries.

Table 5 ERPT coefficients at the good level (SITC classification)

Country	SITC 0&1 Coeff. (*t*-stat)	SITC 2 Coeff. (*t*-stat)	SITC 4 Coeff. (*t*-stat)	SITC 5 Coeff. (*t*-stat)	SITC 6 Coeff. (*t*-stat)	SITC 7 Coeff. (*t*-stat)	SITC 8 Coeff. (*t*-stat)
Belgium	0.989 (6.49)	1.575 (5.92)	0.934 (3.12)	1.342 (5.03)	1.326 (5.52)	1.366 (9.25)	2.098 (7.35)
J-stat	0.146	0.478	0.815	0.278	0.994	0.899	0.162
AR(2)	0.312	0.773	0.913	0.663	0.349	0.343	0.062
No. obs.	684	513	171	456	513	497	456
France	0.711 (10.5)	0.712 (8.20)	1.021 (3.69)	0.654 (4.79)	0.632 (12.8)	0.873 (7.49)	1.413 (16.40)
J-test	0.429	0.899	0.215	0.258	0.995	0.789	0.300
AR(2)	0.574	0.915	0.417	0.246	0.184	0.238	0.074
No. obs.	684	513	171	456	513	513	399
Germany	0.702 (5.85)	0.783 (5.55)	0.655 (3.19)	0.677 (4.16)	0.491 (6.01)	0.904 (6.78)	1.233 (8.85)
J-test	0.791	0.357	0.998	0.296	0.900	0.700	0.739
AR(2)	0.706	0.383	0.093	0.926	0.230	0.046	0.944
No. obs.	684	513	171	456	513	513	456

Notes (a) This table reports the estimated ERPT coefficients from Eq. (1), (b) Corresponding *t*-statistics are given between parentheses, (c) Estimation based on extra-UE trade (i.e., after having removed intra-EU trade), (d) SITC 0 and 1: Food, beverages and tobacco, SITC 2: Crude materials, inedible, except fuels, SITC 4: Animal and vegetable oils, fats and waxes, SITC 5: Chemicals and related products, SITC 6: Manufactured goods, SITC 7: Machinery and transport equipment, SITC 8: Miscellaneous manufactured articles, (e) The null hypothesis of the J-test is the validity of instruments, (f) The null hypothesis of the AR(2) test is the absence of serial autocorrelation of order 2

tariffs increase the ERPT in some cases. However, higher intra-industry trade and, above all, more regional trade reduce the ERPT. Chinese import share, in turn, seems to play no major role in the pricing decisions of foreign exporters.

More in detail, regarding trade openness, the interactive term is positive and significant for goods belonging to SITC 2 (crude materials, inedible, except fuels) for all the countries, SITC 4 (animal and vegetable oils) except in France, and SITC 7 (machinery and transport equipment) for France. In those cases, greater competition is thus associated with higher ERPT, in line with the arguments developed by Dornbusch (1987) and Benigno and Faia (2016).

Turning to intra-industry trade, when significant, the associated coefficient is mostly negative. This means that for the concerned sectors, higher intra-industry trade tends to lower ERPT. This result is consistent with the fact that for sectors characterized by high levels of intra-industry trade, firms react in terms of product differentiation, leading to lower ERPT. A typical example is given by the German case for which machinery and transport equipment sector plays a central role worldwide in the sense that the negative sign reflects the aim of Germany to preserve its market shares.

Our findings also illustrate that Chinese firms' market penetration has not caused a structural change in ERPT, the interactive coefficient being rarely significant, except

Table 6 ERPT and growth in trade openness at the good level

SITC	Belgium		France		Germany	
	θ	θ^C	θ	θ^C	θ	θ^C
	Coeff.	Coeff.	Coeff.	Coeff.	Coeff.	Coeff.
	(t-stat)	(t-stat)	(t-stat)	(t-stat)	(t-stat)	(t-stat)
SITC 0 & 1	0.931 (6.52)	0.008 (1.62)	0.674 (8.20)	−0.009 (−2.51)	0.625 (5.93)	0.006 (1.24)
J-stat	0.134		0.394		0.062	
AR(2)	0.643		0.333		0.803	
SITC 2	1.471 (5.85)	0.008 (2.10)	0.561 (7.08)	0.012 (4.56)	0.669 (5.24)	0.010 (2.70)
J-stat	0.589		0.998		0.302	
AR(2)	0.476		0.549		0.451	
SITC 4	1.603 (4.13)	0.029 (4.73)	0.542 (9.69)	−0.005 (−0.96)	0.994 (4.52)	0.020 (4.71)
J-stat	1.00		0.998		0.988	
AR(2)	0.461		0.243		0.255	
SITC 5	1.261 (3.94)	−0.005 (−0.48)	0.596 (4.53)	0.008 (1.20)	0.785 (9.26)	−0.023 (−2.05)
J-stat	0.325		0.270		0.379	
AR(2)	0.647		0.287		0.476	
SITC 6	1.252 (6.39)	−0.009 (−1.30)	0.516 (5.10)	0.001 (0.19)	0.446 (4.51)	0.002 (0.45)
J-stat	0.995		0.378		0.818	
AR(2)	0.424		0.212		0.096	
SITC 7	1.355 (8.95)	0.005 (0.86)	0.797 (8.16)	0.015 (1.99)	0.764 (6.19)	0.005 (0.99)
J-stat	0.685		0.723		0.608	
AR(2)	0.316		0.240		0.061	
SITC 8	1.853 (5.53)	−0.018 (−1.74)	1.262 (11.60)	−0.012 (−1.88)	1.265 (7.28)	−0.005 (−0.54)
J-stat	0.635		0.998		0.972	
AR(2)	0.190		0.087		0.350	

Notes This table reports the estimated ERPT coefficients from Eq. (2). Corresponding t-statistics are given between parentheses. Estimation based on extra-UE trade (i.e., after having removed intra-EU trade)

in some special cases such as manufactured goods and miscellaneous manufactures in France and Germany, which correspond to sectors in which China is highly competitive.[12] This means that an increase in the share of China tends to weaken ERPT, reflecting a threat from competition with China in these sectors. Those findings may reflect the wish of domestic firms to preserve their market power and protect against foreign competition in these particular sectors.

[12]It is worth mentioning that the manufactured goods sector plays a key role in the Chinese economy since (i) it contributes, with the construction sector, nearly half of China's GDP, and (ii) it is highly competitive and export-oriented.

Table 7 ERPT and growth in intra-industry trade at the good level

SITC	Belgium		France		Germany	
	θ	θ^C	θ	θ^C	θ	θ^C
	Coeff.	Coeff.	Coeff.	Coeff.	Coeff.	Coeff.
	(t-stat)	(t-stat)	(t-stat)	(t-stat)	(t-stat)	(t-stat)
SITC 0 & 1	0.983 (6.34)	−0.009 (−2.29)	0.713 (10.01)	0.002 (0.58)	0.703 (6.06)	−0.002 (−0.56)
J-stat	0.154		0.420		0.065	
AR(2)	0.287		0.584		0.726	
SITC 2	1.700 (6.77)	0.001 (0.31)	0.727 (8.47)	−0.007 (−2.56)	0.798 ((5.18)	−0.006 (−0.89)
J-stat	0.516		0.899		0.338	
AR(2)	0.984		0.964		0.353	
SITC 4	1.248 (4.17)	−0.012 (−2.56)	0.651 (8.24)	0.016 (9.63)	0.835 (3.57)	0.004 (0.83)
J-stat	0.999		0.989		0.999	
AR(2)	0.398		0.411		0.302	
SITC 5	1.432 (6.67)	−0.029 (−4.68)	0.675 (5.96)	0.007 (1.26)	0.662 (4.61)	0.004 (1.76)
J-stat	0.375		0.433		0.357	
AR(2)	0.840		0.642		0.857	
SITC 6	1.331 (5.72)	0.009 (1.33)	0.623 (8.42)	−0.004 (−1.34)	0.516 (8.42)	0.001 (0.09)
J-stat	0.992		0.354		0.835	
AR(2)	0.299		0.229		0.145	
SITC 7	1.379 (9.18)	0.007 (1.69)	0.769 (6.45)	0.006 (0.84)	0.893 (6.88)	−0.027 (−3.64)
J-stat	0.999		0.495		0.752	
AR(2)	0.353		0.245		0.075	
SITC 8	1.900 (5.72)	−0.006 (−2.65)	1.289 (12.50)	−0.003 (−0.96)	1.291 (7.26)	−0.007 (−0.83)
J-stat	0.537		0.994		0.968	
AR(2)	0.190		0.069		0.414	

Notes This table reports the estimated ERPT coefficients from Eq. (2). Corresponding t-statistics are given between parentheses. Estimation based on extra-UE trade (i.e., after having removed intra-EU trade).

However, as an illustration that globalization can act in different ways and has many sides, note the effect of trade tariffs in manufacturing goods (SITC 6): reducing import tariffs clearly increases the ERPT in all the three countries. Contrary to the complementarity hypothesis, what seems to be happening is that exporters who face low tariff rates may also face a low degree of local competition in the markets to which they export and hence will be less limited in passing exchange rate changes onto the prices that they charge. Note that this is also the case in crude materials, inedible, except fuels (SITC 2).

Table 8 ERPT and growth in China's imports at the good level

SITC	Belgium		France		Germany	
	θ	θ^C	θ	θ^C	θ	θ^C
	Coeff.	Coeff.	Coeff.	Coeff.	Coeff.	Coeff.
	(t-stat)	(t-stat)	(t-stat)	(t-stat)	(t-stat)	(t-stat)
SITC 0 & 1	1.125 (10.3)	−0.003 (−2.56)	0.718 (12.5)	0.006 (1.80)	0.719 (8.01)	0.003 (1.49)
J-stat	0.780		0.290		0.744	
AR(2)	0.826		0.242		0.134	
SITC 2	1.875 (11.2)	−0.001 (−0.24)	0.679 (6.27)	−0.001 (−0.65)	0.574 (6.55)	−0.004 (−1.39)
J-stat	0.998		0.091		0.537	
AR(2)	0.432		0.516		0.245	
SITC 5	1.579 (6.03)	0.002 (0.91)	0.839 (5.73)	−0.002 (−1.36)	0.846 (9.95)	0.000 (−0.25)
J-stat	0.912		0.250		0.985	
AR(2)	0.757		0.390		0.183	
SITC 6	1.411 (7.45)	−0.002 (−0.17)	0.663 (10.2)	−0.007 (−2.90)	0.525 (6.52)	−0.005 (−1.42)
J-stat	0.988		0.540		0.906	
AR(2)	0.059		0.199		0.065	
SITC 7	1.202 (8.50)	0.007 (3.87)	0.894 (8.31)	−0.002 (−1.49)	0.928 (6.63)	−0.003 (−1.51)
J-stat	0.151		0.812		0.989	
AR(2)	0.100		0.224		0.140	
SITC 8	1.919 (5.95)	−0.007 (−0.54)	1.327 (12.70)	−0.011 (−1.85)	1.367 (7.74)	−0.010 (−2.36)
J-stat	0.325		0.995		0.974	
AR(2)	0.042		0.041		0.713	

Notes This table reports the estimated ERPT coefficients from Eq. (2). Corresponding t-statistics are given between parentheses. Estimation based on extra-UE trade (i.e., after having removed intra-EU trade)

Finally, even though the effects of globalization in the ERPT are not clear-cut, regional trade decreases ERPT to import prices in a more generalized way. Indeed, when significant, the coefficient of the interaction term is always negative, meaning that a higher presence of intra-EU imports in all sectors, but SITC 5 tends to lower ERPT.

5 Conclusion

Assessing the degree of exchange rate pass-through (ERPT) into import prices in eurozone countries is worthy of investigation. Indeed, import prices being a key channel through which exchange rate changes influence domestic prices and, in

Table 9 ERPT and growth in import tariffs at the good level

SITC	Belgium		France		Germany	
	θ	θ^C	θ	θ^C	θ	θ^C
	Coeff. (*t*-stat)	Coeff. (*t*-stat)	Coeff. (*t*-stat)	Coeff. (*t*-stat)	Coeff. (*t*-stat)	Coeff. (*t*-stat)
SITC 0 & 1	0.993 (5.62)	−0.047 (−2.12)	0.737 (9.29)	−0.001 (−0.07)	0.664 (5.11)	−0.041 (−1.55)
J-stat	0.623		0.880		0.548	
AR(2)	0.600		0.440		0.975	
SITC 2	1.375 (3.73)	−0.099 (−3.95)	0.688 (5.30)	−0.055 (−2.35)	0.836 (3.90)	−0.104 (−2.10)
J-stat	0.999		0.989		0.998	
AR(2)	0.510		0.243		0.956	
SITC 4	0.706 (2.36)	−0.353 (−1.23)	1.242 (5.88)	0.172 (1.96)	0.556 (2.67)	0.003 (0.01)
J-stat	0.998		0.944		0.999	
AR(2)	0.685		0.579		0.093	
SITC 5	1.363 (5.21)	0.001 (0.54)	0.637 (4.48)	0.000 (−0.20)	0.719 (5.87)	0.001 (2.07)
J-stat	0.309		0.238		0.284	
AR(2)	0.617		0.248		0.744	
SITC 6	1.517 (6.12)	−0.126 (−4.54)	0.657 (9.75)	−0.010 (−3.75)	0.529 (7.42)	−0.010 (−3.45)
J-stat	0.806		0.989		0.990	
AR(2)	0.784		0.582		0.134	
SITC 7	1.417 (6.80)	0.013 (0.65)	0.905 (9.83)	−0.090 (−1.55)	0.998 (9.07)	−0.004 (−0.05)
J-stat	0.681		0.996		0.993	
AR(2)	0.335		0.245		0.088	
SITC 8	2.124 (7.20)	−0.081 (−1.22)	1.415 (9.48)	−0.053 (−1.43)	1.256 (8.62)	−0.033 (−1.06)
J-stat	0.213		0.909		0.783	
AR(2)	0.433		0.055		0.996	

Notes This table reports the estimated ERPT coefficients from Eq. (2). Corresponding *t*-statistics are given between parentheses. Estimation based on extra-UE trade (i.e., after having removed intra-EU trade)

turn, inflation and output, evaluating the degree of ERPT is a crucial issue within the context of a monetary union. The same variation in the euro exchange rate change may affect eurozone countries differently, depending on their openness to trade degree. We tackle this issue in the present paper by analyzing ERPT into import prices for three core eurozone countries, namely Belgium, France and Germany, which are characterized by various openness degrees. With protectionism on the rise, the question becomes even more relevant.

Relying on a battery of indicators, we show that globalization plays a limited role in explaining ERPT at the aggregated, country level. The main noticeable exception is Germany, for which higher trade openness and intra-industry trade push up the degree of ERPT. Germany has experienced an important rise in trade openness over

Table 10 ERPT and growth in intra-EU trade at the good level

SITC	Belgium		France		Germany	
	θ	θ^C	θ	θ^C	θ	θ^C
	Coeff.	Coeff.	Coeff.	Coeff.	Coeff.	Coeff.
	(t-stat)	(t-stat)	(t-stat)	(t-stat)	(t-stat)	(t-stat)
SITC 0 & 1	1.057 (7.54)	−0.020 (−2.24)	0.623 (11.4)	−0.422 (−2.85)	0.596 (5.46)	−0.022 (−1.96)
J-stat	0.635		0.973		0.236	
AR(2)	0.239		0.277		0.051	
SITC 2	1.596 (5.62)	−0.016 (−3.70)	0.661 (7.84)	−0.008 (−1.22)	0.790 (6.10)	−0.009 (−0.31)
J-stat	0.995		0.685		0.358	
AR(2)	0.793		0.768		0.702	
SITC 4	1.764 (3.76)	−0.113 (−3.68)	0.605 (5.11)	0.023 (1.19)	0.850 (5.93)	0.006 (0.29)
J-stat	0.989		0.999		1.000	
AR(2)	0.264		0.348		0.220	
SITC 5	1.372 (4.63)	0.028 (1.02)	0.640 (3.61)	−0.009 (−0.81)	0.685 (5.42)	0.013 (0.96)
J-stat	0.291		0.999		0.310	
AR(2)	0.837		0.227		0.982	
SITC 6	1.324 (6.18)	−0.002 (−0.17)	0.555 (11.40)	−0.111 (−2.78)	0.563 (7.11)	−0.014 (−2.38)
J-stat	0.992		0.310		0.851	
AR(2)	0.777		0.215		0.058	
SITC 7	1.355 (8.99)	−0.007 (−2.55)	0.822 (7.59)	−0.044 (−13.20)	0.875 (6.99)	0.020 (1.30)
J-stat	0.673		0.837		0.614	
AR(2)	0.350		0.209		0.058	
SITC 8	1.879 (3.80)	0.020 (1.73)	1.362 (12.00)	−0.049 (−2.28)	1.214 (6.98)	0.014 (0.88)
J-stat	0.999		0.098		0.999	
AR(2)	0.068		0.044		0.048	

Notes This table reports the estimated ERPT coefficients from Eq. (2). Corresponding t-statistics are given between parentheses

the period under study, leading to greater competition coming from an increasing share of foreign products in the market and, in turn, raising ERPT. At the disaggregated, sectoral level ERPT degrees differ across countries, and there is overall sparse evidence that globalization impacts the pass-through. However, increased globalization apprehended through a rise in the degree of trade openness and a reduction in trade tariffs results in higher ERPT in some sectors. Turning to the "Chinese effect", it is very limited but, when significant, it has the expected sign: a rise in the share of China tends to weaken ERPT, illustrating a threat from competition with this country in the concerned sectors. Overall, even though the effects of globalization on ERPT are quite sparse, we show that regional trade decreases the degree of exchange-rate pass-through in a more generalized way.

Most importantly, our paper puts forward the importance, and even the necessity, to control for intra-EU trade when assessing ERPT into import prices in the case of eurozone countries. Specifically, we show that while ERPT is generally found to be incomplete in the literature, this conclusion is strongly called into question when intra-EU trade is accounted for. Indeed, we find evidence that ERPT into import prices dramatically increases when considering only extra-EU trade, at both the country and the good levels. In most cases, ERPT is close to be complete as exchange rate changes tend to be fully reflected in import prices. In this sense, incompleteness generally observed in the literature is in appearance only and not at play for eurozone countries when intra-EU trade is controlled for.

On the whole, our results show that ERPT into import prices is significant and complete in numerous sectors, meaning that exchange rate changes still exert very important pressure on domestic prices in the considered eurozone economies. In addition, the responses of import prices to euro exchange rate variations differ across countries and sectors, a characteristic which has to be taken into account for the conduct of the single monetary policy.

Acknowledgements We would like to thank Anne-Laure Delatte and Sébastien Jean for helpful remarks and suggestions.

6 Appendix

Table 11 SITC classification

Section (Panel) and Division (Panel members)
0. Food and live animals
00 - Live animals
01 - Meat and meat preparations
02 - Dairy products and birds' eggs
03 - Fish (not marine mammals), crustaceans, etc.
04 - Cereals and cereal preparations
05 - Vegetables and fruit
06 - Sugars, sugar preparations and honey
07 - Coffee, tea, cocoa, spices, and manufactures thereof
08 - Feeding stuff for animals (not including unmilled cereals)
09 - Miscellaneous edible products and preparations

(continued)

Table 11 (continued)

1. Beverages and tobacco
11 - Beverages
12 - Tobacco and tobacco manufactures
2. Crude materials, inedible, except fuels
21 - Hides, skins and furskins, raw
22 - Oil-seeds and oleaginous fruits
23 - Crude rubber (including synthetic and reclaimed)
24 - Cork and wood
25 - Pulp and waste paper
26 - Textile fibres
27 - Crude fertilizers and crude minerals
28 - Metalliferous ores and metal scrap
29 - Crude animal and vegetable materials, n.e.s.
4. Animal and vegetable oils, fats and waxes
41 - Animal oils and fats
42 - Fixed vegetable fats and oils, crude, refined or fractionated
43 - Animal or vegetable fats and oils, processed, etc
4. Chemicals and related products, n.e.s.
51 - Organic chemicals
52 - Inorganic chemicals
53 - Dyeing, tanning and colouring materials
54 - Medicinal and pharmaceutical products
55 - Essential oils, etc
56 - Fertilizers
57 - Plastics in primary forms
58 - Plastics in non-primary forms
59 - Chemical materials and products, n.e.s.
5. Manufactured goods classified chiefly by material
61 - Leather, leather manufactures, n.e.s., and dressed furskins
62 - Rubber manufactures, n.e.s.
63 - Cork and wood manufactures (excluding furniture)
64 - Paper, paperboard and articles of paper pulp, of paper or of paperboard
65 - Textile yarn, fabrics, made-up articles, n.e.s., and related products
66 - Non-metallic mineral manufactures, n.e.s.
67 - Iron and steel
68 - Non-ferrous metals
69 - Manufactures of metals, n.e.s.

(continued)

Table 11 (continued)

6. Machinery and transport equipment
71 - Power-generating machinery and equipment
72 - Machinery specialized for particular industries
73 - Metalworking machinery
74 - General industrial machinery and equipment
75 - Office machines and automatic data-processing machines
76 - Telecommunications and sound-recording and reproducing apparatus and equipment
77 - Electrical machinery, apparatus and appliances, n.e.s.
78 - Road vehicles (including air-cushion vehicles)
79 - Other transport equipment
7. Miscellaneous manufactured articles
81 - Prefabricated buildings; sanitary, plumbing, etc.
82 - Furniture, and parts thereof, etc
83 - Travel goods, handbags and similar containers
84 - Articles of apparel and clothing accessories
85 - Footwear
87 - Professional, scientific and controlling instruments and apparatus
88 - Photographic apparatus, equipment and supplies and optical goods, etc
89 - Miscellaneous manufactured articles, n.e.s.

Table 12 ERPT coefficients at the country level. Total trade

	1992Q1–2016Q4		After 2002Q1
	Coeff. (t-stat)	Coeff. (t-stat)	Coeff (t-stat)
Belgium			
θ	0.402 (2.17)	0.427 (2.29)	0.760 (2.11)
Dummy Euro		0.934 (1.10)	
France			
θ	0.196 (0.98)	0.187 (0.93)	0.076 (0.22)
Dummy euro		0.722 (0.67)	
Germany			
θ	0.543 (3.77)	0.545 (3.78)	0.903 (3.70)
Dummy euro		0.516 (0.59)	

Notes This table reports the estimated ERPT coefficients from Eq. (1). Corresponding t-statistics are given between parentheses. Estimation based on total trade (i.e., without excluding intra-EU trade). Dummy euro is a variable that takes the value of 1 for 2002Q1 and all subsequent periods, zero otherwise

References

Anderton, R. (2003). *Extra-euro area manufacturing import prices and exchange rate pass-through*. Working Paper Series 0219, European Central Bank.

Bailliu, J., & Fujii, E. (2004). *Exchange rate pass-through and the inflation environment in industrialized countries: An empirical investigation*. Staff Working Papers 04-21, Bank of Canada.

Ben Cheikh, N., & Rault, C. (2016). Recent estimates of exchange rate pass-through to import prices in the euro area. *Review of World Economics (Weltwirtschaftliches Archiv)*, *152*(1), 69–105.

Benigno, P., & Faia, E. (2016). Globalization, pass-through, and inflation dynamics. *International Journal of Central Banking*, *12*(4), 263–306.

Bernard, A. B., Jensen, J. B., & Schott, P. K. (2006). Survival of the best fit: Exposure to low-wage countries and the (uneven) growth of U.S. manufacturing plants. *Journal of International Economics*, *68*(1), 219–237.

Betts, C., & Devereux, M. B. (2000). Exchange rate dynamics in a model of pricing-to-market. *Journal of International Economics*, *50*(1), 215–244.

Bouakez, H., & Rebei, N. (2008). Has exchange rate pass-through really declined? Evidence from Canada. *Journal of International Economics*, *75*(2), 249–267.

Burstein, A., & Gopinath, G. (2013). *International prices and exchange rates*. NBER Working Papers 18829, National Bureau of Economic Research, Inc.

Campa, J., & Goldberg, L. (2005). Exchange rate pass-through into import prices. *The Review of Economics and Statistics*, *87*(4), 679–690.

Campa, J., & Gonzàlez-Mínguez, J. M. (2006). Differences in exchange rate pass-through in the euro area. *European Economic Review*, *50*(1), 121–145.

Campa, J. M., Goldberg, L. S., & Gonzàlez-Mínguez, J. M. (2005). *Exchange-rate pass-through to import prices in the euro area*. Working Paper 11632, National Bureau of Economic Research.

Caves, R. E. (1998). Industrial organization and new findings on the turnover and mobility of firms. *Journal of Economic Literature*, *36*(4), 1947–1982.

Colantone, I., Coucke, K., & Sleuwaegen, L. (2015). Low-cost import competition and firm exit: Evidence from the EU. *Industrial and Corporate Change*, *24*(1), 131–161.

Deese, D. (2016). *Handbook of the International Political Economy of Trade*. Handbooks of Research on International Political Economy. Edward Elgar Publishing Limited.

Devereux, M. B., Engel, C., & Storgaard, P. E. (2004). Endogenous exchange rate pass-through when nominal prices are set in advance. *Journal of International Economics*, *63*(2), 263–291.

Devereux, M. B., & Yetman, J. (2010). Price adjustment and exchange rate pass-through. *Journal of International Money and Finance*, *29*(1), 181–200.

Dong, W. (2012). The role of expenditure switching in the global imbalance adjustment. *Journal of International Economics*, *86*(2), 237–251.

Dornbusch, R. (1987). Exchange rates and prices. *American Economic Review*, *77*(1), 93–106.

Engel, C. (2002). *The responsiveness of consumer prices to exchange rates and the implications for exchange rate policy: A survey of a few recent new-open-economy models*. Working Paper No. 8725, NBER.

Faruqee, H. (2006). Exchange rate pass-through in the euro area. *IMF Staff Papers*, *53*(1), 63–88.

Gagnon, J. E., & Ihrig, J. (2004). Monetary policy and exchange rate pass-through. This article is a U.S. Government work and is in the public domain in the U.S.A. *International Journal of Finance & Economics*, *9*(4), 315–338.

Gaulier, G., Lahrèche-Révil, A., & Méjean, I. (2008). Exchange-rate pass-through at the product level. *Canadian Journal of Economics*, *41*(2), 425–449.

Gust, C., Leduc, S., & Vigfusson, R. (2010). Trade integration, competition, and the decline in exchange-rate pass-through. *Journal of Monetary Economics*, *57*(3), 309–324.

Hahn, E. (2003). *Pass-through of external shocks to euro area inflation*. Working Paper Series 0243, European Central Bank.

Ihrig, J. E., M. Marazzi, & Rothenberg, A. D. (2006). *Exchange-rate pass-through in the G-7 countries*. International Finance Discussion Papers 851, Board of Governors of the Federal Reserve System (U.S.).

Knetter, M. M. (1989). Price discrimination by U.S. and German exporters. *American Economic Review, 79*(1), 198–210.

Lipsey, R. E. (1976). Intra-industry trade: The theory and measurement of international trade in differentiated products: Herbert J. Grubel and P. J. Lloyd, (John Wiley, New York, 1975) pp. 205. *Journal of International Economics, 6*(3), 312–314.

López-Villavicencio, A., & Mignon, V. (2017). Exchange rate pass-through in emerging countries: Do the inflation environment, monetary policy regime and central bank behavior matter? *Journal of International Money and Finance, 79*, 20–38.

Marazzi, M., & Sheets, N. (2007). Declining exchange rate pass-through to U.S. import prices: The potential role of global factors. *Journal of International Money and Finance, 26*(6), 924–947.

Marazzi, M., Sheets, N., Vigfusson, R., Faust, J., Gagnon, J., Marquez, J. R., Martin, R., Reeve, T. A., & Rogers, J. (2005). *Exchange rate pass-through to U.S. import prices: Some new evidence*. International Finance Discussion Papers 833, Board of Governors of the Federal Reserve System (U.S.).

Milani, F. (2012). Has globalization transformed U.S. macroeconmic dynamics? *Macroeconomic Dynamics, 16*(2), 204–229.

Monacelli, T. (2005). Monetary policy in a low pass-through environment. *Journal of Money, Credit and Banking, 37*(6), 1047–1066.

Schroder, M., & Hufner, F. P. (2002). *Exchange rate pass-through to consumer prices: a European perspective*. ZEW Discussion Papers 02 20. Center for European Economic Research: ZEW.

Taylor, J. B. (2000). Low inflation, pass-through, and the pricing power of firms. *European Economic Review, 44*(7), 1389–1408.

A State-Space Model to Estimate Potential Growth in the Industrialized Countries

Thomas Brand, Gilles Dufrénot, and Antoine Mayerowitz

1 Introduction

The anemic growth rates and rising unemployment rates in the industrialized countries since the Great Financial Crisis have been an important reason for the renewed interest in the issue of potential growth among the economists and policymakers. Can we consider that the economies have gone through a soft patch that will end soon as stronger growth recover? Or does the increasing unemployment rates reflect a downward break in medium-term growth? What is the size of this pull-back? There is no consensus on the answers to these questions. The various strands of theories propose different concepts of potential growth. Additionally, a source of controversies is the uncertainty about how to measure potential growth, since it is an unobservable variable.

The purpose of this paper is to estimate the potential growth of 5 major industrialized countries (France, USA, UK, Canada, and Germany) over the period from 1980 to 2016 by comparing two approaches: (i) an estimate based on a filter where the output-gap equation is enriched with information from several key determinants of the economic cycle, in particular financial variables, (ii) a small semi-structural model where potential growth is estimated simultaneously with other variables: the natural rate of interest and medium-term unemployment rate. Potential growth is a benchmark which serves for growth, fiscal, and monetary policies, but which is

T. Brand
CEPREMAP, 48 bd Jourdan, 75014 Paris, France
e-mail: thomas.brand@cepremap.org

G. Dufrénot (✉)
Aix-Marseille School of Economics and CEPII, Marseille, France
e-mail: gilles.dufrenot@univ-amu.fr

A. Mayerowitz
Paris School of Economics, EHESS, 3 rue d'Ulm, 75005 Paris, France
e-mail: a.mayerowitz@gmail.com

G. Dufrénot and T. Matsuki (eds.), *Recent Econometric Techniques for Macroeconomic and Financial Data*, Dynamic Modeling and Econometrics in Economics and Finance 27, https://doi.org/10.1007/978-3-030-54252-8_3

unobservable. Finding such a benchmark raises two difficulties. The first difficulty is theoretical, while the second is statistical.

Theoretically, there is no consensus among the economists about what is meant by potential growth. The main divergence concerns the opposition on the question as whether the analysis should be demand-based or supply-based.

The Keynesian view of potential growth has its root in the paradigmatic approach initially proposed by Okun (1962). Potential growth can be defined as the highest growth rate attainable without inflationary pressures. When an economy reaches its potential growth, inflation becomes very sensitive to changes in the unemployment rate. This happens because the unemployment rate is expected to gravitate around its 'medium-term' level. In light of this, the flattening of the Phillips curves recently evidenced by the new-Keynesian literature, especially in the European countries, requires a special attention. Flat Phillips curves may signal that the output-gap are wide because demand is capped. This means that policies aiming at reducing or closing the output-gap should be demand-oriented. The economists who share this view refer to several traditions of economic thought. Some authors point that we should look at the conditions in the financial sector as providing us with some explanations on the imbalance between supply and demand, in line with Minsky's vision of the business cycle. Other emphasizes the role of monetary policy in the Wicksellian tradition: what matters is not the policy rate itself, but the discrepancy between this rate and the natural rate of interest (which is also an unobserved variable).

The Neoclassical perspective provides an alternative vision of potential growth as a medium-term attractor toward which the short-term growth adjusts back after an initial deviation. Potential growth is the growth rate of potential output defined as the level of the output that an economy could achieve, if prices were fully flexible. The reason why negative output-gaps are observed is the existence of market imperfections. A modern approach to this view is the so-called new Neoclassical synthesis. Potential output and growth are calculated by taking recourse to DSGE models. In these models, the economies are hit by external shocks (technological, fiscal, and financial shocks) and the propagation mechanisms are described by the agents' intertemporal optimization behaviors. These models contain market imperfections and wage and price rigidities [see, among others, Andrés et al. (2005), Edge et al. (2008), Vetlov (2011)]. A key result of these models is that potential growth fluctuates along the business cycle, much more than in the usual empirical estimates. Potential growth is therefore linked to changes affecting the supply-side of the economy (shocks to the preferences that affect saving and labor supply, technological shocks affecting total factor productivity, fiscal policy reducing the cost of capital and labor, etc.). As a result, closing the output-gap requires lower market inefficiencies.

In this paper, we consider models in line with the theories in which potential growth and GDP are determined from the demand side of the economy. The demand factors are both financial and real. We consider the Minsky (1992)'s hypothesis of a correlation between financial and business cycles, as well as an hypothesis in the tradition of Wicksell according to which monetary policy is a central determinant of short-run and medium-term growths. We first show that the information in the financial variables can change the estimate of potential growth, as compared with a standard HP filter model. We further estimate a simple semi-structural model that consists of an IS curve, a Phillips curve, an Okun law, and an equation linking potential GDP and the natural real rate of interest. Potential growth is defined as the growth rate of potential GDP. To assess the plausibility of a non-neutral influence of finance on potential output and growth, we make two assumptions. We first consider the role of monetary policy on business cycle fluctuations. The central bank is assumed to set its real short-term interest rate with regards to a benchmark which is the natural real interest rate. The short-term interest rate aims at stabilizing inflation and the business cycle. Second, we consider the role of changes in property prices and credit to the private sector, in order to incorporate some ingredients relating output fluctuations to the Minsky'approach of cumulative imbalances in the financial sector.

The second challenge an economist faces when he or she wishes to measure potential growth faces is statistical. Neither the output-gap nor the potential GDP is observable. So their identification is usually done in three ways.

- First approach: one focuses on the estimation of potential GDP and deduces the cycle (output-gap). Production functions are estimated to evaluate potential growth based on its long-term determinants (total productivity, labor and capital productivity, demographic factors, volume of capital, and hours worked).
- Second approach: the cycle and potential GDP are determined jointly using filter identification criteria (choice of smoothing parameters, use of the information contained in their determinants multivariate filters …).
- Third approach: semi-structural models. One estimate reduced forms from a theoretical structural model and uses a state-space approach to estimate simultaneously unobservable variables and the coefficients based on the theoretical relationships .

These three approaches have given rise to alternative methodologies in the literature:

- univariate filters are used when potential output is considered as trend output. The trend is thought of as deterministic or stochastic and can be extracted from the observed GDP using OLS, Hodrick–Prescott filter, unobserved components models, etc. [for a recent paper, see Proietti (2009)];
- one can estimate production functions and examine the contribution of capital and labor to the growth rate of medium-term GDP [see, for instance, Epstein and Macchiarelli (2010)];
- bivariate Kalman filter models have been proposed in which potential GDP is estimated jointly with variables correlated with total factor productivity [see, Clancy (2013)];
- multivariate HP filter models [for recent papers, see Beneš et al. (2010), Andrle (2013)];
- hybrid models combine the growth-accounting approach and multivariate filter approaches (Cheremukhin et al. 2013);
- VAR models (see McNelis et al. 2007);

- State-space models with reduced-forms based on economic theories [see, Laubach and Williams (2003), Laubach et al. (2015), Blagrave et al. (2015), Benati and Vitale (2007)].

The choice between these methods is based on the following trade-off:
(i) Either we consider the estimate of reduced form equations. As stated by Borio et al. (2014), reduced-form models have two main advantages. First, one keeps the dimensionality of the models small, by not considering a large set of theoretical relationships. Secondly, they involve the estimate of a small number of scaling factors, and that makes the estimation of the other parameters easier.
(ii) Or, we want to learn about the mechanisms through which different variables determine both potential growth and the output-gap. In this case, we provide our estimates with more theoretical foundations and we therefore need semi-structural models (reduced-form equations with some theoretical background). This can help us understand why different researchers can find different estimates of potential growth. A downside of semi-structural models is that they can involve specification biases.

This paper proposes both a pure reduced-form model approach and a semi-structural model to estimate the output-gap of our five countries. The paper contributes in three ways to the literature.

First, we find that the so-called leakage effect is not a problem per se. The feature usually required that the output-gap should be of short-term durations so as to reflect the duration of a typical business cycle (between 6 and 32 quarters) does no longer hold when some of the determinants of the cycle have long-term components or are characterized by long waves. When financial variables such as credit or property prices are used for the identification of the trend and cycle components in the GDP, the cycle can have both short- and long-term components (beyond the usual 8 year duration). As a consequence, potential growth can have more variability than usually and can imply trend components that are smoother. What this implies from an economic policy perspective is, for instance, that demand policies that target the economic cycle should be designed at the right 'frequency.' Credit policy is not only a matter of short-term balances but could also be sized in such a way to impact long duration cycles.

Our second result is in line with the first one. A long tradition in economics sees the cycle as fundamentally reflecting short-term dynamics, as affected by transitory shocks. In this regard, the output-gap is thought of as obeying a mean-reverting dynamics. This belief is the backbones of policies suggesting a distinction between short-term demand-oriented policies to tackle the recession and expansion phases of the business cycle, and long-term supply-oriented policies aiming at changing the trend component (long-term or potential growth). We show that such a distinction is tenuous, for the output-gap is found to be $I(1)$. We provide several reasons why this can occur and show that our explanations suit well with the historical examples.

As a third contribution, we discuss the link between the interest rate gap (difference between the observed and natural real interest rate gap) and the changes observed in the growth gap across the years. The countries that have the most reactive monetary policies are those that set their policy rate close to the natural rate.

The remainder of the paper is as follows. In Sect. 2, we present our state-space model with theoretical relationships. Section 3 contains our results. Finally, Sect. 4 concludes.

2 A State-Space Model with Theoretical Relationships

We first present the structure of our general model, within which different sub-models usually considered in the literature are embedded, depending upon some restrictions on the parameters of the equations.

2.1 The General Model

The model embodies several theoretical ideas from the Keynesian tradition of an economic equilibrium determined by demand factors, from the Wicksellian view of a link between the interest rate gap and the business cycle, and Minsky's approach of the role of the financial cycle in determining the business cycle.

A natural interest rate is considered as a benchmark for monetary policy and is determined by both potential growth and short-term determinants. A short-term determinant is, for instance, consumer's rate of preference, while long-run determinants refer, for example, to technological changes or demographic factors, which are both key determinants of potential growth. Unlike the original suggestion by Wicksell, the natural rate is not considered here as a direct determinant of the inflation gap. In the author's reasoning, the economy was assumed to be permanently at its steady-state level and, therefore, any gap between the central bank's nominal interest rate and the natural rate of interest caused inflationary pressure or deflation. As is known, the Wicksell approach was extended by the Stockholm school during the 1930*s* and the natural rate was conceptualized to build a theory of the business cycle. Differences between the nominal and natural rates cause a credit expansion or restriction and this in turn cause increases or decreases in output through changes in private demand. The impact to inflation comes then through a Phillips curve. In some recent models of potential growth, the interest rate gap directly enters as an explanatory variable of the output-gap [see Laubach and Williams (2003)]. Here, we adopt the Minsky's view of cumulative processes in the credit sector as a key source of output fluctuations. Banks and lenders' behavior, through changes in the interest rates of loans tighten or loose credit availability. The interest rate gap also induces speculative investment. Therefore, our model attempts to capture the Minsky's financial instability hypothesis as follows. The output-gap (defined as the gap between the observed GDP and potential GDP) depends upon monetary and financial sector imbalances, and these imbalances are related to the interest rate gap.

We consider a simple model with backward looking expectations. This assumption can be criticized given the importance of inflation expectations in the determination of the real interest rates and with regards to the fact that interest rate rules with forward looking output-gap and inflation-gap have been extensively used in the literature. Further, one may argue that a forward looking Phillips curve better fits the new Keynesian theories of price determination. We are aware of these limits to our framework. Though these views have many advocates, we motivate the use of backward looking expectations models as follows.

We assume bounded rationality as a source of business cycle fluctuations and changes in the inflation rate (this corresponds to Keynes and Minsky's view of strong uncertainty). In terms of forecasting, bounded rationality implies extrapolative or regressive expectations. Unlike the mainstream view of the Neoclassical synthesis, one can adopt the Keynesian view that, (i) agents do not know the future, (ii) their long-term expectations is thus a projection into the future of past trends, and (iii) even if they are aware that their expectations can lead to mistakes, extrapolation is not irrational if they do not know in which direction future changes will occur.

Our IS equation is augmented with variables capturing financial imbalances. A recent literature shows that the financial instability hypothesis (how financial fluctuations affect the economic activity and inflation) based on the hypothesis of bounded rationality matches the real macroeconomic outcomes and explain GDP cycles observed over the past years and explains how the period of Great moderation paved the way to the recent Great recession [see, for instance, Keen (2011)]. To be consistent with the hypothesis of bounded rationality, we assume that the agents do not take into account in their behavior the consequences of the central bank's policy on the economy. The interest rate rule is also backward looking.

The first equation relates the natural real interest rate to potential GDP and other variables:

$$r_t^* = cg_t + z_t, \tag{1}$$

$$g_t = \delta_g g_{t-1} + \overline{g} + \epsilon_t^g, \tag{2}$$

$$z_t = D_z^{p_z}(L)z_{t-1} + \epsilon_t^z, \tag{3}$$

where r_t^* is the natural real interest rate, g_t is potential growth, and z_t represents other determinants of the natural rate. Potential growth is assumed to follow a mean reverting dynamics and z_t follows an $AR(p)$ process with $D_z^{p_z}(L)$ being a lag polynomial of order p. We assume that $\epsilon_t^r \approx N(0, \sigma_r^2)$ and $\epsilon_t^z \approx N(0, \sigma_z^2)$. δ_g and $\overline{g}$ are real parameters.

The second equation is an IS curve:

$$\begin{aligned}(y_t - y_t^*) = \widetilde{y}_t = {} & a_1\widetilde{y}_{t-1} + a_2\widetilde{y}_{t-2} \\ & + D_c^{p_c}(L)(\Delta \text{credit}_t) + D_s^{p_s}(L)(\Delta \text{financial}_t) \\ & + D_w^{p_w}(L)\Delta \text{world}_t + \epsilon_t^{\widetilde{y}}, \quad \epsilon_t^{\widetilde{y}} \approx N(0, \sigma_{\widetilde{y}}^2),\end{aligned} \tag{4}$$

world_t is the real world demand, credit_t is the real credit to the private sector by the banking sector, and financial_t is the real financial variables (real housing prices). The output-gap depends upon its own lags, upon changes in the growth rate of private credit, property prices and changes in world demand. $D_{\tilde{y}}^{p_y}(L)$, $D_c^{p_c}(L)$, $D_s^{p_s}(L)$, $D_w^{p_w}(L)$ are polynomials of respective order p_y, p_c, p_s and p_w.

y_t^* is the logarithm of the potential level of GDP assumed to be $I(1)$:

$$y_t^* = y_{t-1}^* + g_{t-1} + \epsilon_t^{y^*}, \quad \epsilon_t^{y^*} \approx N(0, \sigma_{y^*}^2), \tag{5}$$

where $\epsilon_t^{y^*}$ is a random noise. Our motivation for adding financial variables in the (IS) curve is the expanding literature pointing to the role of financial variables (like credit, stock prices, and property prices) as advanced indicators of business cycle fluctuations (see Drehmann et al. 2012, Schularick and Taylor 2012, Claessens et al. 2012, Borio et al. 2014).

The next two equations relates the financial variables to the interest rate gap:

$$(\Delta\text{credit}_t - \pi_t) = \beta_c(\Delta\text{credit}_{t-1} - \pi_{t-1}) + (1/2)b_c \sum_{j=1}^{2}(r_{t-j} - r_{t-j}^*) + \epsilon_t^c \tag{6}$$

$$(\Delta\text{financial}_t - \pi_t) = \beta_s(\Delta\text{financial}_{t-1} - \pi_{t-1}) + (1/2)b_s \sum_{j=1}^{2}(r_{t-j} - r_{t-j}^*) + \epsilon_t^s \tag{7}$$

where $\epsilon_t^c \approx N(0, \sigma_c^2)$ and $\epsilon_t^s \approx N(0, \sigma_s^2)$.

The interest rate gap is endogenous and described by an interest rate rule

$$\begin{aligned}(r_t - r_t^*) = \lambda_r(r_{t-1} - r_{t-1}^*) + (1/2)\mu_g \sum_{j=1}^{2}(y_{t-j} - y_{t-j}^*) \\ + (1/2)\mu_\pi \sum_{j=1}^{2}(\pi_{t-j} - \pi_{t-1}^*) + \epsilon_t^r, \quad \epsilon_t^r \approx N(0, \sigma_r^2),\end{aligned} \tag{8}$$

where $\tilde{y}_t$ is the output-gap, r_t is the short-term real interest rate, and r_t^* is the natural real interest rate. $\epsilon_{\tilde{y}_t}$ is a random noise. π^* is targeted inflation defined by the central bank and assumed to be exogenous. The endogenous variable is the inflation adjusted short-term rate (which is influenced by the central bank's policy rate). The benchmark interest rate is assumed to be the Wicksellian natural rate of interest.

Some comments are in order here. There are alternative ways of incorporating information about the impact of the financial fluctuations on the business cycle. One way is to set the central bank's interest rate as a function of the output-gap, the inflation-gap and of financial imbalances, and to assume that the output-gap depends upon changes in the interest rate gap. This amounts to assuming that the central bank addresses the build-up of financial imbalances through a macroprudential policy. We know that this was not the case up until the very recent years. Rather, to better fits what was observed in reality, we prefer considering a standard interest rate rule and a causal relationship from the interest rate to financial imbalances (as in Eqs. 6 and 7). The central bank thus aims at stabilizing output around its potential level consistent with a stabilization of inflation around its long-run level π_t^*. It reacts to the output and inflation taking into account an average of their current and lagged values.

The next equation is an augmented Phillips curve with relative price variables. We assume extrapolative expectations:

$$\begin{aligned} \pi_t - \pi_{t-1} &= b_1(\pi_{t-1} - \pi_{t-2}) + b_y(u_{t-1} - u_{t-1}^*) + D_{\text{imp}}^{p_{\text{imp}}}(L)(\pi_t^{\text{imp}} - \pi_t) \\ &+ D_{\text{oil}}^{p_{\text{oil}}}(L)(\pi_t^{\text{oil}} - \pi_t) + \epsilon_t^{\pi}, \quad b_y < 0, \quad \epsilon_t^{\pi} \approx N(0, \sigma_{\pi}^2). \end{aligned} \tag{9}$$

π_t^{imp} and π_t^{oil} are, respectively, logarithmic changes in the import and oil prices. We consider possible lagged effects through the lag polynomials $D_{\text{imp}}^{p_{\text{imp}}}(L)$ and $D_{\text{oil}}^{p_{\text{oil}}}(L)$. π_t^* is the inflation target set by the central bank assumed to be exogenous.

The next two equations define the short-term and medium-term unemployment rates:

$$u_t = u_t^* + \widetilde{u}_t, \quad \text{where } u_t^* = u_{t-1}^* + \epsilon_t^{u^*}, \tag{10}$$

$$\widetilde{u}_t = u_t - u_t^* = -\omega_y(y_{t-1} - y_{t-1}^*) + \epsilon_t^{\widetilde{u}}, \tag{11}$$

where ω_y is a real parameter, $\epsilon_t^{u^*} \approx N(0, \sigma_{u^*}^2)$ and $\epsilon_t^{\widetilde{u}} \approx N(0, \sigma_{\widetilde{u}}^2)$. Equation 11 refers to Okun Law.

The noise variables are assumed to be uncorrelated and to have their covariances equals zero.

2.2 Sub-models and Comparison with Other Models Used in the Literature

By imposing alternative restrictions on the parameters of the model, one can retrieve different types of models used in the literature to estimate potential GDP and growth.

- *Model 1*. Standard HP filter model with two state variables being potential growth and GDP
  It is defined by Eqs. (2), (4) and (5) with the constraint $D_c^{p_c}(L) = D_s^{p_s}(L) = D_w^{p_w}(L) = 0$.
- *Model 2*. Multivariate semi-structural model with no financial variables, Phillips and IS curves and exogenous interest rate gap.

It is defined by Eqs. (2), (4), (5), (9), and (11) with the constraint $D_c^{p_c}(L) = D_s^{p_s}(L) = 0$.

- *Model 3*. Multivariate semi-structural model in which the output-gap and the natural interest rate are determined jointly.
  This is *Model 2* augmented with Eq. (1).
- *Model 4*. Multivariate HP filter with financial variables and Taylor rule with exogenous natural rate
  It is defined by Eqs. (2), (4), (8), (5), (9) and (11)

We, now, compare our general model with four models which have been used with many instances in the literature.
Laubach and Williams (2003)

- no financial variables in their model,
- no Taylor rule,
- instead of Okun law and unemployment rate, the hours worked are signals of cyclical signals in the labor market,
- they consider inflation expectations to define the ex-ante real interest rate in the Phillips curve, similar dynamics for the output-gap and potential GDP.

Beneš et al. (2010)

- No financial variables to account for the role of financial imbalances,
- output-gap linked to both capacity utilization and unemployment gap,
- potential GDP is modeled as a trend component of output using a HP filter (not estimated within the multivariate model),
- long-run unemployment rate defined as a function of the output-gap and of the steady state of the model.

Borio et al. (2013), Borio et al. (2014)

- They use a multivariate HP filter model with real interest rate, real exchange rate, and financial variables,
- the main goal of the authors is to show that financial imbalances are informative about potential growth and GDP,
- no Phillips curve. They motivate their choice not to consider a semi-structural model as a way of avoiding mispecification biases.

Blagrave et al. (2015), Alichi (2015)

- no financial imbalances variables, no role for credit,
- No natural interest rate, the labor market is modelled by labor force participation,
- the non-accelerating inflation rate unemployment (NAIRU) converges to its steady state,
- potential growth is subject to shocks fading gradually with an adjustment to the steady state growth rate,
- their model has a capacity utilization block,
- they make the assumption of rational expectations.

2.3 Estimation Method

Since we have some unobserved variables, the starting point is to write the model using a state-space representation. The parameters and unknown variables are then estimated using Kalman filter method. The state-space representation of the model is as follows:

$$\begin{aligned} X_t &= AX_{t-1} + Z_t + F_t W_t : \text{ state equation} \\ Y_t &= \mu_t + C_t^{'} X_t + V_t : \text{ measurement equation} \end{aligned} \tag{12}$$

X_t is the vector of k_1 state variables (unobserved), Y_t is the vector of k_2 observed variables, A is a $k_1 \times k_1$ matrix, Z_t is a $k_1 \times 1$ vector of deterministic terms, W_t is a $r_1 \times 1$ vector of residuals, F_t is a $k_1 \times r_1$ matrix, μ_t is the product of a $k_2 \times nexpl$ matrix of coefficients by a vector of $nexpl$ explanatory variables. C_t is a matrix of dimension $k_2 \times k_1$ and V_t is a vector of r_2 residual terms.

To estimate the model, we adopt a sequential approach based on five steps.

Step 1: we estimate the model with three state variables: y_t^*, y_{t-1}^*, y_{t-2}^* and three observed variables y_t, π_t, $\widetilde{u}_t$.

The influence of the import prices and oil prices on inflation in Eq. (9) is assumed to be measured by the average impact over two lagged quarters. In Eq. (11) u_t^* is assumed to be a constant (exogenous). In the IS curve, the financial variables are assumed to be exogenous and the influence of world demand is captured by the average of the impacts of current and previous quarters ($D_c^{p_c}(L) = 0.5a_{31}(L + L^2$, $D_s^{p_s}(L) = 0.5a_{32}(L + L^2)$, $D_w^{p_w}(L) = 0.5(1 + L)$). Moreover, we do not consider the influence of the interest rate. Finally potential growth rate is assumed to be a constant (exogenous)

Denoting Θ the set of parameter to estimate, we, therefore, search for $\hat{\Theta}$, $\hat{X}_t$ that minimizes the following loss function:

$$\sum_{t=1}^{T} \left\{ \frac{1}{\sigma_{\widetilde{y}}^2}(y_t - y_t^*)^2 + \frac{1}{\sigma_{y*}^2}(\epsilon_t^{y*})^2 + \frac{1}{\sigma_{\pi}^2}(\epsilon_t^{\pi})^2 + \frac{1}{\sigma_{\widetilde{u}}^2}(\epsilon^{\widetilde{u}})^2 \right\} \tag{13}$$

The estimates depends upon the values of the following weights (scaling factors): $\lambda_1 = \frac{\sigma_{\widetilde{y}}^2}{\sigma_{y*}^2}$, $\lambda_2 = \frac{\sigma_{\widetilde{y}}^2}{\sigma_{\pi}^2}$, $\lambda_3 = \frac{\sigma_{\widetilde{y}}^2}{\sigma_{\widetilde{u}}^2}$. As has been evidenced in the literature, there is no clear guidance about the choice of particular values for these scaling factors. Even the conventional value of 1600 usually chosen for λ_1 would not be appropriate here as the cyclical properties of the output-gap depend upon all the scaling factors. We leave those scaling be estimated by the data. This first-step estimation, with constant g aims at generating some first guess values for the output-gap and the coefficients of the Phillips, IS and Okun equations.

Step 2: Potential growth rate is endogenous and time-varying. Changes in the financial variables are endogenous and vary with the interest rate gap (the natural rate is assumed to be exogenous). We estimate the model with four state variables: y_t^*, y_{t-1}^*, y_{t-2}^*, g_t and 5 observed variables y_t, π_t, $\widetilde{u}_t$, Δcredit_t and $\Delta\text{property}_t$.

We minimize the following loss function:

$$\sum_{t=1}^{T} \frac{1}{\sigma_{\widetilde{y}}^2}(y_t - y_t^*)^2 + \frac{1}{\sigma_{y*}^2}(\epsilon_t^{y*})^2 + \frac{1}{\sigma_{\pi}^2}(\epsilon_t^{\pi})^2 + \frac{1}{\sigma_{\widetilde{u}}^2}(\epsilon^{\widetilde{u}})^2 + \frac{1}{\sigma_g^2}(\Delta y_{t+1}^* - \Delta y_t^*)^2 + \frac{1}{\sigma_c^2}(\epsilon_t^c)^2 + \frac{1}{\sigma_s^2}(\epsilon_t^s)^2 \tag{14}$$

This time, we have 6 scaling parameters. Again, we impose no constraints and estimate them from the data.

Steps 3 and 4: The natural rate is endogenous and varies linearly with potential growth. The model is estimated with the interest rate rule. The long-run unemployment rate is assumed to be constant. The model has nine state variables: y_t^*, y_{t-1}^*, y_{t-1}^*, g_t, g_{t-1},g_{t-2}, z_t, z_{t-1},z_{t-2}. With the interest rate rule, we have six variables that defines the measurement equations: y_t, π_t, $\widetilde{u}_t$, Δcredit_t, $\Delta\text{property}_t$ and r_t. Finally, we consider the long-run unemployment rate to be endogenous and defined by Eq. (11). The loss functions and scaling factors are defined in similar ways as in (13) and (14)

Steps 5: Finally, the medium-term unemployment rate is added to the list of the state variables.

3 Results

We use quarterly data from 1980 to 2015 taken from OECD, trade variables (IMF) and financial variables (BIS). The observable variables are the unemployment rate, GDP, core inflation, import and oil prices, credit as share of GDP, property prices, and the short-term interest rate. World demand for each country is computed as a weighted average of the GDPs of its main trading partners. For each variables, we first compute the logarithmic difference and then quarterly differences are annualized. Kalman filter is used to obtain estimates of the unobservable (or state) variables.

Tables 1 and 2 show the consequences of a situation of secular stagnation: high structural unemployment rates and low core inflation rates. The latter has been steadily declining since the early 1990s, starting in Japan. This country even experienced a situation of deflation in the 2000s. The estimated structural unemployment rates also remained high, specifically in Germany (8%) and France (9%), two important euro area countries.

Table 1 Structural unemployment (%)

	1990–2001	2002–2007	2010–2015
United States	6.7	6.3	6.0
Germany	8.6	7.9	7.3
France	9.0	9.1	9.3
Japan	3.7	4.0	4.3
United Kingdom	7.9	6.8	5.8

Table 2 Core inflation (%)

	1990–2001	2002–2007	2010–2015
United States	3.0	2.1	1.7
Germany	2.4	1.4	1.1
France	1.7	1.0	1.0
Japan	0.9	−0.3	0.2
United Kingdom	2.6	1.4	2.1

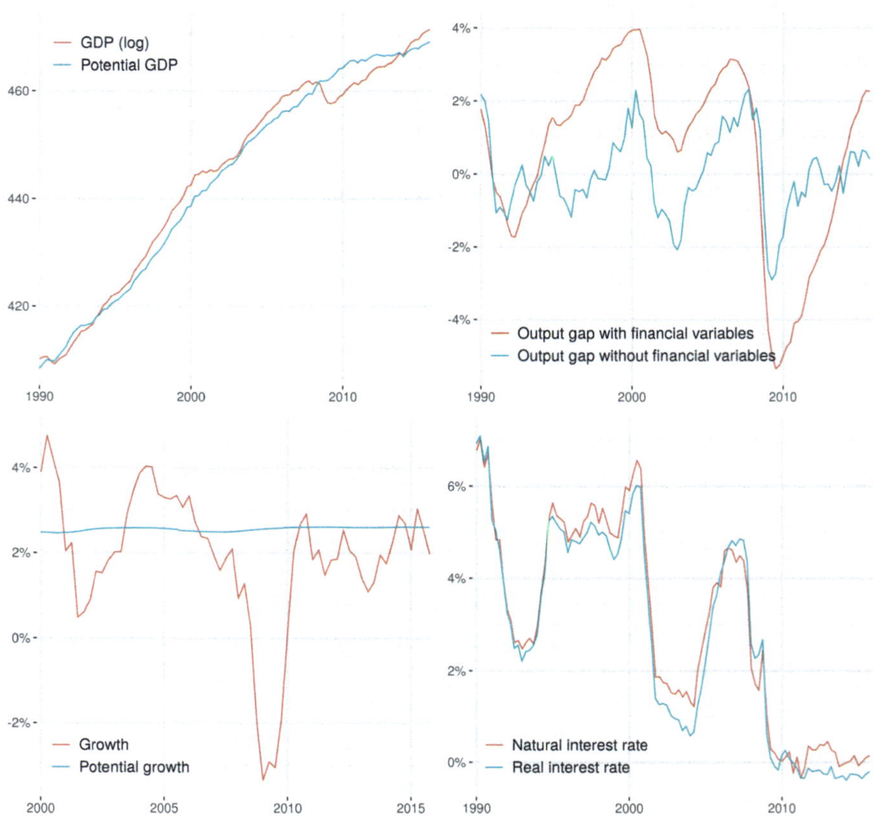

Fig. 1 USA

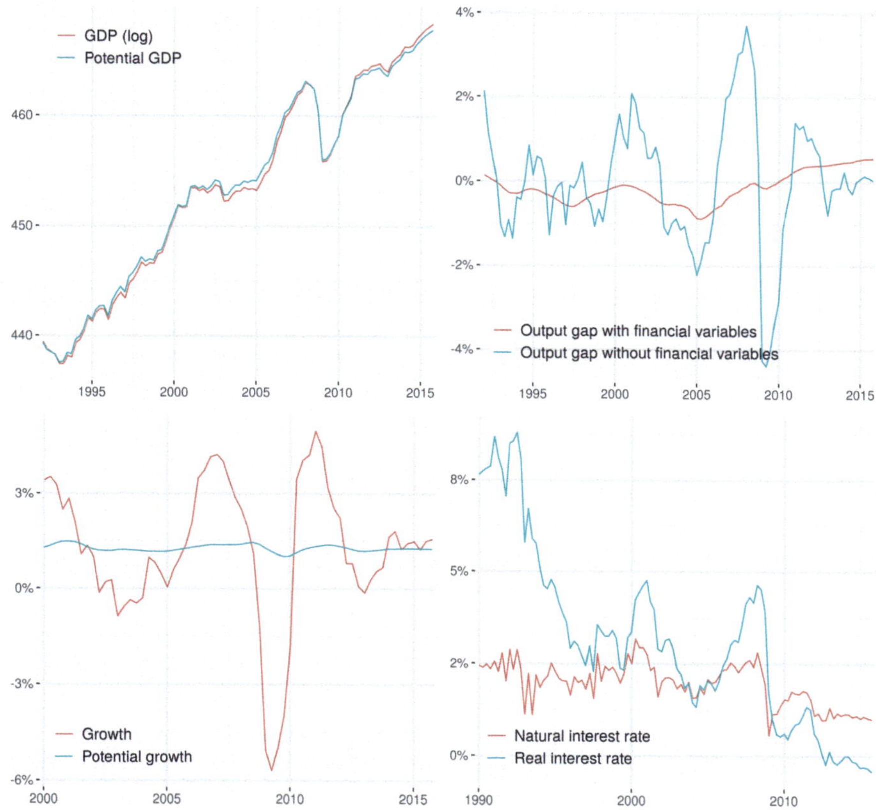

Fig. 2 Germany

Figures 1, 2, 3, 4, and 5 show the following estimated variables: potential GDP, potential growth, and output-gaps—calculated with and without taking into account financial variables—and the natural interest rate. The countries share some common features. We first see a steady decline in natural interest rates that started well before the 2008 financial crisis. In Japan, this rate dropped close to 0% since the mid-1990s, which can be explained by the so-called lost decades. This rate has become negative in the USA since 2008 and in Germany since 2012. It can be used as an indicator of the stance of monetary policies. In the USA and Japan, the real short-term interest rate has often been close to its natural level. In Europe, up until the 2008 financial crises and the public debt crises in 2010, the natural rates have moved below the real interest rates, thereby suggesting that monetary policies have not been sufficiently accommodative. Specifically, in the United Kingdom and Germany, we see that the spread between the two rates has remained significantly positive.

The dynamics of potential growth display various profiles. In the USA and in Germany, it remains stable over time. In France, and even more so in the United Kingdom, the 2008 crisis induced a regime shift with potential growth being on a

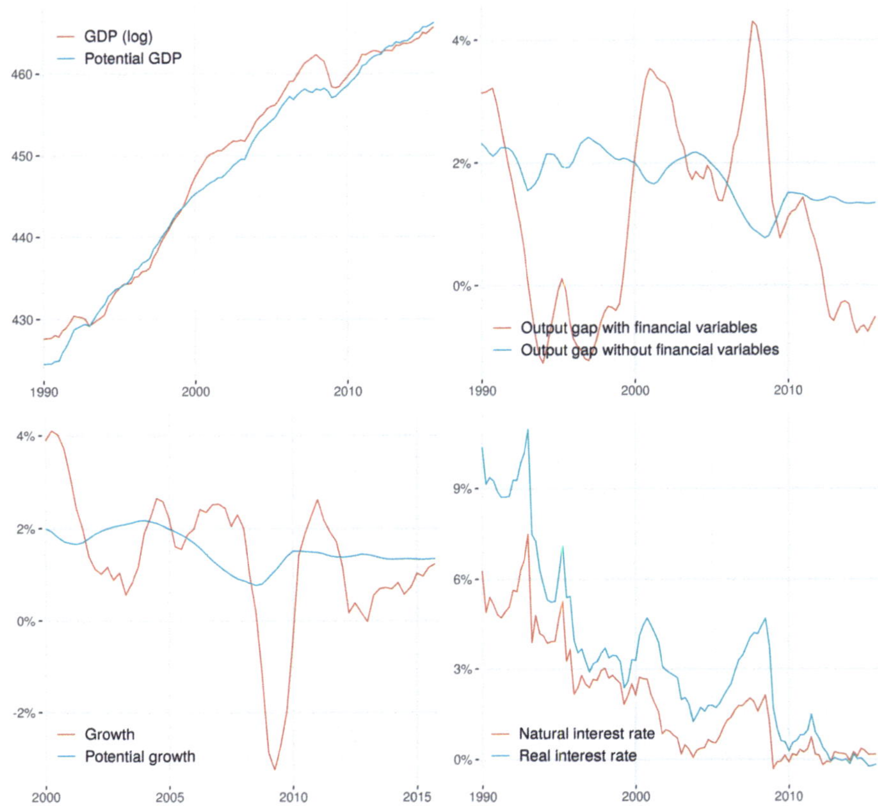

Fig. 3 France

lower trajectory than before the crisis. The United Kingdom also differs from the other countries in that potential growth fluctuates much more than elsewhere, which could be explained by a greater sensitivity to the financial cycle.

To study the influence of the financial cycle, we report the estimates of the IS equation, in Table 3, and then we plot for each country the output-gaps calculated with and without the financial variables. Many coefficients in the table are significant at 1%. In the figures, the inclusion of financial variables leads to business cycles with more pronounced troughs during recession phases and higher peaks during expansion phases. Germany is an atypical case, since the financial cycle leads to smaller fluctuations in the output-gap. Therefore, except the German case, when the role of the financial cycle is ignored, the importance of overheating situations during economic expansion phases could be underestimated and, on the contrary, cycle troughs could be overestimated. The financial cycle thus tends to amplify the real fluctuations during the expansionary regimes, and to cushion recessions.

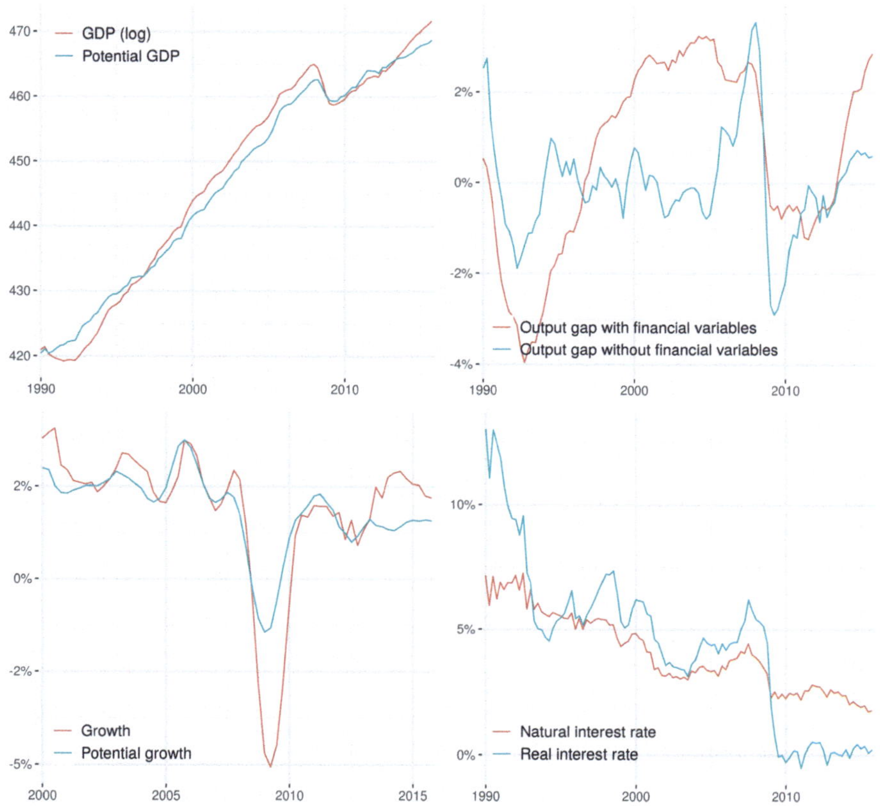

Fig. 4 United Kingdom

These results suggest that two financial factors have been key determinants of the real cycle, potential growth, and natural interest rates: (a) over-indebtedness and deleveraging by private agents, with credit being the driving force of the cycle, and (b) wealth effects resulting from reversals in the real estate cycle.

4 Conclusion

In this article, we have proposed a new method for estimating natural interest rates that explicitly takes into account the financial cycle. The methodology consists of supplementing the information in the measurement equations by adding to the IS and Phillips curve equations, two equations that explain the determination of the credit and real estate cycles. Moreover, the interdependence with monetary policy is taken into account by a Taylor rule where the target real interest rate corresponds to the

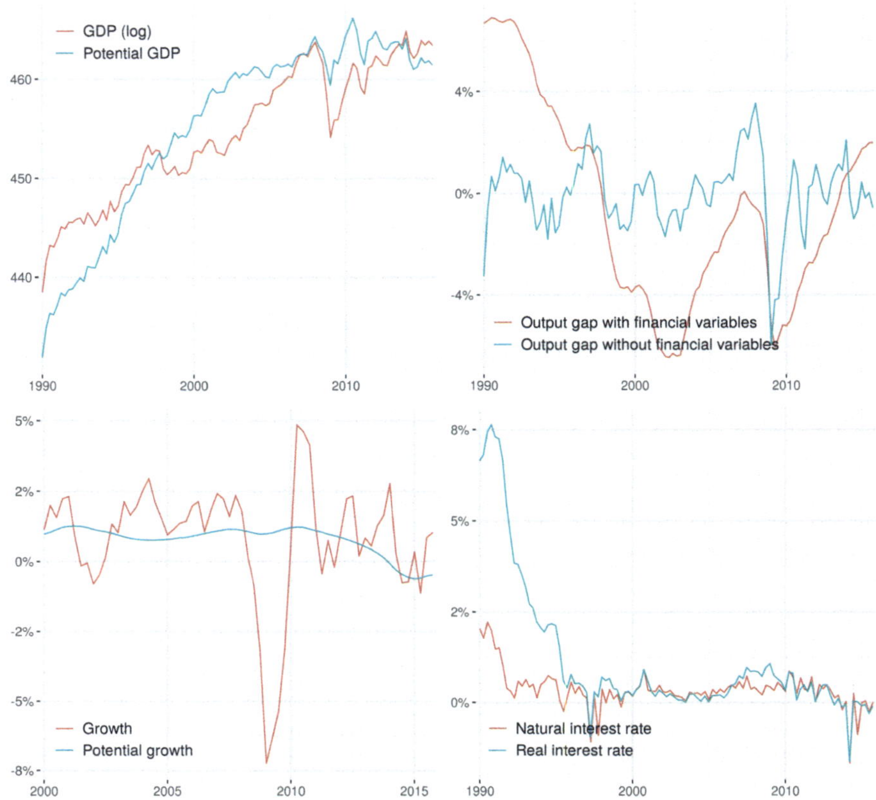

Fig. 5 Japan

Table 3 Estimated coefficients—IS equation

	Domestic credit		Share prices		World demand	
	Coefficient	Std err.	Coefficient	Std err.	Coefficient	Std err.
United States	0.05***	0.01	0.08***	0.013	0.0002**	0.0001
Germany	−0.02	0.03	0.13***	0.04	0.0002*	0.0001
France	0.04***	0.01	0.006	0.006	0.0003***	0.0001
Japan	0.05***	0.02	0.04*	0.02	0.01***	0.002
United States	−0.014	0.01	0.05*	0.008	0.0001	0.0001

natural rate. Thus, the financial cycle is not neutral in the long run, since, in addition to the output-gap, it influences the natural rate and potential growth.

Our results show that, since 1990, the United Kingdom, Germany, the USA, France, and Japan have all experienced a downward trend in their natural interest rates. Contrary to an assumption often made in the literature, this decline is not solely due to real factors (labour productivity, demographics), but to the upward and downward phases of the real estate and credit cycles.

References

Alichi, A. (2015). *A new methodology for estimating the output gap in the United States* (pp. 15–144). International Monetary Fund.

Andrés, J., Lopez-Salido, J. D., & Nelson, E. (2005). Sticky-price models and the natural rate hypothesis. *Journal of Monetary Economics, 52*, 1025–1053.

Andrle, M. (2013). *What is in your output gap? Unified framework & decomposition into observables* (pp. 13–105). International Monetary Fund.

Benati, L., & Vitale, G. (2007). *Joint estimation of the natural rate of interest, the natural rate of unemployment, expected inflation, and potential output.*

Beneš, J., Clinton, K., García-Saltos, R., Johnson, M. P., Laxton, D., Manchev, P. B., & Matheson, T. (2010). Estimating potential output with a multivariate filter. In *IMF Working Papers* (pp. 1–37).

Blagrave, P., Garcia-Saltos, M. R., Laxton, M. D., & Zhang, F. (2015). *A simple multivariate filter for estimating potential output* (pp. 15–79). International Monetary Fund.

Borio, C. E., Disyatat, P., & Juselius, M. (2013). *Rethinking potential output: Embedding information about the financial cycle.*

Borio, C. E., Disyatat, P., & Juselius, M. (2014). *A parsimonious approach to incorporating economic information in measures of potential output.*

Cheremukhin, A. A., et al. (2013). *Estimating the output gap in real time*. Staff Papers.

Claessens, S., Kose, M. A., & Terrones, M. E. (2012). How do business and financial cycles interact? *Journal of International Economics, 87*, 178–190.

Clancy, D. (2013). Output gap estimation uncertainty: extracting the TFP cycle using an aggregated PMI series. *The Economic and Social Review, 44*, 1–18.

Drehmann, M., Borio, C. E., & Tsatsaronis, K. (2012). *Characterising the financial cycle: Don't lose sight of the medium term!*.

Edge, R. M., Kiley, M. T., & Laforte, J. P. (2008). Natural rate measures in an estimated DSGE model of the US economy. *Journal of Economic Dynamics and control, 32*, 2512–2535.

Epstein, N. P., & Macchiarelli, C. (2010). *Estimating Poland's potential output: A production function approach* (pp. 10–15). International Monetary Fund.

Keen, S. (2011). *Debunking economics: The naked emperor dethroned?* Zed Books Ltd.

Laubach, T., & Williams, J. C. (2003). Measuring the natural rate of interest. *The Review of Economics and Statistics, 85*, 1063–1070.

Laubach, T., Williams, J. C., et al. (2015). *Measuring the natural rate of interest redux*. Federal Reserve Bank of San Francisco.

McNelis, P. D., Bagsic, C. B., & Guinigundo, D. C. (2007). *Output gap estimation for inflation forecasting: The case of the philippines*. Center for Monetary and Financial Policy, BSP Working Paper Series.

Minsky, H. P. (1992). *The financial instability hypothesis*. The Jerome Levy Economics Institute Working Paper.

Okun, A. M. (1962). Potential gnp: its measurement and significance. In *Proceedings of the Business and Economic Statistics Section of the American Statistical Association* (pp. 89–104). Washington, DC.

Proietti, T. (2009). Structural time series models for business cycle analysis. *Palgrave handbook of econometrics* (pp. 385–433). Springer.

Schularick, M., & Taylor, A. (2012). Credit booms gone bust: Monetary policy, leverage cycles, and financial crises, 1870–2008. *American Economic Review, 102*, 1029–1061.

Vetlov, I., Hlédik, T., Jonsson, M., Henrik, K., & Pisani, M. (2011). *Potential output in DSGE models.*

Detecting Tranquil and Bubble Periods in Housing Markets: A Review and Application of Statistical Methods

Jun Nagayasu

JEL Classification E1 · G1

1 Introduction

Financial bubbles, that is abnormally high asset and real estate prices, are generally believed to be temporary and recurrent and closely associated with investors' expectations.[1] Because their collapse has catastrophic effects on the standard of living of most people, both policymakers and investors show interest in their identification. However, despite numerous studies,[2] it is difficult to reach an academic consensus on not only the timing and duration of financial bubbles, but also the definition of bubbles. Indeed, there are many common terms in financial research, such as rational/speculative bubbles, irrational bubbles, intrinsic bubbles, and periodically collapsing bubbles. Further, because financial bubbles and investors' expectations are unobservable, previous empirical findings are almost always under scrutiny.

Researchers long recognized the significant role of investors' expectations in asset price determination, and its importance increases during bubble periods. Fama (1970), a Nobel Prize winner in Economic Sciences in 2013, discussed investors' expectations formation using the information in the context of stock price behaviors, known as the efficient market hypothesis, and provided a statistical framework to test investor rationality. Fama proposed three types of expectations: weak, semi-strong, and strong forms of rational expectations. Investors form weak expectations based solely on historical price data, semi-strong expectations using any relevant

[1] Hereafter, financial bubbles refer to both asset and real estate bubbles for convenience because a single economic theory can explain their evolution (see the next section).
[2] See Gurkaynak Gurkaynak (2008) and Shiller (2014) for a survey of the literature.

J. Nagayasu (✉)
Graduate School of Economics and Management, Tohoku University,
27-1 Kawauchi, Aoba-ku, Sendai-city, Miyagi 980-8576, Japan
e-mail: jun.nagayasu.d8@tohoku.ac.jp

G. Dufrénot and T. Matsuki (eds.), *Recent Econometric Techniques for Macroeconomic and Financial Data*, Dynamic Modeling and Econometrics in Economics and Finance 27, https://doi.org/10.1007/978-3-030-54252-8_4

publicly available information, and strong expectations based on private information that can be exclusive to a limited number of investors. Since then, much theoretical and empirical research utilized these concepts of expectations, particularly the weak form. However, more recent research tends to consider the market microstructure as an explanation of how private information transmits to investors and affects financial asset prices over time (Easley et al. 2002).

Today, much research provides evidence against investor rationality. One example is noise traders, who possess limited information and behave irrationally, creating significant noise in the market (De Long et al. 1990). Because institutional investors have more time and financial resources to analyze the market, noise traders are often considered private traders. However, there is no clear empirical evidence to conclude whether private or institutional traders are the main driver of noise and financial bubbles. Nagayasu (2018) showed that the transactions of private traders in real estate markets can explain historically high price movements. Choi and Skiba (2015) concluded that institutional herding helped stabilize securities prices in 41 global markets. In contrast, Choi et al. (2015) showed that private traders tend to engage in contrarian trades in Korea, and Zeng (2016) concluded that institutional traders also create noise in the US market.

Traditionally, financial studies referred to irrational behaviors as a market anomaly, akin to the January effect (Rozeff and Kinney 1976) and the Monday effect (Gibbons and Hess 1981). The January effect refers to the phenomenon of asset price increases in January and the Monday effect to price behaviors on Monday that are similar to those on Friday. Conventional economic theory based on profit maximization behaviors cannot explain these price movements. Similarly, other factors such as weather conditions (Hirshleifer and Shumway 2003) also influence market participants, who react more to bad news than to good news (De Bondt and Thaler 1985). Abreu and Brunnermeier (2003) showed that even rational behaviors create persistent deviations of financial prices from economic fundamentals if investors' coordination fails in selling strategies.

Despite evidence of investor irrationality, most empirical investigations and statistical approaches were proposed within a framework of rational expectations. One popular method is based on the statistical concept of integration. For example, integration studies characterize a weak form of market efficiency as a random walk. Given that asset prices and their determinants (i.e., economic fundamentals) are non-stationary, the lack of cointegration is a traditionally considered evidence of financial bubbles. Typically, unit root and cointegration tests are left-tailed tests designed to detect periods of economic equilibriums, which is reflected in their alternative hypotheses, $\rho < 1$, where $y_t = \rho y_{t-1} + \epsilon_t$ and ϵ_t is a white noise error, as a simple example.

Recently, Phillips et al. (2011) and Phillips et al. (2015) proposed recursive and rolling explosive unit root tests, where they examined the null hypothesis of the unit root ($\rho = 1$) against the alternative of an explosive case ($\rho > 1$). The authors developed these tests to deal with potential statistical problems related to unique price movements characterized as periodically collapsing bubbles (Evans 1991). Since then, explosive tests became a popular analytical tool. However, the advantage of explosive tests may become a potential problem in financial research. Because both the null and alternative hypotheses of explosive tests consider persistent deviations from economic fundamentals, these tests preclude the possibility of tranquil moments that the economy is expected to maintain and in which economists show the most interest.

Against this background, we will critically review recent developments in the statistical methods to investigate financial bubbles. Furthermore, as an extension to traditional approaches, we propose applying a top-down method to a threshold autoregressive distributed lag (T-ARDL) model to study housing markets. This statistical method is not an answer to all questions and criticisms in studies of financial bubbles, but it has interesting features that other popular methods do not share.

2 The Theoretical Concept of Rational Bubbles

The main objective of this study is to review statistical methods, but it is still important to understand the underlying economic theory of financial bubbles in order to specify a composition of statistical models. As discussed, recent research casts doubt on the rationality of investors, but many economic analyses maintain this assumption and it is often explained using the present value model (PVM). The rationality assumption prevails in academic research largely for convenience; it is easier to model rational behaviors than irrational ones. Survey data on investors' expectations are the best source of information about investors' expectations, but in the absence of survey data for a comprehensive number of countries and infrequent dissemination of survey data, we also maintain the rationality assumption.

According to the PVM, rational bubbles are defined as sizable and persistent deviations from economic fundamentals and follow a non-stationary process in a statistical sense. Based on the definition of asset returns or returns on real estate ($r_{t+1} = (P_{t+1} + D_{t+1})/P_t - 1$), the PVM suggests that the contemporaneous prices (P_t) will be determined by the expected value of future economic fundamentals (D) and prices:

$$P_t = E_t\left[\frac{P_{t+1} + D_{t+1}}{1 + r_{t+1}}\right] \tag{1}$$

where t denotes time ($t = 1, \ldots, T$) and E is an expectation operator. D is an economic fundamental, such as dividend payments in equity analyses or rental costs in housing analyses. Solving Eq. (1) forwardly and using the law of iterated expectations, we can obtain Eq. (2):

$$P_t = E_t\left[\sum_{h=0}^{\infty}\left(\prod_{k=0}^{h}\left(\frac{1}{1 + r_{t+k}}\right)\right)D_{t+h} + \lim_{h\to\infty}\prod_{k=0}^{h}\left(\frac{1}{1 + r_{t+k}}\right)P_{t+h}\right] \tag{2}$$

When P and D are cointegrated, and when the transversality condition holds (i.e., $E_t[\lim_{h\to\infty}\prod_{k=0}^{h}(\frac{1}{1+r_{t+k}})P_{t+h}] \to 0$), then the result shows no evidence of bubbles. Therefore, in this case, asset prices tend to move along with the economic fundamentals. On the other hand, when these conditions do not hold, then the results indicate evidence of bubbles. For this reason, prior studies frequently investigated rational bubbles using integration methods.

In the standard setting, r is often assumed to be constant over time. Then, we can simplify Eq. (2):

$$P_t = E_t\left[\sum_{h=0}^{\infty}\left(\frac{1}{1+r}\right)^h D_{t+h}\right] + E_t\left[\left(\frac{1}{1+r}\right)^h P_{t+h}\right] \tag{3}$$

It follows that the current price comprises the present value of the market fundamentals ($P_t^* = E_t[\sum_{h=0}^{\infty}(\frac{1}{1+r})^h D_{t+h}]$) and bubbles ($B_t = E_t[(\frac{1}{1+r})^h P_{t+h}]$). In short,

$$P_t = P_t^* + B_t \tag{4}$$

Furthermore, subtracting a multiple of D from both sides of Eq. (3) and assuming that the transversality condition ($B_t \to 0$) holds,

$$rP_t - D_t = E_t\left[\sum_{h=0}^{\infty}\left(\frac{1}{1+r}\right)^h \Delta D_{t+h}\right] \tag{5}$$

This is probably the most popular theoretical explanation of a financial bubble and has been applied to the analysis of not only housing markets, but also the financial markets of equities and foreign exchange rates. Additionally, this model has been extended in many ways, such as by accommodating a time-varying r. Furthermore, as an extension to this model, Froot and Obstfeld (1991) proposed a concept of intrinsic bubbles that can be born out of rational behaviors and the fundamental determinants of asset prices. Intrinsic bubbles are created by a nonlinear function of the fundamentals. They proposed the following intrinsic bubble equation, which replaces B_t with $P_t^{*\gamma}$ in Eq. (4):

$$P_t = P_t^* + aP_t^{*\gamma} \tag{6}$$

When $a \neq 0$, then a bubble exists in the market. Using dividends as the economic fundamental in the analysis of the UK stock prices, they argued that a nonlinear solution to asset prices was consistent with investors' overreaction to news and showed that this concept offers a better explanation of stock prices than does the standard PVM.

While intrinsic bubbles are an alternative formation of bubbles and provide additional information about sources of rejecting the relationship between asset prices and economic fundamentals, we do not focus on this concept because the most recent statistical models are based on a linear PVM, as in Eq. (5). Furthermore, intrinsic bubbles are sensitive to the specification of the evolution of bubbles using economic fundamentals; moreover, an intrinsic bubble component is very small compared with the regime-switching components in the standard stock price model (Driffill and Sola 1998).

Another drawback of the standard PVM is that it is not possible to distinguish between positive and negative bubbles. Negative bubbles occur during periods of low asset prices and positive bubbles at times of high prices. There is a conceptual problem for negative bubbles because very low prices are not regarded as financial bubbles conventionally.[3] Therefore, it is price levels rather than the level of inflation that should serve as a criterion to determine bubble periods. In fact, distinguishing between negative and positive bubbles was discussed before. Blanchard and Watson (1982) proposed several evolution processes of bubbles, and similarly, Diba and Grossman (1998) described the evolution of bubbles as follows:

$$E(B_t) = (1 + r)B_t$$

which states that bubbles are independent of asset prices and generated by themselves. Then, we can state the expected value of bubbles following Blanchard and Watson (1982) as

$$\begin{aligned} \lim_{i \to \infty} E(B_{t+i}) &= +\infty \quad \text{if } B_t > 0 \\ &= -\infty \quad \text{if } B_t < 0 \end{aligned}$$

It follows that when $B_t > 0$, there is a chance of a bubble. On the other hand, when $B_t < 0$, $E(B_{t+i})$ may become large and negative in the future, which will make P_t negative. Negative prices are not realistic for most financial assets and real estate because in these cases, the solution to the model should be a case of no bubbles (Blanchard and Watson 1982).

3 Econometric Methods

Econometricians proposed many statistical methods, with statistical hypotheses that seem designed to be suitable from their perspectives. Consequently, some approaches were developed to look for tranquil periods, while others investigate financial bubbles. Unlike previous studies, we make a clear distinction between statistical approaches to determine tranquil and bubble periods. This distinction is important since differences in the statistical hypotheses can explain the different results from these two approaches. In this section, we will clarify these two approaches using popular statistical specifications in studies of bubbles.

---

[3] One exception is exchange rate markets, which we can think of as experiencing both negative and positive bubbles, and where the latter refers to currency crises.

To investigate the theoretical model and predictions in Eq. (2) quantitatively, previous studies often focused on a single market utilizing time series methods. These statistical methods are one-tailed tests, but can be broadly categorized into left- and right-tailed approaches according to their alternative hypotheses. The left-tailed test is classical and is designed to look for cointegration between prices and economic fundamentals, and thus revealing tranquil periods. As an extension, we also propose a nonlinear approach that can be classified into a group of left-tailed tests. On the other hand, the right-tailed test, which has become popular, is an approach to identify explosive bubbles.

3.1 Classical Approaches

Left-tailed integration tests are classical approaches. Indeed, the majority of previous analyses used this conventional approach; in particular, the relationship between housing prices and their economic fundamentals was investigated by a cointegration method (Engle and Granger 1987). The concept of cointegration is widely accepted by economists who established a theoretical framework to identify economic equilibrium conditions and led to Prof. Granger receiving a Nobel Prize (2003). Today, many applied studies used this concept to analyze housing markets worldwide (Hendry 1984; Meese and Wallace 2003; McGibany and Nourzad 2004; Gallin 2006; Adams and Fuss 2010; Oikarinen 2012; De Wit et al. 2013). Because most economic and financial data, including real estate prices and their economic fundamentals, follow a non-stationary process (e.g., Nelson and Plosser (1982), cointegration was considered appropriate to test their long-run relationship and bubbles.

The concept of cointegration can be summarized by rewriting it as a dynamic bivariate relationship. More specifically, to derive the long-run relationship between housing prices (y) and covariates (x), for the period ($t = 1, \ldots, T$), consider the following dynamic equation:

$$y_t = \alpha_0 + \rho_1 y_{t-1} + \beta_0 x_t + \beta_1 x_{t-1} + u_t \tag{7}$$

where the residual u is normally distributed ($u_t \sim N(0, \sigma^2)$). Both x and y are in natural the logarithmic form and are assumed to exhibit persistence, in line with many economic and financial variables. Then, we can transform Eq. (7) as follows:

$$\Delta y_t = \alpha_0 + \beta_0 \Delta x_t + (\rho_1 - 1)\left(y_{t-1} + \frac{\beta_0 + \beta_1}{\alpha_1 - 1} x_{t-1}\right) + u_t \tag{8}$$

or simply

$$\Delta y_t = a + b\Delta x_t + c_1(y_{t-1} + f x_{t-1}) + u_t \tag{9}$$

where Δ is the difference parameter and $c_1 = \rho_1 - 1$. When y is a housing price, $\Delta y_t = y_t - y_{t-1}$ represents housing price inflation. Parameters a, b, c_1, and f need to be estimated. The parameter b measures the short-term sensitivity of y to x, and c_1 measures the speed of the return to the long-run path. The parameter f is a vector of cointegrating parameters that summarize the long-run relationship between x and y, and $y_{t-1} + fx_{t-1}$ is the error correction mechanism (ECM). It is stationary, $I(0)$, in the presence of cointegration; in this case, the adjustment parameter c_1 will be $-1 < c_1 < 0$ according to the Granger representation theorem (Engle and Granger 1987). A value of parameter c_1 close to -1 indicates a quick return to the long-run path, and a value close to 0 indicates a slow adjustment. In contrast, when there is no cointegration, c_1 will not lie within this theoretical range, which implies that there are significant deviations of prices from economic fundamentals, which provides evidence of a bubble. Because financial bubbles are unobservable and are considered leftover (i.e., residuals) in Eq. (9), bubble analyses are sensitive to what comprises economic fundamentals.

Despite the importance of defining economic fundamentals, the selection of economic fundamentals remains controversial. The simplest approach used in many studies considers only one economic variable as a proxy for the economic fundamental, such as housing rental costs (Meese and Wallace 1994; Phillips and Yu 2011), mortgage interest rates (McGibany and Nourzad 2004), the price of residential land (Ooi and Lee 2006), or household income (Gallin 2006). The first two are often used to create the price-to-rent ratio and price-to-income ratio. While the former compares financial burdens between different choices of residential type, the latter measures the difficulty of purchasing houses, also known as the affordability ratio. Such investigations become univariate analyses by dropping x from Eq. (9). For y as the price-to-rent ratio, finding stationarity implies that housing prices and rent are cointegrated with a cointegrating parameter equal to unity, and the market is said to be tranquil. Thus, researchers can use unit root tests rather than cointegration tests.

$$\Delta y_t = a + c_1 y_{t-1} + u_t \tag{10}$$

The parameter of interest is c_1, and the interpretation of the parameters remains unchanged. Equation (10) is a specification of the Dicky–Fuller (DF) unit root test and can be extended to include past differenced ratios to create a more general form [i.e., the augmented DF unit root test (ADF)].

$$\Delta y_t = \alpha + c_2 y_{t-1} + \sum_{i=1}^{p} \theta_i \Delta y_{t-i} + \epsilon_t \tag{11}$$

where the residual $\epsilon_t \sim N(0, \sigma_\epsilon^2)$. For the classical unit root tests, the null hypothesis of a unit root ($c_2 = 0$) is tested against the alternative of stationarity ($c_2 < 0$), and thus, can be regarded as a statistical approach to determine a tranquil period. The non-stationary process of y is traditionally regarded as evidence of housing market bubbles (without distinguishing between negative and positive bubbles). The methods described thus far are conventional statistical models for analyzing housing markets and bubbles and are consistent with the traditional definition of bubbles in economics. That is, financial bubbles are phenomena in which housing prices do not move along with economic fundamentals in the long run.

3.2 Explosive Unit Root Tests

Financial bubbles are expected to occur occasionally and be recurrent (Blanchard and Watson 1982); furthermore, housing prices may be more chaotic and integrated of an order higher than one. In these cases, the classical unit root and cointegration tests possess only a weak statistical power for detecting bubbles (Evans 1991). To address these shortcomings, Phillips and Yu (2011) proposed conceptually different statistical methods based on Bhargava (1986) and Diba and Grossman (1998). Their tests are right-tailed and aim to examine a high level of a non-stationary process based on Eq. (11). They are designed to trace the orientation and collapse of bubbles, and thus to find chaotic moments (i.e., explosive bubbles) in financial markets. These statistical tests do not aim to determine tranquil periods.

Phillips et al. (2011) is based on the right-tailed test. Their motivation is (Phillips et al. 2011, p. 206), who state that "In the presence of bubbles, p_t is always explosive and hence cannot comove or be integrated with d_t if d_t is itself not explosive." Here, p_t is the log price, and d_t represents the log economic fundamentals. This is a subtle difference from the view of economists who pay most attention to cointegration between prices and economic fundamentals. Whether or not prices and economic fundamentals follow a unit root or explosive process is not their major interest. Economists claim evidence of tranquility as long as prices and economic fundamentals are cointegrated, regardless of the order of integration for each variable.

This can be seen in the alternative hypotheses of statistical tests. With the same null hypothesis as that of the classical ADF ($c_2 = 0$), Phillips and Yu (2011) suggested evaluating the right-tailed alternative of an explosive unit ($c_2 > 0$). Therefore, compared with the classical unit root tests that define bubbles as $I(1)$ under the null hypothesis, this alternative hypothesis has an implication for stronger bubbles. Thus, the explosive unit root test is conceptually different from the traditional test that looks for cointegration, that is, tranquil periods, and assumes the prevalence of financial bubbles in the market.

Phillips et al. (2011) and Phillips et al. (2015) proposed four types of explosive unit root tests: the right-tailed version of the conventional ADF (ADF), the rolling ADF (RADF), the supremum ADF (SADF), and the generalized SADF (GSADF) tests. The first test is the right-tailed version of the conventional ADF, with its statistic following a nonstandard distribution. Unlike the ADF, which utilizes all observations, the RADF shifts the starting and ending sample data points forward. The SADF is based on the recursive method, and thus the statistic is obtained by fixing the initial point (r_0) equal to the first observation in the data set, but extending the ending point (r_2) one by one for each successive run. The largest statistic obtained from the recursive method is used to evaluate the null hypothesis. Thus, for the period from

0 to r_2, in which r is a fraction of the total time period, the SADF statistic can be expressed as follows:

$$\text{SADF}(r_0) = \sup_{r_2 \in [r_0, 1]} \text{ADF}_0^{r_2}$$

The SADF statistic is consistent if there exists only one bubble, but is problematic if multiple bubbles exist. For this reason, Phillips et al. (2011) proposed the GSADF. It relaxes the SADF such that the initial observation (r_1) in the analysis does not need to be identical to the first observation in the data set. Therefore, the GSADF is considered the most reliable among the right-tailed integration tests. Further, Phillips et al. (2015) proposed a statistical method to identify a period of multiple bubbles, which has reasonable power (Homm and Breitung 2012). The GSADF statistic can be expressed as follows:

$$\text{GSADF}(r_0) = \sup_{\substack{r_2 \in [r_0, 1] \\ r_1 \in [0, r_2 - r_1]}} \text{ADF}_{r_1}^{r_2}$$

Recent studies used these unit root tests. Phillips et al. (2011) applied them to the US markets for housing, crude oil, and bonds, and Phillips and Yu (2011) and Phillips et al. (2015) applied them to the US stock markets (the NASDAQ stock exchange and the S&P500 Index). Kraussl et al. (2016) used these tests to examine bubbles in art markets. In addition to the application of the right-sided tests to different economic areas, researchers made further developments for real-time monitoring of financial bubbles. Phillips and Shi (2018) proposed what they call a recursive reverse-sample regression approach to reduce the delay bias in Phillips et al. (2011).

In short, it is important to note that while explosive tests are suitable for detecting explosive behaviors in data, they are not designed to investigate tranquil periods. Thus, within their framework, we cannot consider normality, a state in which the economy is believed to stay most of the time. This contradicts the choice of economic fundamentals by economists. For example, housing prices in theory are equal to the present value of housing services (or rents) in a steady state of a competitive economy (Blanchard and Watson 1982). There is no scope for the explosive unit root tests to accommodate this possibility.

We therefore examine two types of bubbles and refer to the case in which $c_2 > 0$ as an explosive bubble and the case in which $c_2 = 0$ (unit root) as a mild bubble. We can confirm the existence of a tranquil period when $c_2 < 0$ in Eq. (11). Thus, we can evaluate both conventional (mild) and explosive bubbles and summarize the interpretation of the statistical hypotheses of the classical and explosive ADF tests as follows.

Proposition 1 *Based on classical approaches, there is evidence of mild bubbles if* $c_2 = 0$ *in Eq. (11) and tranquility if* $c_2 < 0$ *in the housing market.*

Proposition 2 *Based on explosive tests, there is evidence of explosive bubbles if* $c_2 > 0$ *and mild bubbles if* $c_2 = 0$ *in the housing market.*

Below, we summarize the statistical hypotheses of the classical and explosive tests. The findings from these two tests do not necessarily need to be consistent, as they consider different data processes under statistical hypotheses.

Test type	Null hypothesis	Alternative hypothesis
Classical test	Mild bubbles ($d = 1, c_2 = 0$)	No bubbles ($d < 1, c_2 < 0$)
Explosive test	Mild bubbles ($d = 1, c_2 = 0$)	Explosive bubbles ($d > 1, c_2 > 0$)

3.3 Panel Approach

A single-country analysis can be extended to a study of financial bubbles in a multivariate context. Panel data estimation approaches often exploit cross-sectional information and increase statistical power. A multi-country analysis may be more appropriate because housing prices are highly correlated, particularly among advanced countries (see next section).

Pavlidis et al. (2016) extended the GADF statistics originally developed for single-country analyses by following Im et al. (2003), who proposed a left-tailed panel unit root test by extending the conventional univariate ADF test. In their approach, test statistics calculated for each country are pooled to create a single statistic that can be used to assess the statistical hypotheses in a panel context. For country k ($k = 1, \dots, K$), we can express the panel data version of Eq. (11) as follows:

$$\Delta y_{k,t} = \alpha_k + c_k y_{k,t-1} + \sum_{i=1}^{p} \theta_{k,i} \Delta y_{k,t-i} + \epsilon_{k,t} \tag{12}$$

where $\epsilon_{k,t} \sim N(0, \sigma^2_{\epsilon_k})$. The null hypothesis is $c_k = 0, \forall k$ against the alternative of explosive behaviors, $c_k > 0$ for some k. The noble feature of this approach is that it allows heterogeneity (i.e., c). However, a conclusion from this test becomes somewhat unclear, as the alternative hypothesis states. In other words, a rejection of the null does not necessarily mean that financial bubbles existed in all countries under investigation, but did in at least one country. To obtain a country-specific conclusion, country-wise analyses are required, as we summarized in the previous subsections.

For example, the panel GSADF can be constructed as the supremum of the panel backward SADF (BSADF). The panel BSADF is in turn obtained as the average of the SADF calculated backwardly for individual countries.

$$\text{Panel GSADF}(\text{r}_0) = \underset{\text{r}_2 \in [\text{r}_0, 1]}{\text{sub}} \quad \text{Panel BSADF}_{\text{r}_2}(\text{r}_0) \tag{13}$$

Given the possible cross-country dependence, we follow the calculation method in Pavlidis et al. (2016) closely and use a sieve bootstrap approach. The panel approach using cross-sectional information may be useful to understand a general trend in real estate prices in global markets.

3.4 T-ARDL Model

Finally, as an extension to the classical approach, we propose the T-ARDL model.[4] The linear ARDL is a classical method used to capture persistence in time series data, and Pesaran et al. (2001) proposed a bounds test to detect cointegration based on the ARDL. An advantage of this method is its ability to determine the presence of cointegration without prior knowledge of the explanatory variables being stationary ($I(0)$) or non-stationary ($I(1)$). This is a useful feature in studies of bubbles, as economies often experience periods of tranquility and mild bubbles.

Pesaran et al. (2001) proposed five specifications of the ARDL with a different combination of deterministic terms. Here, we use the most popular model in financial research, with an unrestricted constant and no trend. For asset prices (y), we can express this as

$$\Delta y_t = a + cy_{t-1} + \boldsymbol{b}\boldsymbol{x}_{t-1} + \sum_{i=1}^{p-1} \boldsymbol{d}_i' \Delta \boldsymbol{z}_{t-i} + \boldsymbol{f}' \Delta \boldsymbol{x}_t + u_t \quad (14)$$

where a, c, $\boldsymbol{b}$, $\boldsymbol{d}$, and $\boldsymbol{f}$ are the parameters to estimate by the ordinary least squares (OLS) for time ($t = 1, \ldots, T$), and u_t is the residual ($u_t \sim N(0, \sigma^2)$). $\boldsymbol{x}$ is a matrix of explanatory variables and $\boldsymbol{z} = [y, \boldsymbol{x}]$. The appropriate lag length (p) is determined such that it captures the data generating process of y. We can study the cointegrated relationship between y and $\boldsymbol{x}$ by analyzing the time series properties of $cy_{t-1} + \boldsymbol{b}\boldsymbol{x}_{t-1}$, known as the ECM. We can test the null hypothesis of no ECM ($c = 0$ and $\boldsymbol{b} = \boldsymbol{0}$) by the F test or $c = 0$ the t test.

As the conventional asymptotic distribution is invalid here, Pesaran et al. (2001) provided critical values based on Monte Carlo simulations for a different dimension of $\boldsymbol{x}$. Because economic variables may be $I(0)$ or $I(1)$, the critical values for these tests have both lower and upper bounds. For the F tests, the lower bound is determined when the data are $I(0)$, and the latter when they are $I(1)$. Test statistics above the upper bound imply evidence of cointegration, and those below the lower bound suggest the absence of cointegration. Test statistics between these bounds are inconclusive. For the t tests, the lower bound is determined when the data are $I(1)$, as the test statistics are expected to be negative. The upper bound is designed for $I(0)$ data.

However, this bounds testing approach is inappropriate for a study of bubbles because it investigates the possibilities of both negative and positive bubbles. That is, like the standard unit root tests, it considers bubbles even when housing prices

[4]This model is the simplest version of a regime-switching model and can be extended to a Markov-switching model.

are low. To treat bubbles as high price phenomena, we introduce nonlinearity into the ARDL as follows:

$$\Delta y_t = \alpha_1 g + \alpha_2 \widetilde{g} + c_1 g y_{t-1} + c_2 \widetilde{g} y_{t-1} + b_1 g \boldsymbol{x}_{t-1} + b_2 \widetilde{g} \boldsymbol{x}_{t-1} + \sum_{i=1}^{p-1} \boldsymbol{d}_i^{'} \Delta z_{t-i}$$
$$\boldsymbol{f}^{'} \Delta \boldsymbol{x}_t + u_t \tag{15}$$

where g and $\widetilde{g}$ are dummy variables that distinguish the regimes, and Eq. (15) has two (upper (g) and lower ($\widetilde{g}$)) regimes. When these regimes are determined by a certain threshold point (w), the dummies are defined as follows:

$$g = \begin{cases} 1 & \text{if } y > w \\ 0 & \text{otherwise} \end{cases} \quad \text{and} \quad \widetilde{g} = \begin{cases} 1 & \text{if } y \leq w \\ 0 & \text{otherwise} \end{cases} \tag{16}$$

To analyze the price-to-rent ratio when y already includes the economic fundamental variables, it becomes a univariate analysis by dropping $\boldsymbol{x}$ from (15). Furthermore, in the next section, we propose a top-down strategy for the T-ARDL model in order to determine the threshold point endogenously. To summarize, we can state the conditions of mild bubbles for this test in Proposition 3.

Proposition 3 *For housing prices that follow an I(1) process, evidence of mild bubbles exists if prices are higher than the historical average (or predictions from economic fundamentals) and if there is no cointegration relationship between prices and economic fundamentals (i.e., the F test statistic $<$ ucv, where ucv is an upper bound critical value, and/or the t test statistic $>$ lcv, where lcv is a lower bound critical value).*

4 Empirics

Using housing market data from advanced countries and the abovementioned statistical methods, we analyze the presence of tranquility and bubbles in real estate markets. We focus on advanced countries because the economic theory, for example, the link between housing prices and rents, assumes perfect and competitive markets, for which advanced countries provide a better proxy than do developing countries. Furthermore, longer sample period observations are available from advanced countries only.

4.1 Data

We obtain quarterly data on housing price-to-rent ratios from the OECD for the Euro area, Japan, the UK, and the USA (Table 1). The maximum sample period is from 1968 to 2018 (nearly 200 observations for each country), and the base year of the data is 2015. Here, the Euro area consists of 15 countries: Greece, France, the Slovak Republic, Italy, Spain, Belgium, Luxembourg, Germany, Portugal, the Netherlands, Finland, Ireland, Austria, Slovenia, and Estonia. (Hereafter, we refer to the Euro area as a country for convenience.) Because financial bubbles are traditionally considered infrequent phenomena, we chose countries with more than 195 observations. In terms of the standard deviations of this ratio, the US housing market is most stable, and the Japanese market is most volatile.

We also obtain housing price indices from the OECD. Housing prices appear to have a positive relationship with the price-to-rent ratios for all countries and exhibit more stable movements compared to prices (Fig. 1). Furthermore, while there are some similarities in the ratios across four countries, they have a declining trend in Japan during the "Lost Decades" (i.e., after 1990). This trend indicates relatively higher inflation in rental properties than houses in Japan and is indeed attributable to the deflation in housing prices according to this figure. This result indicates a weak demand for house purchases during the weak economic conditions of this period. In contrast, there is an increasing trend in the UK ratio from the late 1990s, indicating a housing market boom.

Table 2, which summarizes the correlation coefficients of the price-to-rent ratios and housing prices, also indicates the uniqueness of the Japanese market. On the one hand, housing prices are highly and positively correlated among countries. High correlation coefficients indicate that individual housing markets are highly integrated into global housing markets and are in similar stages of the business cycle, so housing markets are expected to respond similarly to global economic shocks. On the other

Table 1 Descriptive statistics: price-to-rent ratios

	Euro area	Japan	UK	US
Mean	95.009	128.039	73.532	98.310
Median	96.886	128.456	66.606	96.344
Maximum	117.140	187.226	115.279	126.591
Minimum	68.600	90.983	48.394	87.139
Std. Dev.	11.831	25.638	20.350	9.151
Skewness	−0.334	0.268	0.530	1.215
Kurtosis	2.561	2.088	1.866	4.206
Observations	195	195	202	195
Sample period	1970q1–2018q3	1970q1–2018q3	1968q2–2018q3	1970q1–2018q3

Notes Data source: OECD (2019), Housing prices (indicator). doi: 10.1787/63008438-en (accessed on 17 February 2019). 2015=100

Note: Sample period: 1970Q1-2018Q3. The Euro area consists of Euro 15 countries. OECD (2019), Housing prices (indicator). doi: 10.1787/63008438-en (Accessed on 17 February 2019).

Fig. 1 Housing prices and the price-to-rent ratios

Table 2 Correlations among the price-to-rent ratios

Variables	Euro area	Japan	UK	USA
Price-to-rent ratios				
Euro area	1.000			
Japan	−0.370	1.000		
UK	0.733	−0.655	1.000	
USA	0.534	−0.551	0.709	1.000
Housing prices				
Euro area	1.000			
Japan	0.289	1.000		
UK	0.974	0.119	1.000	
USA	0.985	0.255	0.982	1.000

Notes Sample period: 1970q1–2018q3. Data source: OECD (2019), Housing prices (indicator). doi: 10.1787/63008438-en (accessed on February 17, 2019). 2015=100

hand, although the price-to-rent ratios are also highly correlated among countries in general, there is a peculiarity in the Japanese market, in that it is negatively associated with the other markets. This result makes the Japanese market very unique among global housing markets.

We conduct a more formal analysis of the spillover effects between global housing markets using advanced statistical methods. Among the many statistical methods to measure spillovers, Diebold and Yilmaz (2009) proposed a method based on the vector autoregression (VAR) model to decompose aggregate spillovers into disaggregate and directional spillovers. Barunik and Krehlik (2018) extended their method to decompose spillovers using different frequencies of data (see the Appendix for a discussion of their method). Given that there may be time lag in spillovers, frequency analyses are interesting and give us a means to identify sources of spillovers. The short-term spillover is often believed to be influenced more by investors' expectations and the long-term spillovers by changes in the real economy (i.e., business cycles).

Here, we study spillovers across housing markets using the volatility of the price-to-rent ratios. Volatility is often measured by the squared or absolute value of returns (i.e., x^2 or $|x|$, where x is the price) in the finance literature. Using the latter proxy of volatility and VAR(2), we estimate the overall and frequency spillovers using OLS. We plot the results in Fig. 2. Furthermore, we decompose the overall spillovers over the short (within three months), middle (four months to one year), and long (over one year) time ranges. This frequency division is somewhat arbitrary, but shows that spillovers are sensitive to the frequency range, and spillovers seem to increase in the second half of our sample period (i.e., from this century). Furthermore, spillovers are more prominent over the medium to long term. In contrast, we observe no significant spillovers in housing markets at a high frequency. The differing connectedness by data frequency appears to be related to the characteristics of housing markets; residences

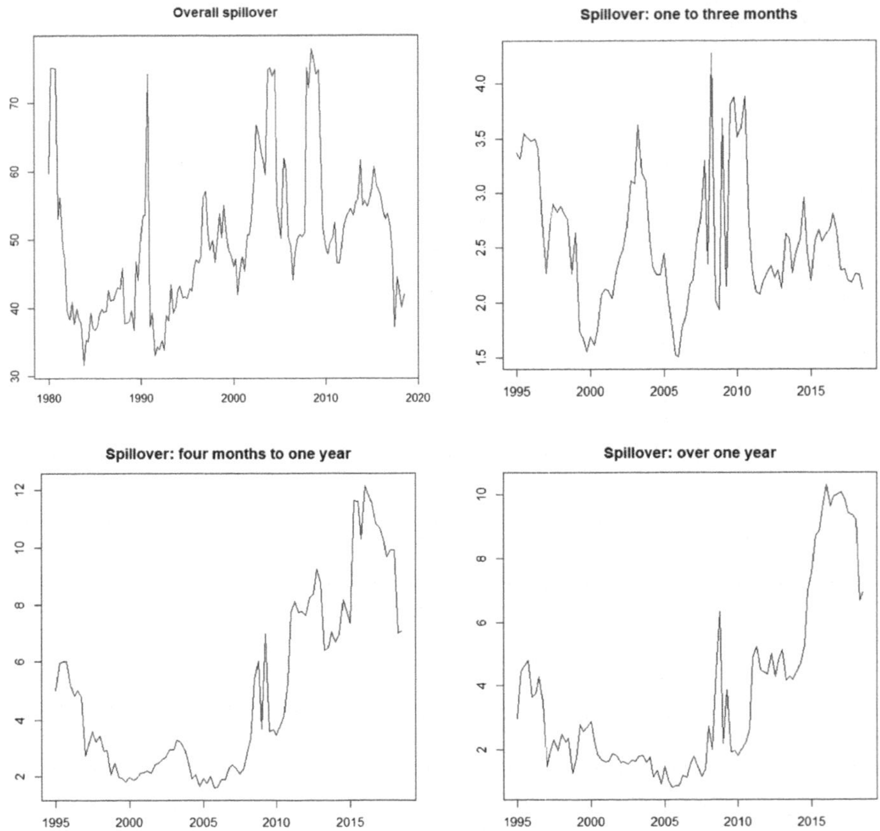

Note: Overall and frequency spillovers using price-to-rent ratios based on Diebold and Yilmaz (2012) and Barunik and Krehlik (2018) .

Fig. 2 Spillover

are traded much less frequently than financial assets are. In this regard, we infer that business cycle channels may be more relevant to explain the size of spillovers in housing markets.

4.2 Classical Test Approaches

We use three left-tailed unit root tests (the ADF, Phillips–Perron, and DF-GLS) that are popular univariate tests in economic and financial research. These tests investigate the null hypothesis that the price-to-rent ratio in levels follows the unit root process ($I(1)$), and a rejection of this null provides evidence of stationarity in this ratio, and thus cointegration between housing prices and rents. Therefore, a failure to reject

Table 3 Standard unit root tests for the price-to-rent ratios

Tests	Level	Difference	Level	Difference	1% cv	5% cv
	Euro area		Japan			
ADF	−2.495	−3.648	−1.565	−4.335	−3.465	−2.877
PP	−2.296	−4.592	−1.432	−4.486	−3.464	−2.876
DF-GLS	−0.174	−3.581	−1.213	−1.792	−2.577	−1.943
	UK		US			
ADF	−1.782	−5.243	−3.061	−4.461	−3.465	−2.877
PP	−1.227	−6.269	−1.967	−8.235	−3.464	−2.876
DF-GLS	−0.759	−5.223	−2.577	−2.053	−2.577	−1.943

Notes Sample period: 1970q1–2018q3. The null hypothesis of the augmented Dicky–Fuller (ADF), Phillips–Perron (PP), and Dickey–Fuller-generalized least squares (DF-GLS) tests examine the null hypothesis of a unit root process against the alternative of stationarity. The lag length is determined by the AIC with a maximum of four lags

this null hypothesis indicates that rents cannot explain the long-term housing price movements, thereby suggesting the presence of mild bubbles. We conduct these tests for the ratios in levels and first differences in order to check the order of integration.

Table 3 summarizes the test statistics for the Euro area, Japan, the UK, and the USA. The results suggest that these ratios follow the unit root process. Using the 5% critical values, we often fail to reject the null hypothesis for the data in levels, but can do so for the differenced data. Therefore, we conclude that mild bubbles existed in all countries, suggesting that rental increases are not always associated with housing price inflation, and there must be some periods when housing prices deviate substantially from the trend in rentals. Obviously, these tests preclude a possibility of explosive bubbles, and moreover, we need to pay attention to the composition of economic fundamentals. However, these outcomes are consistent with our expectations that all housing markets experienced chaotic moments during our sample period.

The potential non-stationary periods identified by the classical method are shaded in Fig. 3. We present two graphs for each country, and the upper figures (denoted as OLS estimates) are obtained from the classical method. As explained earlier, a drawback of this approach is the lack of statistical power to differentiate between hypotheses and that it allows for negative bubbles. For consistency with the standard phenomenon of financial bubbles, we should consider only the positive bubbles (above the horizontal line) as relevant to financial bubbles. Thus, only positive bubble periods are highlighted in gray in this figure and are potential bubble periods because these classical tests are not designed to identify the exact periods of bubbles while they give us evidence of mild bubbles during the sample period.[5]

[5]The identification of mild bubble periods can be made more clearly using a recursive or rolling method, which will be discussed in the T-ARDL.

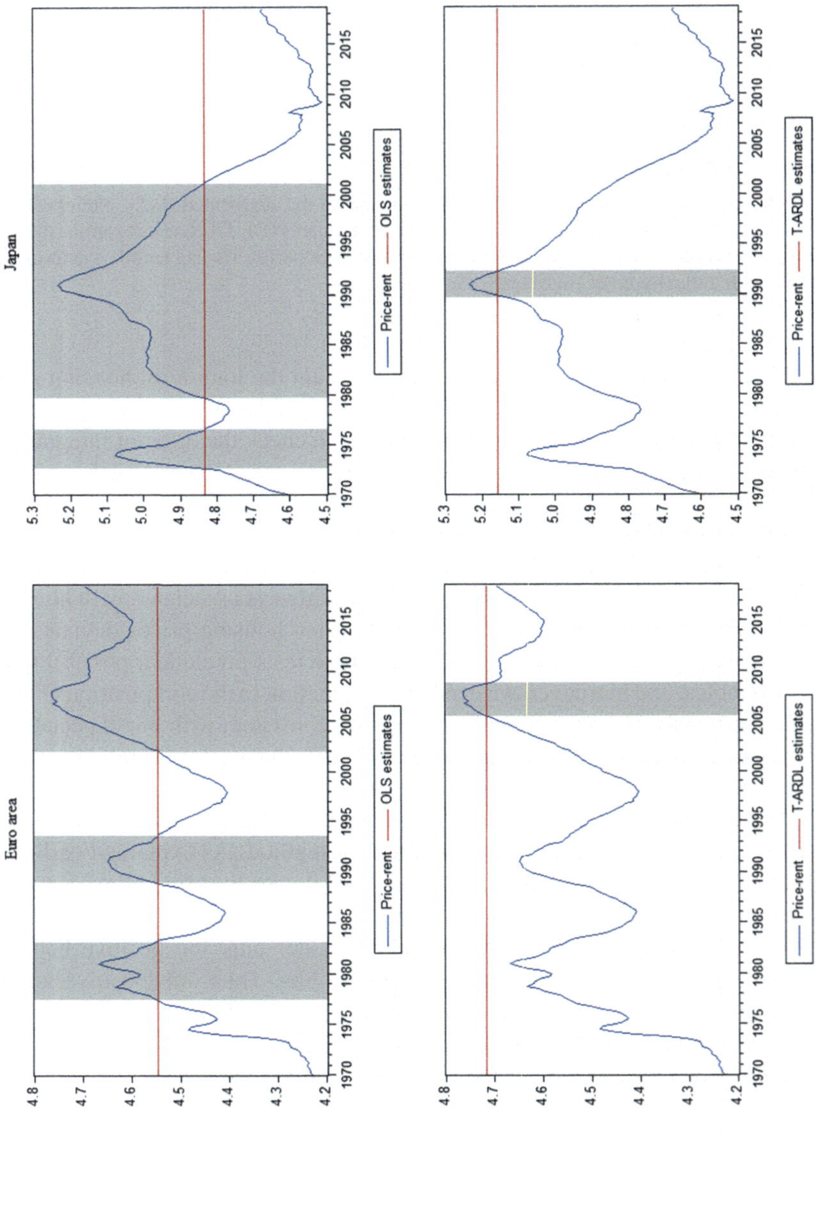

Fig. 3 OLS and threshold estimation results

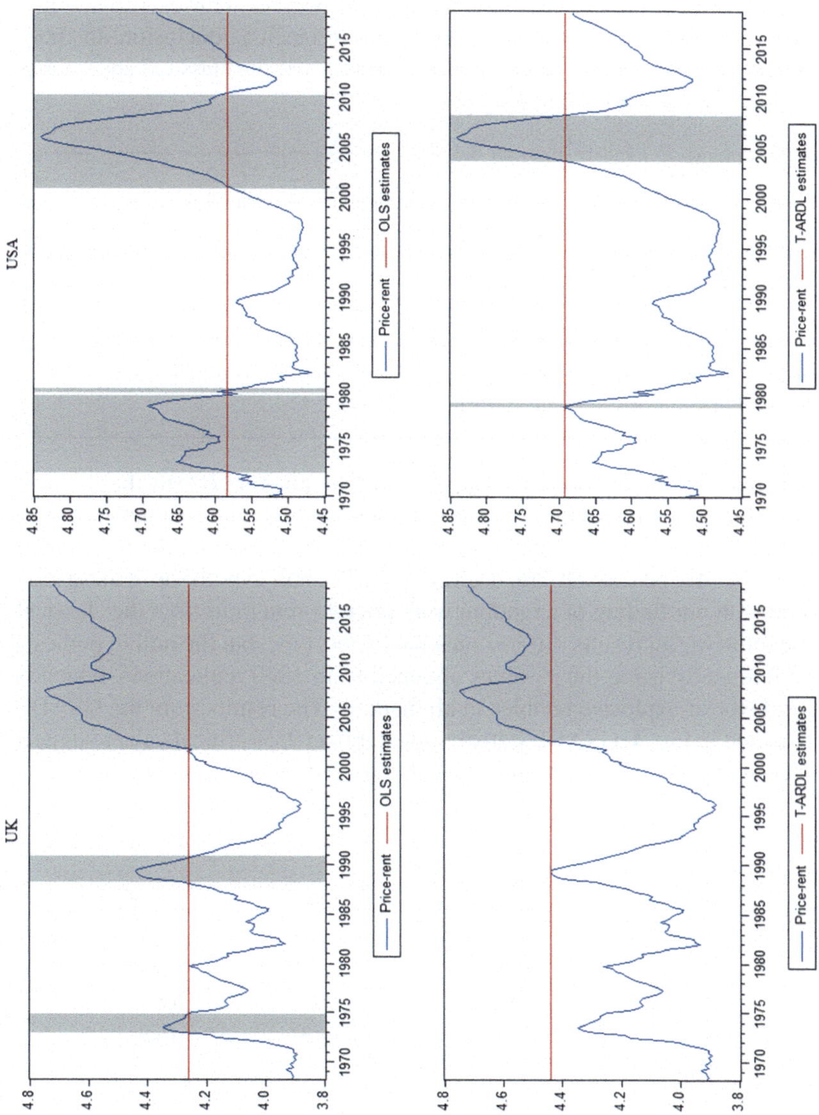

Sample period: 1970Q1-2018Q3. The Euro area consists of 15 Euro countries.

Fig. 3 (continued)

4.3 Explosive Test Approaches

Next, we conduct explosive unit root tests for each country and for a group of countries. From the classical tests, we already know that the price-to-rent ratio is non-stationary and, in fact, implies the presence of mild bubbles. However, when it is not known, we propose the following general steps to reach a conclusion. In short, explosive unit root tests should be conducted if and only if the classical approaches show the data (y) to be a non-stationary process.

A general approach to identify financial bubbles

1. Use the classical approaches to check for the presence of tranquil periods
2. If the null hypothesis of $I(1)$ cannot be rejected, then go to Step 3; otherwise, conclude that the housing market is tranquil.
3. Conduct the explosive bubble tests. If the null hypothesis of these tests is rejected, then conclude the presence of explosive bubbles; otherwise, conclude the presence of mild bubbles in the housing market.

Failing to reject the null hypothesis that price-to-rent ratios are $I(1)$ by the classical tests, we eliminated the possibility of market tranquility and thus conduct explosive unit root tests for each market. Table 4 summarizes the results from the right-tailed tests (RADF, SADF, and GSADF) for each country. The null hypothesis of these tests is consistent with our finding of a random walk price-to-rent ratio from the classical tests. The explosive test results differ somewhat by test type, but the null hypothesis is rejected frequently using the p-values obtained from 1000 replications, which is evidence in favor of explosive bubbles in all markets. The results from the GSADF are also depicted in Fig. 4. GSADF statistics greater than 95% critical values suggest the presence of explosive bubbles, which are also shaded in this figure, and generally

Table 4 Explosive unit root tests for the price-to-rent ratios

	Statistic	p-value	Statistic	p-value
	Euro area		Japan	
RADF	3.808	0.000	9.827	0.000
SADF	0.714	0.214	0.201	0.436
GSADF	4.045	0.000	10.575	0.000
	UK		US	
RADF	3.251	0.000	9.539	0.000
SADF	0.878	0.166	2.422	0.000
GSADF	4.162	0.000	15.336	0.000

Notes p-values obtained from 1000 replications. The rolling ADF (RADF), the supremum ADF (SADF), and the generalized SADF (GSADF) are explosive unit root tests of the null hypothesis of a unit root against the alternative of an explosive process

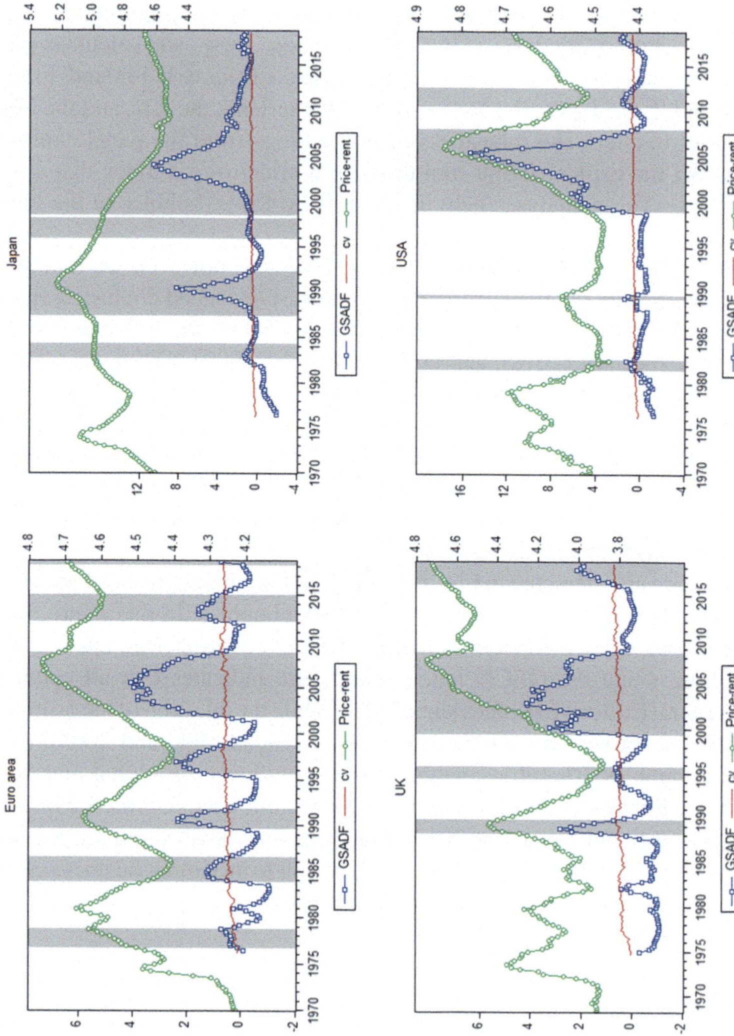

Note: Sample period: 1970Q1-2018Q3. Euro area consists of Euro 15 countries.

Fig. 4 GSADF results

identify explosive bubble periods when housing prices are high. The timing and duration of explosive bubbles differ among countries, but many countries seemed to experience explosive bubbles before the Lehman Brothers collapse in September 2018. The presence of real estate markets is consistent with the results from the classical approaches, but here we have evidence of explosive bubbles, which the classical approach cannot capture.

Given the cross-sectional dependence in international housing markets, we also conduct a multi-country analysis using two methods. First, we calculate panel GSADF statistics using the data of 12 countries: Canada, Denmark, Finland, France, Germany, Ireland, Italy, Japan, the Netherlands, Switzerland, the UK, and the USA. Second, in order to check the robustness of the findings from the panel explosive tests, we conduct the explosive unit root tests for a price-to-rent index that covers OECD countries.[6] These analyses help us identify explosive bubbles in the global housing market.

The country coverage differs slightly in these analyses; however, we find many similarities in the results obtained from these groups of advanced countries. In particular, we observe that explosive bubbles existed just before the Lehman Brothers collapse in 2008, recently (2018), and at the end of the 1980s. The panel GSADF and GSADF statistics for the group of OECD countries are 1.882 and 3.157, respectively, which are greater than the 95% critical values of 0.349 and 2.169. This result confirms that explosive bubbles prevailed in the global housing market in the past decades. It follows that, despite some peculiarities of the Japanese market based on the correlation coefficients, there are similarities in terms of the timing of the evolution of explosive bubbles across countries.

These results are also depicted in Fig. 5. The GSADF statistics greater than critical values (cv, shaded) indicate the presence of explosive bubbles. In this figure, panel indicates the results using data from individual OECD countries (the upper graph) and OECD those using the OECD index. The figure indicates that all countries experienced several explosive bubbles since 1970. While the duration of the estimated bubbles differs by market, in some instances (e.g., around 1990 and 2005), many countries faced bubbles. Therefore, we confirm the spillovers in global real estate markets, and the global spillovers resemble regional spillovers in domestic housing markets (Montagnoli and Nagayasu 2015). Our frequency analysis in the previous section suggests that global market spillovers may result from similar business cycles, and common shocks occurred in the group of countries.

[6]The OECD countries are Greece, Italy, Finland, Korea, Switzerland, France, Japan, Poland, Belgium, Denmark, Chile, Estonia, Germany, Spain, Slovenia, Israel, the UK, Luxembourg, Norway, the Netherlands, the Slovak Republic, the USA, Mexico, Austria, Australia, Lithuania, Sweden, Portugal, Latvia, Ireland, the Czech Republic, Hungary, New Zealand, Canada, Turkey, and Iceland.

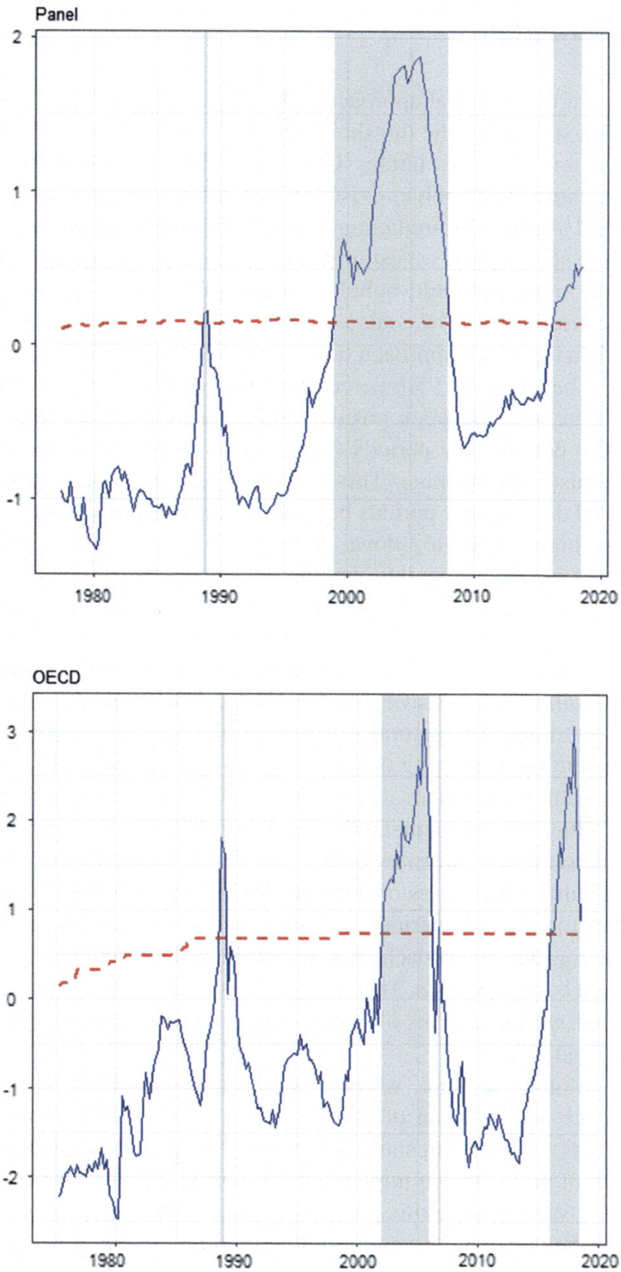

Note: Sample period: 1970Q1-2018Q3.

Fig. 5 Multi-country analyses

4.4 T-ARDL Approach

As an extension of the classical approaches that focused on individual markets, we apply a top-down strategy to the threshold ARDL (T-ARDL) method to distinguish between positive and negative bubbles. Choosing a threshold point requires careful consideration because mild bubbles exist only when housing prices are much higher than the historical average (or predictions from economic fundamentals). Moreover, like the explosive approaches, we estimate this method by changing the sample.

To choose the timing of mild bubbles endogenously, we suggest a theoretical threshold range within which we calculate the test statistics for each percentile of housing prices in a top-down approach (i.e., moving down on the y-axis). We use the theoretical range between the 95th percentile and the historical average. The latter allows us to differentiate between positive and negative bubbles. Importantly, Eq. (15) regards even deflationary periods ($\Delta y < 0$) as potential bubbles if prices are higher than the historical average. This is a more realistic assumption of bubbles than excluding all deflationary periods because real estate markets are volatile, even when prices are high. In the top-down approach, we estimate the T-ARDL from high to low prices and continue until we find evidence against the null in the upper regime. Finding a significant ECM points to tranquil periods, during which economic fundamentals can explain the trend in real estate prices.

This top-down approach has several advantages over the conventional approach, which relies on rolling or recursive methods (i.e., moving sideways on the x-axis), and is often used to detect structural shifts. First, the top-down approach does not require that we trim the beginning or end of the sample periods, which we need for the initial estimation. Researchers are often interested in an extreme sample periods. Second, like a rolling estimation method, we need not fix an unknown rolling window period to calculate the test statistics over time. Third, unlike rolling and recursive methods that require calculations of the test statistics at each point in time, further analyses to determine bubble periods are unnecessary once we find evidence of no bubble from the top-down approach. A y greater than the estimated threshold point ($\bar{y}$) indicates mild bubble periods. Therefore, this approach is more efficient than the conventional approaches in terms of computational time. In this way, mild bubbles can be detected endogenously.

To summarize our procedure, we propose the following top-down approach to study the presence and duration of bubbles based on the t statistics. Initially, we conduct the bounds test at the threshold point of the 95th percentile of the data. Failing to reject the null of no ECM, we implement this test again, this time decreasing the threshold by 1%. We continue this process until we find evidence against the null hypothesis (i.e., the t statistics become smaller than lcv) in the upper regime where $i = 0$–45 (see Fig. 6).

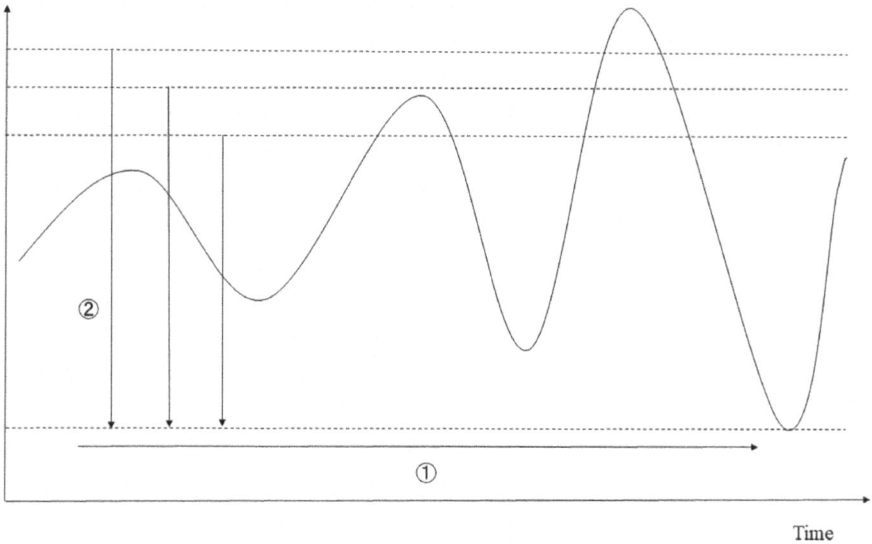

Note: ① indicates the standard (i.e., recursive or rolling) method to detect structural breaks, and ② represents a top-down method.

Fig. 6 Image of the top-down approach

Top-down strategy

For i (i=0 to 45):

1. Determine the initial threshold point using a percentile of y (say, 95%)
2. Estimate the T-ARDL excluding the top 5% (=100-95+i) values of y
3. Evaluate the null hypothesis of $b_1 = 0$ (i.e., no ECM) in the upper regime
4. Return to Step 2 if the test statistic fails to reject the null; otherwise, end.

Figure 7 plots the sequence of test statistics, along with the 5% critical value, and shows that we can reject the null hypothesis after 2, 1, 27, and 5 iterations of the test for the Euro area, Japan, the UK, and the USA, respectively. As we start with at the 95th percentile of the data, we can regard the 93rd, 94th, 68th, and 90th percentile points as the thresholds for these countries. We report the estimation results in Table 5, though do not show the results for the parameters ds in Eq. (15) as they are insignificant and were removed from the model.

Figure 3 plots mild bubbles (shaded areas) predicted from the T-ARDL. As expected, possible periods of mild bubbles are sharpened now; there is evidence of shorter bubble periods from the T-ARDL compared with the standard classical test. This is noticeable for all countries. Compared with the OLS estimates, the evi-

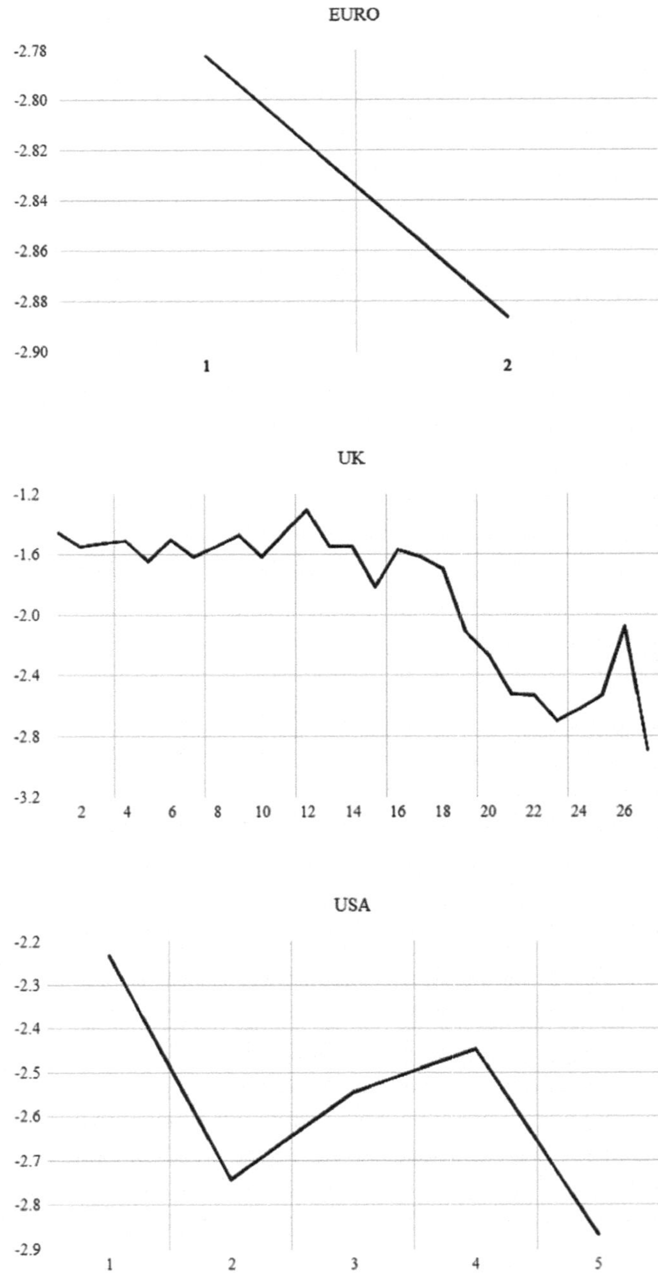

Note: Sample period: 1970Q1-2018Q3. Euro area consists of Euro 15 countries.

Fig. 7 Sequence of t-statistics from the threshold estimation

Table 5 T-ARDL estimation results

	Coefficient	*t*-statistic	Coefficient	*t*-statistic
	Euro area		Japan	
Threshold point	93 percentile		94 percentile	
a_2	0.103	2.901	0.105	4.290
b_2	−0.022	−2.886	−0.022	−4.311
	UK		USA	
Threshold point	68 percentile		90 percentile	
a_2	0.204	2.927	0.136	2.883
a_2	−0.049	−2.891	−0.030	−2.867

Notes Full example. The threshold points are determined by the top-down approach. The model specification is Eq. (15)

dence of bubbles immediately after the Lehman Shock (September 2008) disappears when we use the T-ARDL, and this T-ARDL result is consistent with an underling method that allows us to focus only on the upper regime.

4.5 Discussion

In this study, we considered three states of an economy; none, mild, and explosive bubble periods, and classified statistical approaches into two groups by their research objectives. In classical methods, non-tranquil periods can be considered mild bubble periods, and non-explosive bubble periods are equivalent to mild bubble periods in the explosive bubble analyses. Therefore, identifying explosive bubbles is of primary interest in the recent development in statistical approaches, while economists and policymakers are probably more interested in tranquil periods that may be identified by economic fundamentals (i.e., economic theories) and the classical approaches. The results are naturally sensitive to the test types; therefore, researchers need to decide the research objectives first and adopt an appropriate research framework accordingly.

Table 6 summarizes the possible mild and explosive bubble periods identified by the classical and explosive statistical approaches, which we explained using figures in the previous section. We determined the mild bubble periods from the classical approaches, which we denote as the OLS and the T-ARDL results in the table. These periods are equivalent to ones in which the price-to-rent ratio deviates substantially from the economic fundamental, suggesting no long-run relationship between housing and rental prices. The results from the OLS and T-ARDL differ slightly, and the latter tends to identify shorter periods for mild bubbles. This outcome shows that the threshold approach focuses on only housing prices that are historically expensive, while the OLS considers both expensive and inexpensive periods as potential bubble periods. The explosive bubbles obtained from the GSASF suggest a greater number

Table 6 Periods of mild and explosive bubbles

Mild OLS	Explosive GSADF	Mild T-ARDL	Common	Mild OLS	Explosive GSADF	Mild T-ARDL	Common
Euro area				Japan			
1977q3–1983q1	1976q4–1978q4	2005q3–2008q4	2005q3–2008q4	1972q4–1976q3	1982q4–1984q2	1989q4–1992q2	1989q4–1992q2
1989q1–1993q3	1984q1–1986q3			1979q4–2001q1	1987q3–1992q2		
2002q1–2018q3	1989q4–1991q4				1996q1–1998q1		
	1995q4–1998q4				1998q3-2018q3		
	2002q1–2008q4						
	2012q2–2015q1						
	2018q2–2018q3						
UK				USA			
1973q1–1974q4	1988q2–1989q4	2002q4–2018q3	2002q4–2008q3	1972q3–1980q1	1981q3–1982q3	1979q1–1979q2	2003q4–2008q1
1988q2–1990q4	1994q3–1995q4		2016q2–2018q3	1980q3–1980q4	1989q3–1989q4	2003q4–2008q1	
2001q4–2018q3	1999q3—2008q3			2001q1–2010q2	1999q1–2008q1		
	2016q2–2018q3			2013q3–2018q3	2010q3–2012q3		
					2017q2–2018q3		

Notes Full sample. Mild bubble periods (Mild) are those from the classical approaches (i.e., OLS and T-ARDL), and explosive bubble periods (Explosive) are from the explosive unit root tests (GSADF)

of shorter periods for bubbles overall. As expected, the three tests predict different bubble periods, but the predictions show a lot of overlap. These overlapping periods are also reported as common periods in this table and are the mostly strongly supported bubble periods. In short, we reported several results from different statistical approaches and objectives in financial bubble studies.

5 Conclusion

Housing prices, especially abnormally high prices, are of interest to many people and policymakers because the residence is necessary, but can be very expensive. Consequently, property prices are a long-standing, popular research area. However, recent research appears to be diversified in terms of objectives with respect to detecting financial bubbles and tranquil periods. The methodological approach using statistics to identify explosive bubbles gives only limited economic implications from the data analysis. Given that such research does not give any scope for markets being in equilibrium, it is of less interest to economists and policymakers, who are concerned more with tranquil periods and are interested in providing economic explanations of how the economy changes over time.

Furthermore, we proposed applying a top-down method to the T-ARDL to analyze rational bubbles. While we presented a very simple example and further research is necessary to define the economic fundamentals, this top-down strategy is more efficient than the traditional one is and can be applied to other statistical tests (e.g., unit root and cointegration tests) that attempt to analyze rational bubbles.

Finally, future developments in statistics will be useful, particularly if the null and alternative hypotheses could be better designed to make both economic and statistical sense. An example of such a research direction may be to develop statistical tests evaluating the null hypothesis of no bubbles against the alternative of bubbles of any type. Another example, which is more radical and the most useful direction of statistical development, may be described as two-sided tests. Research along this line would be quite challenging, but useful in synthesizing economic and statistical methodologies.

Acknowledgements The earlier version of this paper was presented at the annual meeting of the Nippon Finance Association. I modified and extended the conference paper (Nagayasu 2016) substantially. I would like to thank Naoya Katayama and the conference participants for constructive comments. However, all remaining errors are mine.

Appendix

In deregulated markets, both domestic and foreign shocks are expected to influence economic development. Therefore, it is important to understand co-movements in global markets, even in an analysis of housing markets, which represent non-tradable goods. Among others, Diebold and Yilmaz (2009, 2012) and Barunik and Krehlik (2018) contributed recent methodological developments. The former authors proposed measurements for the total and directional spillovers, and the latter incorporated a frequency domain in these spillovers. These techniques were developed based on the decomposition method known as generalized impulse response functions, which are invariant to the order of variables in the VAR (Pesaran and Shin 1998).

Consider a stationary N-variate VAR(p) with the errors $\epsilon \sim (0, \sigma)$.

$$x_t = \sum_{i=1}^{p} \Phi_i x_{t-i} + \epsilon_t$$

The moving average representation of this VAR is

$$x_t = \sum_{i=0}^{\infty} A_i \epsilon_{t-i}$$

where $A_i = \Phi_1 A_{i-1} - \Phi_2 A_{i-2} + \cdots + \Phi_p A_{i-p}$. The proportion of the kth variable to the variance of the forecast errors of element j, for the h time horizon, is

$$(\theta_H)_{j,k} = \frac{\sigma_{k,k}^{-1} \sum_{h=0}^{H} ((A_h \sigma)_{j,k})^2}{\sum_{h=0}^{H} (A_h \sigma A_h')_{j,j}}$$

The row of θ_H does not need to be equal to one, and it can be normalized by the row sum.

$$(\tilde{\theta}_H)_{j,k} = \frac{(\theta_H)_{j,k}}{\sum_{k=1}^{N} (\theta_H)_{j,k}}$$

$(\tilde{\theta}_H)_{j,k}$ measures a pairwise spillover from k to j, and by construction, $\sum_{k=1}^{N} (\tilde{\theta}_H)_{j,k} = 1$ and $\sum_{j,k=1}^{N} (\tilde{\theta}_H)_{j,k} = N$. We can obtain a total spillover index by aggregating each pair. That is,

$$C_H = \frac{\sum_{j \neq k} (\tilde{\theta}_H)_{j,k}}{\sum \tilde{\theta}_H} \times 100$$

Barunik and Krehlik (2018) defined a scaled generalized variance decomposition on a specific frequency band, $d = (a, b)$, where $a < b$ and $a, b \in (-\pi, \pi)$, which is from a set of intervals D.

$$(\tilde{\theta}_d)_{j,k} = \frac{(\theta_d)_{j,k}}{\sum_k (\theta_\infty)_{j,k}}$$

Then, we can define the frequency spillover on the frequency band d as follows:

$$C_d = \left(\frac{\sum \tilde{\theta}_d}{\sum \tilde{\theta}_d} - \frac{\mathrm{Tr}\{\tilde{\theta}_d\}}{\sum \tilde{\theta}_d} \right) \times 100$$

Each component of C_d can be obtained by defining the generalized caution spectrum over frequencies $\omega \in (-\pi, \pi)$ as (Barunik and Krehlik 2018):

$$(f(\omega))_{j,k} \equiv \frac{\sigma_{kk}^{-1} \left| (\Psi(e^{-i\omega}) \Sigma)_{j,k} \right|^2}{(\Psi(e^{-i\omega}) \Sigma \Psi'(e^{+i\omega}))_{j,j}}$$

where $\Psi(e^{-i})$ is the Fourier transform of the impulse response. It is a proportion of the spectrum of variable j in response to shocks in variable k. Then, we can specify the components of C_d as follows:

$$(\theta_{\infty})_{j,k} = \frac{1}{2\pi}\int_{-\pi}^{\pi} \Gamma_j(\omega)(f(\omega))_{j,k}\mathrm{d}\omega$$

and

$$(\theta_{d})_{j,k} = \frac{1}{2\pi}\int_{d} \Gamma_j(\omega)(f(\omega))_{j,k}\mathrm{d}\omega$$

where

$$\Gamma_j(\omega) = \frac{(\Psi(\mathrm{e}^{-\mathrm{i}\omega})\Sigma\Psi'(\mathrm{e}^{+\mathrm{i}\omega}))_{j,j}}{\frac{1}{2\pi}\int_{-\pi}^{\pi}(\Psi(\mathrm{e}^{-\mathrm{i}\lambda})\Sigma\Psi'(\mathrm{e}^{+\mathrm{i}\lambda}))_{j,j}\mathrm{d}\lambda}$$

References

Abreu, D., & Brunnermeier, M. K. (2003). Bubbles and crashes. *Econometrica, 71*(1), 173–204. https://doi.org/10.1111/1468-0262.00393.

Adams, Z., & Fuss, R. (2010). Macroeconomic determinants of international housing markets. *Journal of Housing Economics, 19*, 38–50.

Barunik, J., & Krehlik, T. (2018). Measuring the frequency dynamics of financial connectedness and systemic risk. *Journal of Financial Econometrics, 16*(2), 271–296.

Bhargava, A. (1986). On the theory of testing for unit roots in observed time series. *Review of Economic Studies, 53*(3), 369–384.

Blanchard, O. J., & Watson, M. W. (1982). *Bubbles, rational expectations and financial markets* (Working Paper 945). National Bureau of Economic Research.

Choi, J. J., Kedar-Levy, H., & Yoo, S. S. (2015). Are individual or institutional investors the agents of bubbles? *Journal of International Money and Finance, 59*, 1–22.

Choi, N., & Skiba, H. (2015). Institutional herding in international markets. *Journal of Banking & Finance, 55*, 246–259.

De Bondt, W. F. M., & Thaler, R. (1985). Does the stock market overreact? *Journal of Finance, 40*(3), 793–805. https://doi.org/10.1111/j.1540-6261.1985.tb05004.x.

De Long, J. B., Shleifer, A., Summers, L. H., & Waldmann, R. J. (1990). Noise trader risk in financial markets. *Journal of Political Economy, 98*(4), 703–738.

De Wit, E. R., Englund, P., & Francke, M. K. (2013). Price and transaction volume in the dutch housing market. *Regional Science and Urban Economics, 43*, 220–241.

Diba, B. T & Grossman, H. I. (1988). Explosive rational bubbles in stock prices? *American Economic Review, 78*(3), 520–530.

Diebold, F. X., & Yilmaz, K. (2009). Measuring financial asset return and volatility spillovers, with application to global equity markets. *Economic Journal, 119*(534), 158–171. https://doi.org/10.1111/j.1468-0297.2008.02208.x.

Diebold, F. X., & Yilmaz, K. (2012). Better to give than to receive: Predictive directional measurement of volatility spillovers. *International Journal of Forecasting, 28*(1), 57–66.

Driffill, J., & Sola, M. (1998). Intrinsic bubbles and regime-switching. *Journal of Monetary Economics, 42*(2), 357–373.

Easley, D., Hvidkjaer, S., & O'Hara, M. (2002). Is information risk a determinant of asset returns? *Journal of Finance, 57*(5), 2185–2221. https://doi.org/10.1111/1540-6261.00493.

Engle, R. F., & Granger, C. W. J. (1987). Co-integration and error correction: Representation, estimation, and testing. *Econometrica, 55*(2), 251–276.

Evans, G. (1991). Pitfalls in testing for explosive bubbles in asset prices. *American Economic Review*, *81*(4), 922–30.
Fama, E. F. (1970). Efficient capital markets: A review of theory and empirical work. *Journal of Finance*, *25*(2), 383–417.
Froot, K. A., & Obstfeld, M. (1991). Intrinsic bubbles: The case of stock prices. *American Economic Review*, *81*(5), 1189–1214.
Gallin, J. (2006). The long-run relationship between house prices and income: Evidence from local housing markets. *Real Estate Economics*, *34*(3), 417–438.
Gibbons, M. R., & Hess, P. (1981). Day of the week effects and asset returns. *Journal of Business*, *54*(4), 579–596.
Gurkaynak, R. S. (2008). Econometric tests of asset price bubbles: Taking stock. *Journal of Economic Surveys*, *22*(1), 166–186.
Hendry, D. (1984). *Econometric modelling of house prices in the United Kingdom*. Oxford: Basil Blackwell.
Hirshleifer, D., & Shumway, T. (2003). Good day sunshine: Stock returns and the weather. *Journal of Finance*, *58*(3), 1009–1032. https://doi.org/10.1111/1540-6261.00556.
Homm, U., & Breitung, J. (2012). Testing for speculative bubbles in stock markets: A comparison of alternative methods. *Journal of Financial Econometrics*, *10*(1), 198–231.
Im, K. S., Pesaran, M. H., & Shin, Y. (2003). Testing for unit roots in heterogeneous panels. *Journal of Econometrics*, *115*(1), 53–74.
Kraussl, R., Lehnert, T. & Martelin, N. (2016). Is there a bubble in the art market? *Journal of Empirical Finance, 35*(C), 99–109.
McGibany, J. M., & Nourzad, F. (2004). Do lower mortgage rates mean higher housing prices? *Applied Economics*, *36*(4), 305–313. https://doi.org/10.1080/0003684041000l674231.
Meese, R., & Wallace, N. (1994). Testing the present value relation for housing prices: Should I leave my house in San Francisco? *Journal of Urban Economics*, *35*(3), 245–266.
Meese, R., & Wallace, N. (2003). House price dynamics and market fundamentals: The Parisian housing market. *Urban Studies*, *40*, 1027–1045.
Montagnoli, A., & Nagayasu, J. (2015). UK house price convergence clubs and spillovers. *Journal of Housing Economics, 30*(C), 50–58.
Nagayasu, J. (2016). A top-down method for rational bubbles: Application of the threshold bounds testing approach. In *Nippon Finance Association Proceedings*.
Nagayasu, J. (2018). *Condominium prices and inflation: the role of financial inflows and transaction volumes in Japan*, DSSR Discussion Papers 76. Graduate School of Economics and Management, Tohoku University, Japan.
Nelson, C. R. & Plosser, C. I. (1982). Trends and random walks in macroeconmic time series: some evidence and implications. *Journal of Monetary Economics, 10*(2), 139–162.
Oikarinen, E. (2012). Empirical evidence on the reaction speeds of housing prices and sales to demand shocks. *Journal of Housing Economics*, *21*, 41–54.
Ooi, J., & Lee, S.-T. (2006). Price discovery between residential & housing markets. *Journal of Housing Research*, *15*(2), 95–112.
Pavlidis, E., Yusupova, A., Paya, I., Peel, D., Martínez-García, E., Mack, A., et al. (2016). Episodes of exuberance in housing markets: In search of the smoking gun. *Journal of Real Estate Finance and Economics*, *53*(4), 419–449.
Pesaran, H., & Shin, Y. (1998). Generalized impulse response analysis in linear multivariate models. *Economics Letters*, *58*(1), 17–29.
Pesaran, M. H., Shin, Y., & Smith, R. J. (2001). Bounds testing approaches to the analysis of level relationships. *Journal of Applied Econometrics*, *16*(3), 289–326.
Phillips, P. C., & Shi, S.-P. (2018). Financial bubble implosion and reverse regression. *Econometric Theory*, *34*(4), 705–753.
Phillips, P. C., & Yu, J. (2011). Dating the timeline of financial bubbles during the subprime crisis. *Quantitative Economics*, *2*, 455–491.

Phillips, P. C. B., Shi, S., & Yu, J. (2015). Testing for multiple bubbles: Historical episodes of exuberance and collapse in the S&P 500. *International Economic Review*, *56*(4), 1043–1078.
Phillips, P. C. B., Wu, Y., & Yu, J. (2011). Explosive behavior in the 1990s Nasdaq: When did exuberance escalate asset values? *International Economic Review*, *52*(1), 201–226.
Rozeff, M. S., & Kinney, W. R. (1976). Capital market seasonality: The case of stock returns. *Journal of Financial Economics*, *3*(4), 379–402.
Shiller, R. J. (2014). Speculative asset prices. *American Economic Review*, *104*(6), 1486–1517.
Zeng, Y. (2016). Institutional investors: Arbitrageurs or rational trend chasers. *International Review of Financial Analysis*, *45*, 240–262.

An Analysis of the Time-Varying Behavior of the Equilibrium Velocity of Money in the Euro Area

Mariam Camarero, Juan Sapena, and Cecilio Tamarit

JEL Classification E4 · E5 · C54 · C87

1 Motivation

The European Central Bank (ECB) seems to have been successful anchoring inflation expectations since the launching of the euro up to the Great Recession (GR). However, one can ask if this has been just as a result of good luck or good policy. To reach its final target in terms of inflation, the ECB monitors and controls some intermediate variables that work as medium targets. The ECB's approach to organizing, evaluating and cross-checking the information relevant for assessing the risks to price stability is based on two analytical perspectives, referred to as the "two pillars": economic analysis and monetary analysis. The historical importance of money growth on inflation changes has been well established in the literature (Friedman and Schwartz 1963; Crowder 1998; Haug and Dewald 2012). This is the reason why the study of the stability of the money demand function is of such a great importance. Nevertheless, for this function to be stable, the velocity of circulation of money should not change or, at least, its deviations from its long-run value should not be permanent.

M. Camarero
INTECO, University Jaume I, Valencia, Spain

J. Sapena
Catholic University of Valencia, Valencia, Spain

C. Tamarit (✉)
INTECO, University of Valencia, Avda. Tarongers, 46071 Valencia, Spain
e-mail: cecilio.tamarit@uv.es

G. Dufrénot and T. Matsuki (eds.), *Recent Econometric Techniques for Macroeconomic and Financial Data*, Dynamic Modeling and Econometrics in Economics and Finance 27, https://doi.org/10.1007/978-3-030-54252-8_5

In the euro area (EA), the decoupling between interest rates, money supply and inflation started from the very first moment of its creation. However, recent developments in M3 velocity in the EA have seriously questioned the reliability of M3 growth as a pillar of the ECB's monetary policy strategy. In fact, since 2001, M3 growth has systematically exceeded its reference value without a clear effect on inflation. Thus, the ECB has actually been an inflation targeter in deeds and a monetary targeter in words only. Overall, inflation has not diverged much from its 2% target, if anything, has been below and currently is struggling to return to 2%. Since 2000–2001, the income velocity of M3 has declined at a stable rate, close to 3.5%, and this fall has accelerated from 2015 with the more recent enhanced asset purchase programs (APPs). These results tend to show that fundamental changes in M3 velocity trends relative to its historical patterns have occurred and we argue that this should be made explicit in the derivation of the reference value. During the mid-70s and mid-80s, the key challenge for central banks in OECD countries was very high inflation, not too low. Since then, countries that progressively adopted an inflation-targeting strategy experienced a reduction in the level of inflation that was accompanied by a decline in the volatility of inflation together with an improved anchoring of inflation expectations. However, after the global financial crisis, this approach has received criticism.

The most salient feature of monetary policy in recent years is that its main classical instrument, the short rate interest rate, has been set at the zero lower bound (ZLB) and has in practice become inactive, leading the ECB to implement monetary policy differently, mainly using forward guidance and QE programs. Although these new unconventional tools might be necessary, they are not exempt of risk, and therefore, the prominent view is that they should only be implemented temporarily. The first argument against a large balance sheet is the classical monetarist view establishing that a high level of liquidity could result in rapid credit creation and ultimately in an acceleration of inflation. However, the economy's response so far to this expansionary monetary policy has been a stable inflation. Persistently low inflation since the great financial crisis has led to question the ability of the ECB to deliver on their mandate.

In any case, as briefly described above, the practice of monetary policy in the EA has lately been much more complex than any particular theory, questioning our common beliefs on this matter. First, it seems that the relationship linking monetary supply to inflation is not being fulfilled, challenging the grounds of the quantitative approach to monetary policy; second, at the ZLB the monetary multiplier linking money base and monetary aggregates seems to be also failing, which jeopardize the capacity of the monetary policy to follow any ruled-based approach to control aggregate demand. In our opinion, both elements materialize into a shocking evolution of the velocity of circulation of money that questions the theoretical grounds (and practical capacity) of the monetary policy as a stabilization tool.

In this paper, we follow a purely empirical approach to analyze the drivers of the dynamics of the velocity of circulation of money in the EA, both at an aggregate level and across countries, to undercover possible heterogeneity. We start our analysis by examining the statistical properties of the log-velocity of M3 and its stability. We then estimate a model of equilibrium velocity that factors in the opportunity cost of M3, along the lines suggested by Orphanides and Porter (2000), Brand et al. (2002), Leão (2005) and Biefang-Frisancho et al. (2012). To overcome the difficulty of capturing the recently enhanced policy intricacy, our approach is to ascertain the computation of the elements driving the evolution of the income velocity of circulation of money, an indicator that may help to assess the overall monetary policy stance and the sentiment of the public.

To fully take the dynamics of the income velocity into account, we use a very flexible and comprehensive state-space framework. Based on Camarero et al. (2019), our research extends the simple canonical model usually employed in the state-space literature into a panel data time-varying parameter framework, combining fixed (both common and country-specific) and varying components. Regarding the transition equation, our specification allows for the estimation of different autoregressive alternatives and includes control instruments, whose coefficients can be set up either common or idiosyncratic. This is particularly useful to detect asymmetries among individuals (countries) to common shocks. Using this framework has two advantages. First, the panel dimension allows exploiting data of individual countries from the core and the periphery of the EA (Austria, Belgium, France, Germany, Italy, Portugal and Spain) that were part of the European Monetary Union (EMU) since its inception. Second, the presence of a time-varying term allows capturing nonlinearities and estimating the determinants of income velocity.

Turning to the answer to our research question, our findings confirm the role of the permanent income per capita to explain the trend in M3 velocity across the Eurozone, showing country heterogeneity and an asymmetric impact of the business cycle position that determines the evolution of velocity. These results are in line with recent evidence stressing the role of the decline of both potential output and uncertainty as the most relevant explanatory elements of the "inflation puzzle."

The remainder of the paper is structured as follows. Section 2 explains the theoretical and empirical issues related to stability of the money demand function and, particularly, the evolution of the money velocity. Section 3 explains the methodology and the econometric specification. Section 4 reports the results. Finally, Sect. 5 concludes.

2 The Inflation Puzzle and Money Velocity in the EA: Theoretical and Empirical Issues

The quantity theory of money (QTM) predicts a positive relationship between monetary growth and inflation. However, even if inflation truly were a monetary phenomenon in the long run, as stated in Friedman (1963) seminal book, a conventional vector cointegration approach does not necessarily identify the long-run relation between money and prices correctly, because it neglects the structural development of the velocity of money. Since excess supply and demand of money are captured by transitory movements of velocity in a world where real and nominal rigidities prevail, the identification of excess liquidity that endangers price stability is tied to the identification of equilibrium velocity.

In general, evidence from cross-country studies strongly supports the one-to-one correlation of average money growth and average inflation. However, the impact of money on prices is hard to identify within one country. De Grauwe and Polan (2005) have argued that the long-run link between nominal money growth and inflation might be much looser than commonly assumed in countries which have operated in moderate inflation environments as it is the case of the EA.

Flexible inflation targeting implies that the central bank attempts to reach the target gradually in the medium term and not in the immediate period. As recently stressed by Cochrane (2017), existing theories of inflation make straight predictions. The Keynesian school argues that velocity is a highly fluctuating variable which is significantly affected by economic policies. As a result, changes in velocity could nullify the effects of monetary policy. They stress that the velocity of money is severely affected by demand management policies; hence, it is a non-stationary variable. Furthermore, they argue that the movements of velocity are the opposite of the movement of money supply. Interest rate is also regarded as one of the variables influencing velocity. The opposite forecast is made by monetarist models, who predict that, as provided the velocity is "stable" in the long run, a massive increase in reserves must lead to galloping inflation. Yet, none of these predictions have proved right. This issue was already central to the debates of the 1950s and 1960s between Keynesians and monetarists. Keynesians thought that at the zero rates of the Great Depression, money and bonds were perfect substitutes, so monetary policy could do nothing, and advocated fiscal stimulus instead. On the contrary, monetarists held that additional money, even at zero rates, would be stimulative; therefore, the failure to provide additional money was the big monetary policy mistake of that time. The view that inflation is always a monetary phenomenon has a long tradition based on the quantity theory of money. In its simplest form, the QTM states that changes in money supply growth are followed by equal changes in the inflation rate and, through the force of the Fisher effect, in the nominal interest rate.

According to the monetarist doctrine:

$$MV = PY \tag{1}$$

where M stands for money and P for the price level. Keeping V and Y (respectively, velocity and output) fixed, then a proportional increase in M leads to a proportional increase in P. These identities are transformed into a theory, the quantity theory, by the following two propositions. First, in the long run, there is a *proportionality* relation between inflation and the growth rate of money; i.e., in a regression of inflation on money growth, the coefficient of money is estimated to be 1. Second, over a sufficiently long period of time, output and velocity changes are orthogonal (*super-neutrality proposition*) to the growth rate of the money stock. This means that a permanent increase in the growth rate of money leaves output and velocity unaffected in the long run.

According to the monetarists, income velocity is independent from government active policies; hence, it is a function of real as well as institutional variables with negligible fluctuations in the short run. Therefore, as a result of small changes in these factors, velocity is regarded as a stationary variable in both the short run and long run. In this same vein, the New Classical School is of the opinion that the velocity of money (due to the stability of the money demand function) is a stationary variable in the long run. Income velocity is a measure of the rate of the use of money or the average number of transactions per unit of money.[1]

The Cambridge School modified the former equation by placing emphasis on cash balance holdings used in facilitating expenditures

$$M = kY \tag{2}$$

where $k = 1/v$ which represents average cash balances as a fraction of nominal income. This equation shifted emphasis to the determinants of the demand for money rather than the effects of changes in the supply of money. According to Keynesians, economic agents hold a constant fraction of their incomes in cash balances. They argue that the medium of exchange role was only one of the motives of holding money, stressing that liquidity preference could be influenced by yields or alternative financial assets. As a result, velocity could change due to expectations about future interest rates or risk. Also, changes in money stock alone could affect the velocity of money through interest rates.

The equation of exchange can be also written in growth terms (using logarithm differences) as, for example, Orphanides and Porter (2001) do.[2] Thus, the identity is formulated in terms of money growth and inflation. By doing so, they emphasize the importance of accounting for changes in velocity to assess whether money growth is an adequate anchor for inflation.

Equation (1) can be written in logarithmic form and taking differences as:

$$\mu + \Delta v = \pi + \Delta y \tag{3}$$

where $\pi = \Delta p$ is inflation and $\mu = \Delta m$ is money growth. As previously, this relation is an identity and should hold at any horizon. However, it could be useful to decompose the growth of both output and velocity into their cyclical and long-run equilibrium components. We can denote Y^* the potential output, whereas the cyclical component can be represented by the rate of growth of the output gap, $(\delta y - \delta y^*)$. Similarly, we can define the velocity growth gap, $(\Delta v - \Delta v^*)$, to capture the cyclical component of velocity growth. In the long run, these two components would tend to zero.

[1] It is a flow concept which is measurable but not visible.

[2] We base what follows in their decomposition of the logarithmic version of the equation of exchange.

We can rewrite Eq. (3) in terms of the long-run and cyclical components:

$$\mu + \Delta v^* + (\Delta v - \Delta v^*) = \pi + \Delta y^* + (\Delta y - \Delta y^*) \tag{4}$$

Also, we can express the equation above in terms of inflation:

$$\pi = \mu - \Delta y^* + \Delta v^* - (\Delta y - \Delta y^*) + (\Delta v - \Delta v^*) \tag{5}$$

This decomposition separates a cyclical component from a component determined by money growth adjusted by potential output and changes in equilibrium velocity. This adjusted money growth can be denoted as μ^*:

$$\mu^* \equiv \mu - \Delta y^* + (\Delta v^* - \Delta v) \tag{6}$$

The two cyclical terms can be denoted $\eta = -(\Delta y - \Delta y^*) + (\Delta v - \Delta v^*)$, so that Eq. (5) can be expressed as:

$$\pi = \mu^* + \eta \tag{7}$$

Cyclical factors tend to zero in the medium to long run, so that if the central bank wants to achieve its long-run inflation objective, it must ensure that money grows at the same rate ($\pi^* = \mu^*$). This type of equation has been the basis for monitoring the growth of monetary aggregates for a long time.

However, even if the relation between inflation and money growth is an identity, many times this fact is forgotten. Indeed, in the short run, cyclical factors are relevant. Moreover, if one does not consider the role of potential output and the need to adjust velocity in the same terms, the relation seems not to hold.

"Monetary targeting" has remained the prominent framework for the long-run implementation of monetary policy by the ECB since the euro launching. The question of whether the money multiplier is stable and predictable provides the theoretical ground for the adoption of this monetary strategy. This assumption is, however, violated in the presence of marked regime shifts in liquidity conditions of the domestic money market, particularly in the banking sector that are projected on income money velocity instabilities.[3] One of the major issues in monetary policy is the stability and predictability of the money multiplier. Yet, the empirical verifications of such long-term stability have produced conflicting outcomes.

Figure 1 displays the evolution of the M3 velocity and the short-run real interest rate for the EA. We can see a consistent decrease in M3 velocity under a monetary regime that has been causing inflation to be $I(0)$. According to Benati (2019), the money demand literature has routinely interpreted deviations from the long-run

[3]The money multiplier equation indicates that the higher the supply of base money, the higher the money supply. The size of the money multiplier is determined by the behavior of the public in their preference for holding cash relative to bank deposits; the monetary policy stance of the central bank in determining the required reserve ratio; and the incentives for holding excess reserves by the commercial banking system. The lower these ratios are, the larger the money multiplier is.

Fig. 1 Euro area velocity and real interest rate

equilibrium between the short rate and velocity (or money balances as a fraction of GDP) as signaling possible future inflationary pressures, although there seems to be an equilibrium for the EA as a whole. If we compare the M3 velocity evolution for individual EMU countries, we can see heterogeneous situations.

More specifically, it seems rather clear that one monetary policy does not fit all the member countries giving rise to tensions at some concrete moments, as for instance, during the debt crisis that triggered asymmetric capital movements across the EA. The 2008 crisis pushed the ECB to use monetary policy as a discretionary tool to soften the adverse effects of the economic contraction. Monetary policy ceased to operate in the conventional way and entered uncharted territory. The policy rate was lowered repeatedly until it reached the ZLB.[4] The sharp cuts in nominal interest rates might be expected to have led to sharp movements in inflation but they did not, giving birth to an "inflation puzzle" situation (Bobeica and Jarociński 2017). As noticed by Cochrane (2017), the consensus linking interest rates and inflation appears to starkly fail to explain the data after 2009. Prior to the financial crisis, the performance of inflation targeting was in practice widely considered a success.[5] But the situation changed with the crisis. The period after 2007 should offer a good test of monetarism. Keynesians posit that at ZLB, or even with negative interest rates, money

[4]Given the present low level of inflation, the risk of interest rates reaching the effective lower bound (ELB) too often is a real threat. Until recently, the conventional wisdom was that the lower bound of interest rates was zero, the so-called ZLB. However, as demonstrated in recent years, the ELB can be below zero, determined by the cost of holding cash instead of remunerated assets, including deposits.

[5]See, for instance, Svensson (2010).

and short-term bonds become perfect substitutes. A negative nominal interest rate on the deposit facility means that commercial banks must pay for parking their deposits with the ECB. It incentivizes credit creation and a greater velocity of circulation of excess reserves held at the central bank. Velocity becomes a correspondence, not a function of interest rates. In other words, $MV = PY$ becomes $V = PY/M$; therefore, velocity passively adjusts to whatever split of debt between money and reserves the central bank chooses (Fig. 2).[6]

Empirically, the money multiplier is not a mechanical relationship and has not been stable over time. In particular, since 2007 and the significant injections of liquidity into the system by the ECB, first through its refinancing operations and later through its asset purchases, the multiplier has fallen considerably as the two variables clearly decoupled.[7] After asset purchases began and expanded greatly in 2015 with the inclusion of sovereign assets, the increase in base money was entirely supply-driven and induced mechanically by the creation of reserves by the ECB to pay for its asset purchases.

The question is how to test empirically these hypotheses in an econometric tractable way. There is a vast empirical literature concerning the long-run relation between money growth and inflation. In many cases, the empirical approach has been to analyze the long-term quantity theory using data in the frequency domain and confirming the proportionality prediction of the quantity theory (Lucas 1980; Fitzgerald 1999), although their methodology has been criticized by McCallum (1984). More recently, researchers have tested the time series properties of inflation, output and money in vector autoregression models. In this context, the empirical approach using the P-star model[8] has been applied by Gerlach and Svensson (2003) to the EA case.

A conventional vector cointegration approach does not necessarily identify the long-run relation between money and prices correctly, because it neglects the structural development of the velocity of money. Note that a potential bias may arise if the independent variables money and output are correlated with the error term (velocity). Velocity is traditionally viewed as an analogue of the demand for real money

[6]The financial system is perfectly happy to hold arbitrary amounts of reserves in place of short-term bonds. However, below a certain threshold, called by some economists the *reversal rate*, policy rates going further into negative territory may become counterproductive. This would happen because of the detrimental effect of negative interest rates on bank profitability and lending, especially in an environment in which banks are largely financed with retail deposits at nonnegative nominal interest rates.

[7]As explained by the European Central Bank (2017), the increased provision of central bank reserves before the crisis was in fact demand-driven and mirrored the increase in broad money because of the rise in the supply of credit to the non-financial sector that was taking place at the time. Yet, the increase in money base after 2007 was of a different nature. From 2007 to 2012, it was related to an increase in the banks' demand for reserves in refinancing operations, not because they were increasing credit (quite the opposite), but because they were seeking to insure themselves against liquidity shortfalls when short-term money markets were dysfunctional.

[8]The P-star model may be regarded as a modern monetarist approach to modeling inflation. It starts by defining the price gap as the difference between the price level and the long-run equilibrium price level, which is implied by the long-run quantity relation. The model then specifies a direct effect from the lagged price gap and the current price level.

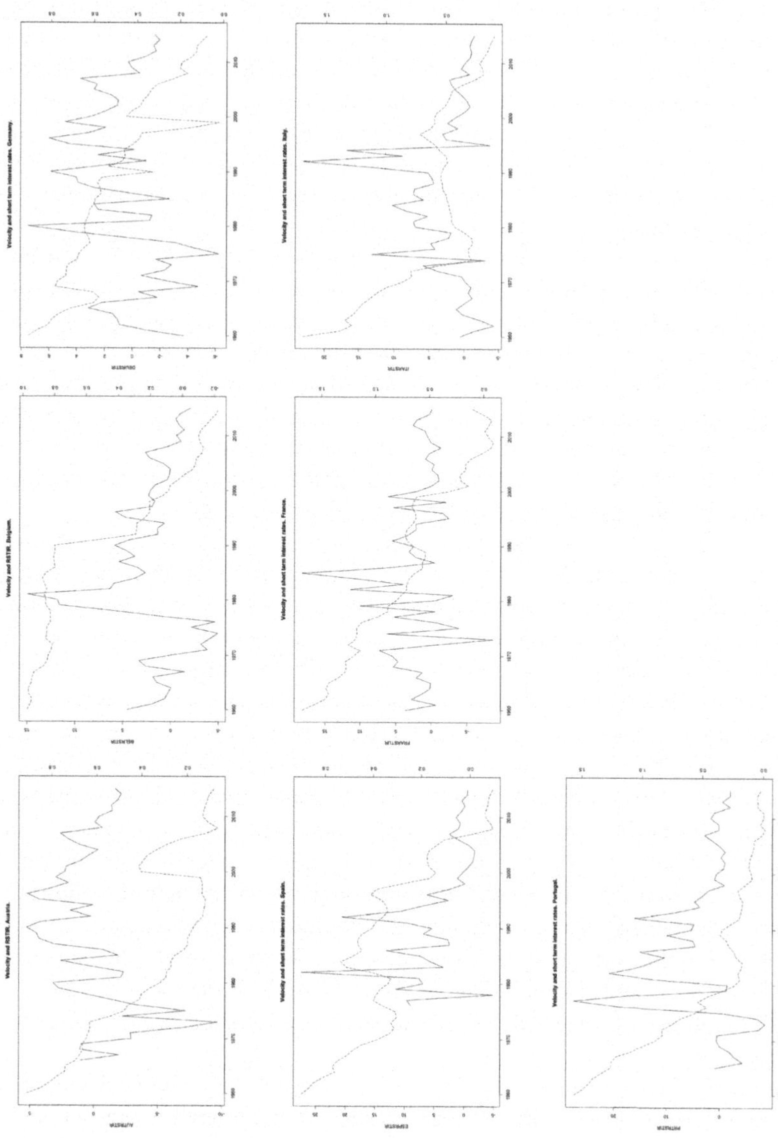

Fig. 2 Velocity and nominal short-run interest rates. Euro area countries

balances. Consequently, it is treated as a function of income (or permanent income) and an interest rate that serves as a proxy for the opportunity cost of holding money. In addition to its traditional determinants, other approaches consider that velocity can be a function of institutional changes in the financial system (Gravy 1959; Bordo and Jonung 1987, 1990; Bordo et al. 1997) or the business cycle (Leão 2005). Since excess supply and demand of money are captured by transitory movements of velocity in a world where real and nominal rigidities prevail, the identification of excess liquidity that endangers price stability is tied to the identification of equilibrium velocity. Thus, in our paper we try to investigate the behavior of velocity in more detail, to capture more information that might be relevant for the determination of future inflation. A key issue in the identification of the relation between money and prices is the identification of the *money overhang*, that is, when movements of money supply are decoupled from money demand, causing velocity to drift away from its equilibrium. When long samples are covered, movements of the equilibrium of velocity have proven to be a key obstacle to the identification of excess liquidity that is essential in the new P-star literature, where a standard Phillips curve framework is extended by including a measure of excess liquidity through the opportunity costs of holding money (Orphanides and Porter 2000) or applying some structural filters to velocity (Bruggeman et al. 2005). There is a related strand of literature that deals with the nonlinearities of money demand (Calza and Zaghini 2009) or that accounts for the role of wealth and asset prices (Dreger and Wolters 2009).

In this paper, we follow an alternative approach proposed by El-Shagi and Giesen (2010) and El-Shagi et al. (2011) who consider any persistent movement to reflect a change of equilibria. The key empirical innovation of this study over previous empirical investigations on the stability of the EA money demand lies in the introduction of a multivariate state-space framework capturing with a time-varying parameter money velocity changes induced by changes in observed and unobserved variables. We adopt a multivariate decomposition of velocity that allows the identification of the long-run equilibrium velocity while applying less restrictive assumptions on specific driving forces of velocity. We merely have to assume that an equilibrium exists, where deviations can be eroded by the growth of money, prices or production. Essentially, we do not only test whether money velocity exhibits a tendency to return to a long-run equilibrium velocity or not, but also through which channels this adjustment occurs. Persistent changes of any potential driving force of velocity are by construction attributed to the persistent velocity component. Thus, a change in velocity that is caused by a persistent change of income is correctly identified as non-transitory.[9] This high flexibility of our model allows a parsimonious specification in terms of further controls.

Additionally, lack of support for the quantity theory of money in the short run could be attributed to nonlinearity in pass-through from money growth to inflation. According to Morgan (1993), the idea of asymmetry arose during the Great Depression, when an expansionary monetary policy was found to have little effect during

[9]The same holds true for developments that are caused by institutional change as financial innovation, wealth and other factors that are discussed in the corresponding literature.

a recession, compared to a tight monetary policy during an economic expansion. Despite the fact that a large number of studies have investigated the relation between money supply and inflation (from the seminal work Friedman and Schwartz (1963) to recent research, like Benati (2009), among others), surprisingly there is only a limited, mainly old literature, which investigates the asymmetry in pass-through from money growth to inflation. More recently, Reynard (2006) argues that very low interest rates (i.e., the ZLB) can generate nonlinearity due to changes in financial market participation, which would then induce relatively high growth rates in monetary aggregates not followed by high inflation.

We contribute to the literature in several ways: First, we distinguish changes in velocity of money that are due to institutional developments and thus do not induce in inflationary pressure and changes that reflect transitory movements in money demand. This is achieved with a newly developed multivariate unobserved component decomposition. Second, we use our model to illustrate the consequences of the monetary policy that has been employed to mitigate the impact of the financial crisis.

3 Methodology

3.1 State-Space Models and Time-Varying Parameter Models

State-space representation of a linear system constitutes a statistical framework for modeling the dynamics of a $(n \times 1)$ vector of variables observed at regular time intervals t, y_t, in terms of a possibly unobserved (or state) $(r \times 1)$ vector ξ_t.[10] The origin of state-space modeling is intimately linked with the Kalman filter, a recursive algorithm for generating minimum mean square error forecasts in state-space models.

The measurement equation models the dynamics of the observable variables y_t, possibly measured with noise, that are assumed to be related to the state vector, providing information on ξ_t. It takes the following general form:

$$\underset{(n\times 1)}{y_t} = \underset{(n\times k)}{\mathbf{A}^\top}\underset{(k\times 1)}{x_t} + \underset{(n\times r)}{\mathbf{H}^\top}\underset{(r\times 1)}{\xi_t} + \underset{(n\times 1)}{w_t} \tag{8}$$

where y_t represents an $(n \times 1)$ vector of variables that are observed at date t and x_t represents a $(k \times 1)$ vector of exogenous determinants, their coefficients being included in the $(k \times n)$ matrix A. H is an $(r \times n)$ matrix of coefficients for the $(r \times 1)$ vector of unobserved components ξ_t. Finally, the measurement or observational error, w_t, is an $(nx1)$ vector assumed to be i.i.d. $N(0, R)$, independent of ξ_t and ν_t and for $t = 1, 2, \ldots$, where

[10] Excellent textbook treatments of state-space models are provided in Harvey (1989, 1993), Hamilton (1994a, b), West and Harrison (1997) or Kim and Nelson (1999), among others. They all use different conventions, but the notation used here is based on James Hamilton's, with slight variations.

$$E(w_t w_t^\top) = \underset{(n\times n)}{R} \tag{9}$$

and variance covariance equal to

$$E(w_t w_\tau^\top) = \begin{cases} \text{R} & \text{for } t = \tau \\ 0 & \text{for } t \neq \tau \end{cases} \tag{10}$$

The state-transition equation describes the evolution of the underlying unobserved states that determine the time series behavior, generated by a linear stochastic difference representation through a first-order Markov process, such as in 11:

$$\underset{(r\times 1)}{\xi_{t+1}} = \underset{(r\times r)}{F}\underset{(r\times 1)}{\xi_t} + \underset{(r\times s)}{B}\underset{(s\times 1)}{Z_{t+1}} + \underset{(r\times 1)}{\nu_{t+1}} \tag{11}$$

where F denotes an $(r \times r)$ state-transition matrix, which applies the effect of each system state parameter at time $t - 1$ on the system state at time t, ξ_t, and Z_t is a $(s \times 1)$ vector containing any control inputs, either deterministic (drift and/or deterministic trend) or stochastic. If present, control inputs affect the state through the $(r \times s)$ control input matrix, B, which applies the effect of each control input parameter in the vector on the state vector.

The introduction of stochastic control inputs is common practice in the literature on control engineering where this concept was coined. Basically, the idea is to simulate the effect of changes in the control variable on a system, namely the state vector.[11] Despite their many potential uses, empirical economic research generally has employed simple state-transition equations, where the unobserved component evolves as a random walk process and no control inputs are present.

Finally, ν_t represents the $(r \times 1)$ vector of serially uncorrelated disturbances containing the process noise terms for each parameter in the state vector. ν_t is assumed to be i.i.d. $N(0, Q)$, where

$$E(\nu_{t+1}\nu_{t+1}^\top) = \underset{(r\times r)}{Q} \tag{12}$$

and variance–covariance equal to

$$E(\nu_{t+1}\nu_{\tau+1}^\top) = \begin{cases} \text{Q} & \text{for } t = \tau \\ 0 & \text{for } t \neq \tau \end{cases} \tag{13}$$

Once both measurement and transition equations have been specified, (14) represents the initial condition of the system:

$$\xi_1 \sim N\left(\xi_{1|0}, P_{1|0}\right). \tag{14}$$

[11] A very simple example proposed by Faragher (2012) is what happens to the trajectory of a rocket when fuel injection is activated during flight.

Writing a model in state-space form means imposing certain values (such as zero or one) on some of the elements of F, Q, B, A, H and R, and interpreting the other elements as particular parameters of interest. Typically, we will not know the values of these other elements, but need to estimate them on the basis of observation of $(y_1, y_2, \ldots, y_T)$ and $(x_1, x_2, \ldots, x_T)$. In its basic form, the model assumes that the values of F (and even B), Q, A, H and R are all fixed and known, but (some of them) could be functions of time.

State-space representation of dynamic models is particularly useful for measuring expectations that cannot be observed directly. If these expectations are formed rationally, there are certain implications for the time series behavior of the observed series that can help to modelize them. According to Hamilton (1994a), it does not exist a unique representation of a state-space formulation of a model. That is why the state variables obtained internally in the system have to be specified according to the nature of the problem with the ultimate goal of containing all the information necessary to determine the behavior of the period-to-period system with the minimum number of parameters.

Time-varying parameter regression models constitute an interesting application of the state-space representation, as:

$$\underset{(n\times 1)}{y_t} = \underset{(n\times k)}{\mathbf{A}^\top} \times \underset{(k\times 1)}{x_t} + \underset{(n\times r)}{\mathbf{H}^\top(x_t)} \times \underset{(r\times 1)}{\xi_t} + \underset{(n\times 1)}{w_t} \tag{15}$$

where A represents a matrix of fixed parameters $\bar{\beta}$. Compared to the general model where the elements of the matrices F, Q, A, H and R are treated as constants, in this model H depends on the observed regressors, as $[H(x_t)]^\top = x_t$. As stated in Hamilton (1994b), assuming that the eigenvalues of F in (11) are all inside the unit circle, the coefficients can be interpreted as the average or steady-state coefficient vector, and the measurement equation can be written as follows:

$$y_t = x_t^\top \bar{\beta} + x_t^\top \xi_t + \omega_t \tag{16}$$

where the vector of unobserved coefficients, $\xi_t = \left(\beta_t - \bar{\beta}\right)$, evolves along time according to the expression:

$$\left(\beta_{t+1} - \bar{\beta}\right) = F\left(\beta_t - \bar{\beta}\right) + \underset{(r\times s)}{B}\underset{(s\times 1)}{Z_t} + \upsilon_{t+1} \tag{17}$$

Equation (17) represents a simple transition equation to be estimated through the Kalman filter, where $\left(\beta_t - \bar{\beta}\right) = \xi_t$ is the unobserved component of our time-varying parameter, while the fixed component is also included at the measurement equation as $\bar{\beta}$.

3.2 A Panel Time-Varying State-Space Extension

In this subsection, we extend the previous time-varying parameter model to a panel setting. Our main goal is to explore the use of the state-space modelization and the Kalman filter algorithm as an effective method for combining time series in a panel. This flexible structure allows the model specification to be affected by different potential sources of cross-sectional heterogeneity. This approach can be a superior alternative to the estimation of the model in unstacked form, commonly employed when there is a small number of cross sections.

The general model can be written as follows:

$$y_{i,t} = x_{i,t}^{\top}\bar{\beta} + x_{i,t}^{\top}\xi_{i,t} + \omega_t \tag{18}$$

or in matrix form:

$$\underset{(n\times t)}{y} = \underset{(n\times n*k)}{\mathbf{A}^{\top}} \times \underset{(n*k\times t)}{x} + \underset{(n\times r)}{\mathbf{H}^{\top}(x)} \times \underset{(r\times t)}{\xi} + \underset{(n\times t)}{w} \tag{19}$$

representing the measurement equation for a $y_t \in \mathbb{R}^n$ vector containing the dependent variable for a panel of countries. $x_t \in \mathbb{R}^{k\times n}$ is a vector of k exogenous variables, including either (or both) stochastic or deterministic components. The unobserved vector $\xi_t \in \mathbb{R}^r$ influences the dependent variable through a varying $H^{\top}(x_t)$ $(n \times r)$ matrix, whose simplest form is $H^{\top}(x_t) = x_t$. Finally, $w_t \in \mathbb{R}^n$ represents the $(n \times 1)$ vector of N measurement errors.

The specification of the model in Eq. (19) relies on a mean-reverting-type modelization of the measurement equation, which also allows for the inclusion of fixed parameters, in matrix **A**. Each of the fixed parameters can be modeled, either as a common parameter for all the agents in the panel, $\bar{\beta}$, or, alternatively, as a country-specific coefficient, $\bar{\beta}_i$. The model also includes time-varying parameters (ξ_t) for some of the regressors that eventually can be interpreted as deviations from the mean parameters ($\left(\beta_{it} - \bar{\beta}_i\right) = \xi_t$).

The measurement equation for each ith element in the tth period $\left(y_{i,t}\right)$ in the vector of the dependent variable can be expressed as follows:

$$\begin{aligned} y_{i,t} = & \sum_{\mathrm{ks}=\mathrm{ksmin}}^{\mathrm{ksmax}} \bar{\beta}_{\mathrm{ks},i} x_{\mathrm{ks},i,t} + \sum_{\mathrm{kc}=\mathrm{kcmin}}^{\mathrm{kcmax}} \bar{\beta}_{\mathrm{kc}} x_{\mathrm{kc},i,t} \\ & + \sum_{\mathrm{kv}=\mathrm{kvmin}}^{\mathrm{kvmax}} \xi_{\mathrm{kv},it} x_{\mathrm{kv},i,t} + h_i \xi_{r,it} + \underset{(n\times 1)}{w_{it}} \end{aligned} \tag{20}$$

In Eq. (20), the measurement equation for each of the individuals (countries) included in the panel allows for the potential inclusion of both fixed (mean) and/or time-varying parameters for the regressors included in the model to be estimated. For the fixed-parameter case, different partitions can be considered. First, one can choose a subset of ksnum = ksmax − ksmin + 1 regressors $(x_{\text{ksmin}}, \ldots, x_{\text{ks}}, \ldots, x_{\text{ksmax}})$ in the interval $[0, k]$, whose coefficients will be modeled as country-specific or idiosyncratic. A second subset of kcnum = kcmax − kcmin + 1 regressors $\left(x_{\text{kcmin}}, \ldots, x_{\text{kc}}, \ldots, x_{\text{kcmax}}\right)$, also defined in the interval $[0, k]$, is also possible, but related in this case to the dependent vector through a common coefficient for all the countries included in the panel. Last, the varying parameters are associated with a third subset of kvnum = kvmax − kvmin + 1 regressors $(x_{\text{kvmin}}, \ldots, x_{\text{kv}}, \ldots, x_{\text{kvmax}})$, also belonging to the interval $[0, k]$.

Regressors with varying parameters can eventually enter the measurement equation with fixed parameters if any of the following conditions hold:

$$(x_{\text{kvmin}}, \ldots, x_{\text{kv}}, \ldots, x_{\text{kvmax}}) \bigcap (x_{\text{ksmin}}, \ldots, x_{\text{ks}}, \ldots, x_{\text{ksmax}}) \neq \varnothing \tag{21}$$

$$(x_{\text{kvmin}}, \ldots, x_{\text{kv}}, \ldots, x_{\text{kvmax}}) \bigcap (x_{\text{kcmin}}, \ldots, x_{\text{kc}}, \ldots, x_{\text{kcmax}}) \neq \varnothing \tag{22}$$

Condition at Eq. (21) presents a combination of varying and fixed idiosyncratic parameters, while Eq. (22) presents the combination of varying parameters with common ones.[12]

The panel specification presented in Eq. (20) can also be enriched to allow for the potential inclusion, as in Broto and Perez-Quiros (2015), of a dynamic common factor in the measurement equation driving the dependent variable vector, $y_{i,t}$. This common factor is modeled simply as an additional unobserved state in the state vector (whose number of rows ups now until $n * \text{kv} + 1$) that enters each country's measurement equation with a country-specific loading parameter, h_i.[13]

Finally, some restrictions can be imposed to the $(n \times n)$ variance–covariance matrix R of the measurement vector:

$$\underset{(n \times n)}{R} = \begin{bmatrix} \sigma_{w_1}^2 & 0 & \cdots & 0 \\ 0 & \sigma_{w_2}^2 & 0 & \vdots \\ \vdots & 0 & \ddots & 0 \\ 0 & 0 & 0 & \sigma_{w_n}^2 \end{bmatrix} \tag{23}$$

The general assumption implies uncorrelated errors with idiosyncratic variances, and hence it to be a diagonal matrix with idiosyncratic components, as in Eq. (23). To reduce the number of hyper-parameters, one could impose a common variance for the uncorrelated errors, $\sigma_{w_1}^2 = \cdots = \sigma_{w_n}^2 = \sigma_w^2$.

---

[12]Camarero et al. (2019) also consider the possibility of the fixed parameters being affected by a deterministic time trend.

[13]Note that our Gauss code allows for multiple common factors as well as the inclusion of potential restrictions on them.

Regarding the state-transition equation, the $\xi_t \in \mathbb{R}^r$ vector of unobserved state evolves according to Eq. (24):

$$\underset{(r\times 1)}{\xi_{t+1}} = \underset{(r\times r)(r\times 1)}{F \quad \xi_t} + \underset{(r\times s)(s\times 1)}{B \quad Z_t} + \underset{(r\times 1)}{\nu_t} \tag{24}$$

In Eq. (24), the vector of unobserved components ξ_t follows an autoregressive process where F denotes an $(r \times r)$ state-transition matrix:

$$\underset{(r\times r)}{F} = \begin{bmatrix} \phi_{1,1} & 0 & 0 & \cdots & 0 & 0 \\ 0 & \phi_{2,1} & 0 & 0 & 0 & 0 \\ 0 & 0 & \ddots & 0 & \vdots & \vdots \\ \vdots & 0 & 0 & \phi_{\mathrm{kv},1} & 0 & 0 \\ 0 & \vdots & 0 & 0 & \ddots & 0 \\ 0 & 0 & \cdots & 0 & 0 & \phi_{\mathrm{kv},n} \end{bmatrix} \tag{25}$$

$Z_{i,t}$ represents the vector containing any control inputs affecting the state through the control input matrix B, which applies the effect of each control input on the state vector. These control variables are frequently employed in engineering but are not so commonly applied to state-space models in economics. Their use could be interpreted as the "coefficient drivers" of the second-generation TVP models, described in Swamy and Tavlas (2001) and related work. As stated in Gourieroux (1997) with the introduction of an input in the "transition equation" or in the "measurement equation," all the formulae of the filter remain valid with the exception of the introduction of the variable in the update equation.

Finally, ν_t represents the $(r \times 1)$ vector containing the process noise terms for each parameter in the state vector and is assumed to be i.i.d. $N(0, Q)$.

Each one of the $(r = \mathrm{kv} \times n)$ first components of ξ_{t+1} is driven by the following expression:

$$\xi_{\mathrm{kv},i,t+1} = \phi_{\mathrm{kv},i} \cdot \xi_{\mathrm{kv},i,t} + \sum_{\mathrm{js}=\mathrm{jsmin}}^{\mathrm{jsmax}} \bar{\mu}_{\mathrm{js},i} \cdot z_{j1,i,t} + \sum_{\mathrm{jc}=\mathrm{jcmin}}^{\mathrm{jcmax}} \bar{\mu}_{\mathrm{jc}} \cdot z_{jc,i,t} + \nu_{\mathrm{kv},i,t+1} \tag{26}$$

Equation (26) describes the autoregressive process followed by every unobserved component whose coefficients $\phi_{\mathrm{kv},i}$ are to be estimated. The state component is also influenced by the evolution of a vector containing s observed variables or control instruments, z_t.

Our frame allows to define the parameter for each control instrument in the state transition equation, either as common for all individual in the panel or, alternatively, as specific for every individual considered. Thus, a subset of $\mathrm{jsnum} = \mathrm{jsmax} - \mathrm{jsmin} + 1$ control inputs $\left(z_{\mathrm{jsmin}}, \ldots, z_{\mathrm{js}}, \ldots, z_{\mathrm{jsmax}}\right)$ in the interval $[0, s]$, whose coefficients $\bar{\mu}_{\mathrm{js},i}$ are modeled as country-specific or idiosyncratic. Similarly, for a second subset of $\mathrm{jcnum} = \mathrm{jcmax} - \mathrm{jcmin} + 1$ regressors $\left(z_{\mathrm{jcmin}}, \ldots, z_{\mathrm{jc}}, \ldots, z_{\mathrm{jcmax}}\right)$, also defined in the interval $[1, s]$, the coefficient $\bar{\mu}_{\mathrm{jc}}$ is estimated as common for all the countries included in the panel. This specification is particularly helpful to detect heterogeneities or asymmetries among countries when estimating the impact of any particular variable.

If dynamic common factor is introduced in the model, its transition equation can be conveniently restricted, so it does not require the inclusion of potentially idiosyncratic control variables.

The autoregressive parameters in the state-transition equation are estimated, in contrast to most of the literature, that restricts their process to follow a random walk. According to Hamilton (1994b), if the eigenvalues of matrix F remain inside the unit circle, then the system is stable and the vector process defined by (24) is stationary. In this case, the inclusion of both fixed and varying parameters for the regressors can be interpreted as a mean-reverting model. The fixed parameter, either idiosyncratic ($\bar{\beta}_i$) or common ($\bar{\beta}$), should be interpreted as the average or steady-state coefficient vector, and the varying parameter as the deviation from this mean in a mean-reverting framework:

$$\xi_{\mathrm{kv},i,t} = \left(\beta_{\mathrm{kv},i,t} - \bar{\beta}_{\mathrm{kv},i}\right) \tag{27}$$

Equation (27) presents the case of a mean-reverting parameter with specific fixed mean for each individual.

$$\xi_{\mathrm{kv},i,t} = \left(\beta_{\mathrm{kv},i,t} - \bar{\beta}_{\mathrm{kv}}\right) \tag{28}$$

On the contrary, in Eq. (28) the mean-reverting parameter exhibits a common mean for all the individuals in the panel.

As for the measurement equation, the above specification of the transition equation can be adapted to introduce different restrictions. Regarding the variances of the process noise for the varying components, they can be restricted to be null or identical for all countries in the panel and/or regressors with varying parameters in the model. It can be also defined a value for the signal-to-noise ratio (ratio between the two variances).[14]

Another set of restrictions has to deal with the autoregressive parameters in the diagonal of F that can also be restricted to one (the random walk) or their value to be identical among all countries in the panel, and/or regressors included in the model with varying parameters. Further restrictions can also be introduced on any of the components of the hyper-parameter vector.

[14]Increasing signal-to-noise ratio would weigh the observation heavier in the correction equations of the Kalman filter.

3.3 A Time-Varying Parameter Model for the M3 Velocity

In modern economies, neglecting what happens to money velocity leads to large relative errors in estimating inflation and output. Moreover, velocity, or its twin sibling, the demand for money, turns out to be highly volatile, difficult to model and hard to measure. Hence, movements in P end up being dominated by unexplained movements in V rather than in M.

Traditional theories of money demand identify income as the principal determinants of velocity. As highlighted in Friedman and Schwartz (1963), if money demand elasticity to income is greater than one, then economic growth would induce a secular downward trend in velocity, inflation and interest rates. The theoretical literature (see Orphanides and Porter 2000) also posits that velocity fluctuates with the opportunity cost of money, driven by inflation and interest rates.

A benchmark regression representing the traditional theories of money demand is presented in Bordo and Jonung (1987), updated in Bordo and Jonung (1990) and revisited using cointegration techniques by Bordo et al. (1997). This formulation is described in Hamilton (1989) using an equation such as:

$$\log V_{i,t} = \beta_{0,i} + \xi_{i,t} + \lambda_i f_t + \beta_{1,i} i_t + \beta_{2,i}\pi_t^e + \beta_{3,i}\log Y_{pc_{i,t}} + \beta_{4,i}\log Y_{pc_{i,t}}^{P} + \varepsilon_{i,t} \tag{29}$$

The above model expresses the log of velocity (V_t) as a function of the opportunity cost of holding money balances in terms of an appropriate nominal interest rate (i_t), expected inflation (π_t^e), proxied by the fitted values of a univariate autoregression for actual inflation, the log of real GNP per capita (Y_{pc_t}) and its smoothed version ($Y_{\mathrm{pc}_t}^{P}$) interpreted as permanent real GNP per capita. The velocity formulation is strongly based on economic theory of permanent income hypothesis (Friedman and Schwartz 1963). We expect a positive sign for permanent income as any increase in it will rise the number of transactions in the economy affecting the velocity positively. Transitory income with a positive coefficient but less than one would indicate that velocity moves pro-cyclically, which would be in line with Friedman's permanent income hypothesis. Over the cycle, the transitory income would increase the demand for money, because cash balances serve as buffer stock, and therefore, in the long run these transitory balances would disappear, returning the coefficient to unity. As for the real interest rate, it is expected to have a positive sign as an increase in it would decrease the demand for real money balances and thus a raise in the velocity for a given level of income. Finally, the impact of inflation on velocity is ambiguous depending upon its relative influence on money balances and income growth.

Bordo and Jonung (1987) add to the regression (29) a battery of variables representing the degree of sophistication and stability of the financial sector. These variables include the ratio of currency to a broader monetary aggregate, the share of the labor force in agriculture, the ratio of total nonbank financial assets to total financial assets and a six-year moving average of the standard deviation of Y_t. According to their hypothesis, the monetization of the economy causes velocity to fall over time during the early phases of financial development. In the later phases of financial development, however, increasing financial sophistication leads to the invention of a variety of money substitutes, causing velocity to rise over time. A similar finding is obtained by Mele and Stefanski (2019) that argue that monetary velocity declines as economies grow, due to the process of structural transformation and the subsequent rising income.

If institutional change is a plausible explanation of at least some of the ongoing changes in the long-run velocity function in a number of countries over the last decade (as stressed by both Wicksell and Fisher long ago, and more recently by Bordo and Jonung (1987)), then it might be impossible to obtain a stable relationship. Moreover, this would imply that policy rules that attempt to impose some sort of fixed targets on the monetary aggregates to ensure monetary stability are unlikely to be successful. In the same vein, Raj (1995), and also Bordes et al. (2007), finds some evidence of instability in equilibrium velocity.

Our time-varying parameter specification for the determinants of M3 velocity (lnM3V) is represented as in the following equation:

$$\begin{aligned} \mathrm{lnM3V}_{i,t} = \beta_{0,i} + (\beta_{1,i} + \xi_{1,i,t}) \cdot \mathrm{lnGNIpc}^{p}_{i,t} \\ + \beta_{2,i} \cdot \mathrm{lnM3V}_{i,t-1} + \beta_{3,i} \cdot \mathrm{baaspread}_t \\ + \beta_4 \cdot \pi^{e}_{i,t} + \beta_5 \cdot \mathrm{rstir}_{i,t-1} + \omega_{i,t} \end{aligned} \tag{30}$$

where $\mathrm{lnM3V}_{i,t}$ is the logarithm of M3 velocity, $\mathrm{lnGNIpc}^{p}_{i,t}$ is the log of permanent component of real GNI per capita that enters the equation with country-specific parameters, estimated using Hamilton (2018) as an alternative to Hodrick–Prescott filtering, and the lagged dependent is represented by $\mathrm{lnM3V}_{i,t-1}$.

A novelty in our model is the introduction of a measure of global risk aversion, as literature suggests that changes in money velocity are inversely associated with confidence. As in Bernoth and Erdogan (2012) or Camarero et al. (2019) among others, we use $\mathrm{baaspread}_t$, the yield spread between low-grade US corporate bonds (BAA) and the 10-year treasury bonds as an empirical proxy for this overall investors' risk attitude ("BAAS"). This variable has been obtained from the Federal Reserve Economic Database (FRED), provided by the Federal Reserve Bank of St. Louis.[15]

Our measurement equation also includes the real short-term interest rate $\mathrm{rstir}_{i,t}$ and expected inflation, $\pi^{e}_{i,t}$, proxied by the fitted values of a univariate autoregression for actual inflation, whose parameter is defined as common for the panel.

In addition to the fixed coefficient $\beta_{1,i}$, $\xi_{1,i,t}$ stands for the varying component of the parameter for $\mathrm{lnGNIpc}^{p}_{i,t}$. In a mean-reverting framework, this varying compo-

[15]An alternative variable for global risk aversion is the CBOE volatility index, known by its ticker symbol VIX; this variable was not available for the whole period. Note that CBOE VIX measures the stock market's expectation of volatility implied by S&P 500 index options, calculated and published by the Chicago Board Options Exchange. Moreover, although the literature finds a relevant role for both proxies, while BAA spread measures risk appetite, a variation in implied volatility on a market may stem from a change in the quantity of risk on this market and not necessarily from a change in the investor's risk aversion.

nent can be interpreted as the deviations from the mean parameter, $\left(\beta_{1,i,t} - \bar{\beta}_{1,i}\right)$. Hence, our state-space model incorporates heterogeneity by allowing for a country-specific slope in the measurement equation. The measurement equation also includes a uncorrelated error disturbance vector $\omega_{i,t}$, with zero mean and variances $\sigma^2_{w,i}$.

Then, the model is estimated by maximum likelihood using the Kalman filter algorithm, where each of the m components (one in our model) of the unobserved vector, $\xi_{i,t}$, follows a stochastic process defined as:

$$\xi_{m,i,t+1} = \phi_m \cdot \xi_{m,i,t} + \mu_{m,1,i} \cdot \text{GNIpc}^{c+}_{i,t} + \mu_{m,2,i} \cdot \text{GNIpc}^{c-}_{i,t} + \upsilon_{i,t+1} \qquad (31)$$

where $\upsilon_t \sim N\,(0, Q)$, and the transition of the unobserved vector includes, in addition to the autoregressive component ϕ_m, two control instruments $\text{GNIpc}^{c+}_{i,t}$ and $\text{GNIpc}^{c-}_{i,t}$ that capture the impact of positive and negative deviations from the trend of GNI per capita, calculated as the ratio of the logarithm of GNI to GNI trend. These control instruments enter as country-specific parameters, $\mu_{m,1,i}$ and $\mu_{m,2,i}$. Finally, the state equation includes a error disturbance $\upsilon_{i,t+1}$, with mean zero and variance $\sigma^2_{\upsilon,i}$.

4 Results

In this section, we estimate the empirical specification described in Sect. 3.3 using monthly data for the period 1961–2016 in a panel of 7 countries.

Prior to the estimation of our model using TVP, we analyze the univariate properties of the series using both single and panel unit root tests, allowing for both cross-country dependence and structural breaks. This is an important feature given that unit root tests can lead to misleading conclusions if the presence of structural breaks is not accounted for when testing the order of integration (Perron 1989). Therefore, the first stage of our analysis has focused on a pretesting step of the income velocity that aims to assess whether the time series is affected by the presence of structural breaks regardless of its order of integration. This pretesting stage is a desirable feature, as it provides an indication of whether we should then apply unit root tests with or without structural breaks.

4.1 Univariate Properties of the Data

Trend breaks appear to be prevalent in macroeconomic time series, and unit root tests therefore need to make allowance for these if they are to avoid the serious effects that unmodeled trend breaks have on power.[16] Consequently, when testing for a unit

[16]See, *inter alia*, Stock and Watson (1996, 1999, 2005) and Perron and Zhu (2005).

Table 1 Perron and Yabu (2009) test for structural breaks

Variable	Model 1	Break	Model 2	Break	Model 3	Break
$velaut_t$	29.08***	2000	0.8843	1984	28.80***	1984
$velbel_t$	26.95***	1992	−0.04	1984	30.74***	1991
$velger_t$	12.72***	2000	−0.29	1985	12.72***	1971
$velesp_t$	1.20	1983	0.15	1997	1.67	1982
$velfra_t$	0.43	1987	0.03	2002	0.81	1988
$velita_t$	2.43	1971	44.38***	1974	47.51***	1978
$velpo_t$	0.08	1971	36.23***	1977	34.91***	1977
$veleuro_t$	9.62***	1990	0.09	1976	10.03***	1983

Note ***, ** and * denote rejection of the null of no structural break at 1%, 5% and 10% levels of significance, respectively. The critical values for Model 1 (model with a break in the level) are 1.36 (10%), 1.74 (5%) and 3.12 (1%). For Model 2 (model with a break in the trend), the critical values are 1.13 (10%), 1.67 (5%) and 3.06 (1%). In the case of Model 3 (break in both the level and the trend), the critical values are 2.48 (10%), 3.12 (5%) and 4.47 (1%)

root it has become a matter of regular practice to allow for this kind of deterministic structural change.

In order to avoid this pitfall, we run tests to assess whether structural breaks are present in the series. This testing problem has been addressed by Perron and Yabu (2009), who define a test statistic that is based on a quasi-GLS approach using an autoregression for the noise component, with a truncation to 1 when the sum of the autoregressive coefficients is in some neighborhood of 1, along with a bias correction. For given break dates, one constructs the F-test (Exp $-$ W_{FS}) for the null hypothesis of no structural change in the deterministic components. The final statistic uses the Exp functional of Andrews and Ploberger (1994). Perron and Yabu (2009) specify three different models depending on whether the structural break only affects the level (Model I), the slope of the trend (Model II) or the level and the slope of the time trend (Model III). The computation of these statistics, which are available in Table 1, shows that we find more evidence against the null hypothesis of no structural break with Model III.

The analysis shows instabilities in the money velocity for all the countries with two exceptions, Spain and France. Therefore, in a second step, we have computed the unit root test statistics in Carrion-i Silvestre et al. (2009). The unit root tests in Carrion-i Silvestre et al. (2009) allow for multiple structural breaks under both the null and alternative hypotheses which make especially suitable for our purpose, since we have obtained evidence in favor of the presence of structural breaks regardless of their order of integration. The results of all these statistics are reported in Table 3. As can be seen, the unit root tests proposed by Carrion-i Silvestre et al. (2009) led to the non-rejection of the null hypothesis of a unit root in most of cases at the 5% level of

Table 2 Carrion-i Silvestre et al. (2009) unit root test with structural breaks

Variables	P_T^{GLS}	MP_T^{GLS}	ADF	Z_α	MZ_α^{GLS}	MSB^{GLS}	MZ_t^{GLS}	$\hat{T}_1$	$\hat{T}_2$	$\hat{T}_3$
$velaut_t$	14.05**	13.23**	−4.46**	−29.40	−21.36	0.15	−3.24	1978	1991	2004
$velbel_t$	23.29**	19.80**	−3.12	−15.82	−12.28	0.19	−2.42	1981	2002	2009
$velger_t$	17.61**	16.79**	−3.77	−22.47	−17.72	0.16	−2.97	1971	1986	1993
$velesp_t$	19.31**	17.40**	−3.65	−21.38	−17.12	0.17	−2.92	1969	1981	1989
$velfra_t$	19.59**	17.65**	−3.27	−18.97	−15.76	0.17	−2.72	1968	1979	1987
$velita_t$	18.02**	16.78**	−3.67	−22.08	−17.66	0.167	−2.94	1971	1991	2000
$velpo_t$	24.58**	21.76**	−2.96	−15.06	−12.18	0.19	−2.39	1970	1985	2006
$veleuro_t$	14.07**	13.57**	−3.73	−31.25	−21.90	0.15	−3.28	1968	1984	1992
$baas_t$	13.76**	14.03**	−4.076**	−24.61**	−18.01**	0.16**	−2.978**	1974	1979	2007

Notes ** and * denote rejection of the null hypothesis of unit root at the 5 and 10% levels of significance, respectively

significance.[17] Our conclusion is that the income velocity variable for the countries considered has unit roots with breaks both in levels and in most of the cases.

Now, we focus on the analysis of the stochastic nature of the income velocity for the EA (aggregate) and the rest of the variables explaining the evolution of the income velocity over time. As the variables have been defined at country level, we can construct a panel consisting of the different individuals and test for unit roots using Bai and Carrion-I-Silvestre (2009) panel unit root test. In the case of $baas_t$, we apply the univariate unit root tests proposed by Carrion-i Silvestre et al. (2009). Both, the panel and the univariate tests, as shown in Table 2, allow for multiple and unknown structural breaks.

Bai and Carrion-I-Silvestre (2009) propose a set of panel unit root statistics that pool the modified Sargan–Bhargava (hereafter MSB) tests (Sargan and Bhargava 1983) for individual series, taking into account the possible existence of multiple structural breaks[18] and cross-sectional dependence modeled as a common factors model.[19] The common factors may be non-stationary, stationary or a combination of both. The number of common factors is estimated using the panel Bayesian information criterion proposed by Bai and Ng (2002).

The possible presence of non-stationary variables, as the potential structural changes, is non-trivial features. A non-stationary-dependent variable, without further restrictions, could be most likely driven by a non-stationary state, such as one following a random walk. An important part of the literature on time-varying parameter models would make this choice. However, as we will see in the next subsection, in our case we do not assume that the transition equation follows a random walk, but let it be estimated as a component of the hyper-parameter vector. On the other hand, the presence of the structural breaks could, in fact, be an indication of a time-varying relationship for the model.

---

[17]It should be mentioned that the ADF test statistic rejects the null hypothesis of unit root for one time series, although at the 10% level of significance.

[18]Adapting Bai and Perron (2003) methodology to a panel data framework.

[19]Following Bai and Ng (2004) and Moon and Perron (2004).

Table 3 Bai and Carrion-I-Silvestre (2009) panel unit root test with common factors and structural breaks. Period 1980–2016. Model 2 (change in the constant and trend)

Variables	Z	P_m	P	Z^*	P_m^*	P^*	T	N	m	fr
$\ln M3V_{it}$	−0.77	−0.68	12.14	2.43	−1.00	10.32	56	8	3	4
$\ln GNI_{pc}^{p}$	−1.14	1.34	23.58	−1.31	0.29	17.68	56	8	3	4
$rstir_{it}$	0.10	−0.75	11.75	0.93	−0.94	10.67	56	8	3	4
π_{it}^{e}	−0.90	−0.41	13.63	−0.90	−0.42	13.63	56	8	3	4

Note Z, P and P_m denote the test statistics proposed by Bai and Carrion-I-Silvestre (2009). Z and P_m follow a standard normal distribution, and their 1%, 5% and 10% critical values are 2.326, 1.645 and 1.282, respectively. P follows a chi-squared distribution with n (breaks + 1) degrees of freedom and critical values 46.46, 43.19 and 37.48, at 1%, 5% and 10%, respectively. The number of common factors is chosen using the panel Bayesian information criterion proposed by Bai and Ng (2002). Z^*, P^* and P_m^* refer to the corresponding statistics obtained using the p-values of the simplified MSB statistics. One, two and three asterisks detone the rejection of the null hypothesis of a unit root at 10%, 5% and 1% significance levels, respectively, when the statistic is greater than the upper level

Concerning the panel unit root tests, we have allowed for a maximum of 4 breaks, determined using the Bai and Perron (1998) procedure.[20] In Table 3, we present the unit root results for the variables $\ln M3V_{i,t}$, $\ln GNIpc_{i,t}^{p}$, $\pi_{i,t}^{e}$ and $rstir_{i,t-1}$ estimated for the panel.

The panel unit root tests have a constant and a trend and allow for structural changes in both (Model 2, trend break model). In the case of $\ln M3V_{i,t}$, the evidence points to a non-rejection of the unit root hypothesis and a similar outcome is found for $\ln GNIpc_{i,t}^{p}$, $\pi_{i,t}^{e}$ and $rstir_{i,t-1}$. Neither the P-tests nor the Z-tests allow us to reject the null hypothesis of a unit root at different levels of significance.

The position of the structural breaks found for each of the panel dimension variables is reported in Table 4. The structural breaks are distributed along the whole period. We can find however, some patterns in the position of the breaks for the EA associated with the first and second oil crises (1973/1979) and the changes in monetary policy due to the Great Recession. In the case of Spain, notably, the break coincides with its EU membership.

Therefore, although with some mixed results, the null hypothesis of a unit root (with structural breaks) cannot be rejected for all the series at the 5% level of significance. Accordingly, we can conclude that the variables in Tables 2 and 3 are $I(1)$ with structural breaks.

4.2 M3 Velocity Panel TVP Model Estimation

In this subsection, we present the results from the estimation of the model for the time-varying determinants of M3 velocity. The estimates have been obtained by using

[20] See Bai and Carrion-I-Silvestre (2009) for details.

Table 4 Bai and Carrion-I-Silvestre (2009). Structural breaks

Country	$\ln M3V_t$	$\ln GNI^{p}_{pct}$	$rstir_t$	π^e_t
Austria	1990	1979		
	2001			
Belgium		1976		1974
				1982
Germany		1981		
Spain	1973	1976		
	1985	1987		
		2008		
France		1975	1986	
		2008		
Italy	1975	1972		
	2000	2005		
Portugal	1975			
EA		1975		

a Gauss code that extends the traditional approach by Hamilton (1994b) and includes all the elements of the model presented in Sect. 3.2. The results for the maximum likelihood estimation of the elements of the hyper-parameter vector are reported in Table 5, for both the measurement equation and the state-transition equations.

The first part of Table 5 displays the hyper-parameters for the "measurement equation" in Eq. (30), estimated for our panel including the following Eurozone members: Austria, Belgium, Germany, Spain, France, Italy and Portugal. The last column presents the results for the model when estimated for the Eurozone as a whole. The estimated country-specific fixed parameters are reported for the measurement equation, where β_{0i} is the fixed-mean intercept, β_{1i} is the fixed-mean parameter for the log of the permanent component of real GNI per capita ($\log GNIpc^p$), β_{2i} is the fixed coefficient for the lagged dependent variable, ($\ln M3V_{i,t-1}$), and β_{3i} is the mean parameter for $baaspread_t$, our measure of global risk aversion. The table also displays the estimated common mean fixed parameters β_4 and β_5, for the expected inflation ($\pi^e_{i,t}$), and the real short-term interest rate, ($rstir_{i,t}$), respectively.

The rest of the table includes the estimated hyper-parameters for the "state-update equation" in Eq. (31) that contribute to explain the transition of the country-specific varying parameter vector, $\xi_{1,i,t}$, which stands for the varying component of the parameter for $\ln GNIpc^p_{i,t}$. Each unobserved component follows an autoregressive process estimated with a common autoregressive parameter ϕ_1. Finally, our state-space equation also includes control instruments that drive the varying components of both parameters: $GNIpc^{c+}_{i,t}$ and $GNIpc^{c}_{i,t}$. They capture the asymmetric impact of positive and negative deviations from the trend of GNI per capita, calculated as the logarithm of the GNI-to-GNI trend ratio. These control instruments enter as country-specific.

Table 5 M3 velocity function 1981–2016. Selected European countries

Measurement equation. Dependent variable: $\ln M3V_{i,t}$								
	AUT	BEL	DEU	ESP	FRA	ITA	POR	Eurozone
intercept	0.1588	2.5832***	3.3618***	0.8454***	0.9287*	0.6886***	0.6388***	2.3651***
(std error)	*(0.2347)*	*(0.7951)*	*(1.0251)*	*(0.2798)*	*(0.5831)*	*(0.2245)*	*(0.2237)*	*(0.8488)*
$\ln M3V_{-1}$	0.7767***	0.6760***	0.2949*	0.7141***	0.7340***	0.9333***	0.3613**	0.3812**
(std error)	*(0.1213)*	*(0.0960)*	*(0.1702)*	*(0.0926)*	*(0.0833)*	*(0.0835)*	*(0.1655)*	*(0.1546)*
$\ln GNIpc^{p}$	−0.0362	−0.7683***	−0.9373***	−0.2385***	−0.2387	−0.1963***	−0.1938***	−0.6799***
(std error)	*(0.0698)*	*(0.2335)*	*(0.2940)*	*(0.0858)*	*(0.2047)*	*(0.0745)*	*(0.0729)*	*(0.2554)*
baaspread	−0.0045	0.0100	0.0068	−0.0438***	−0.0473**	−0.0250**	−0.0217**	−0.0613***
(std error)	*(0.0186)*	*(0.0161)*	*(0.0292)*	*(0.0106)*	*(0.0195)*	*(0.0123)*	*(0.0103)*	*(0.0215)*
π^{e}	0.0001							0.0124
(std error)	*(0.0007)*							(0.0086)
rstir	0.0002							−0.0069
(std error)	*(0.0006)*							*(0.0079)*
$\sigma_{w,n}$	0.0283	−0.0322	−0.0761	0.0000	0.0486	0.0000	−0.0099	−0.0523
(std error)	*(0.0083)*	*(0.0082)*	*(0.0107)*	*0.0000*	*(0.0077)*	*0.0000*	*0.0000*	*(0.0062)*
State-transition equation. Dependent variable: $\xi_{1,i,t+1}$								
	AUT	BEL	DEU	ESP	FRA	ITA	POR	
tvp for ln(gnipc2010(p))								
$\xi_{1,i,t}$	0.4767***							0.8129***
(std error)	*0.0865*							(0.0690)
$GNIpc^{c+}$	0.1194	0.1249	0.0379	0.1229	−0.2425*	0.1874	0.1909*	0.6617***
(std error)	*(0.2901)*	*(0.1380)*	*(0.2621)*	*(0.1154)*	*(0.1342)*	*(0.1402)*	*(0.1049)*	*(0.2143)*
$GNIpc^{c-}$	−0.1023	−0.3019	−0.1700	−0.0548	0.1753*	−0.1155	0.0196	−0.1734*
(std error)	*(0.1934)*	*(0.1836)*	*(0.1993)*	*(0.0753)*	*(0.0907)*	*(0.1082)*	*(0.0674)*	*(0.1023)*
σ_{v}	0.0096							0.0000
(std error)	*(0.0008)*							*(0.0000)*
Number of observations: 252 (7 countries)								36
Value of log-likelihood: 587.26550								73.845113

Notes t-tests in parentheses. * $p < 0.10$, ** $p < 0.05$, *** $p < 0.01$

We should emphasize the importance of considering the two equations jointly. Whereas the first one (the measurement equation) is the equivalent to many OLS models reported in the empirical literature, the state-update equation represents the behavior of the state components that determine the observed vector of M3 velocity across the countries analyzed.

To ease the interpretation of the results, we complement the estimated coefficients presented in Table 5 with the graphs of the time-varying parameters shown in Fig. 3 that plots the time-varying parameter for $\ln gGNIpc^{p}$ and its mean.

The results confirm the role of permanent income per capita to explain the declining trend in M3 velocity across the Eurozone. Country-specific coefficients range from −0.937 in the case of Germany to −0.193 for Portugal. Only in the cases of Austria and France, the coefficients are nonsignificant. We also find that M3 velocity has a highly persistent nature, as the coefficient of lagged velocity lies between 0.295 for Germany and 0.933 for the Italian case. These results are in line with those stressed by Benati (2019). According to him, velocity fluctuations are systemati-

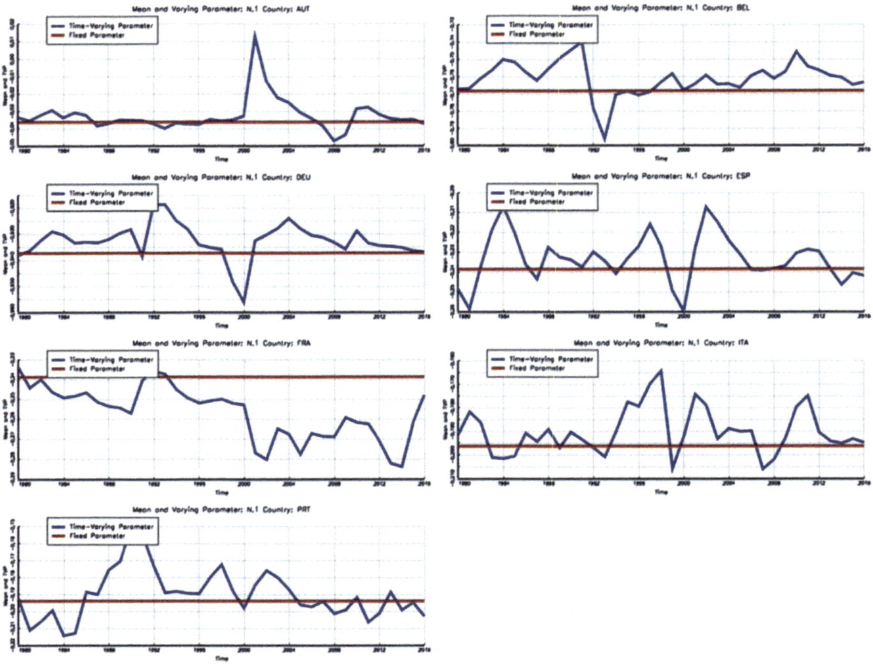

Fig. 3 Time-varying parameter estimation (trend component of real net income per capita), 1961–2016

cally strongly correlated with time-varying parameter estimate of trend real GDP growth.[21]

A salient result is that the explanatory power on the evolution of the M3 velocity according to the estimated parameters of the variable baaspread is different depending on the individual countries. In the case of the peripheral countries (Spain, Portugal, Italy, including France), the coefficients are negative and highly significant whereas in the case of the core countries (Austria, Belgium and Germany), these coefficients are not significant. As the baaspread is a commonly used variable to capture global risk appetite, our results reveal first, a negative link between confidence and money velocity, as suggested by the literature, and second, that this relationship might affect differently to peripheral and core EMU countries. This coefficient is again significant, when the model is re-estimated for the EA as a whole, with a negative parameter (−0.06).

On the contrary, the coefficients for $\pi^e_{i,t}$ and $\text{rstir}_{i,t}$ estimated as a common parameter of the panel appear as nonsignificant. Therefore, the uncertainty and the flight-to-safety effects in the peripheral countries, especially during turmoils, tend to take

[21] However, our interpretation differs from Benati (2019) in that, as we use a broader definition of money, velocity is subject to short-term movements that stem from financial innovation, uncertainty and global risk.

in other explanatory variables (i.ex. baaspread_t) signaling the opportunity cost of holding liquid assets.

Regarding the transition equation, it includes the estimated values of the autoregressive parameter in matrix F. The autoregressive parameter for the permanent component of the net income per capita, ϕ_1, is 0.4767. It exhibits a rather persistent pattern, but clearly lower than one. This means that velocity does not behave as a random walk, a finding that we would have considered an unsatisfactory modelization of the process.

Moreover, according to Hamilton (1994a), if the eigenvalues of the matrix F that contains the autoregressive parameters ϕ are all inside the unit circle, as it happens in our case, the varying coefficient can be considered a deviation from the mean parameter that is estimated in the measurement equation. This pattern can be clearly observed in Fig. 3 that shows the heterogeneity of the estimated coefficients across countries, when the fixed and varying components are aggregated, for the variable $\text{lnGNIpc}^p_{i,t}$.

Concerning the parameters of the cyclical (negative and positive) deviations from the GNI trend per capita, our results suggest an asymmetric influence of the cycle, in particular for the case of France, as periods of recession and expansion seem to affect differently the state-transition of the varying component of the parameter for the $\text{lnGNIpc}^p_{i,t}$.

Summing up, our findings suggest that the downward trend of M3 velocity is mainly explained by the evolution of permanent income, proxied by the trend component of the per capita income, and also exhibits a high persistence. A second relevant result is the asymmetric impact of business cycle fluctuations in the evolution of the unobserved state that drives the varying parameters, causing the instability of velocity. Third, the estimated model emphasizes the role of changes in uncertainty and risk premia with heterogeneous and asymmetric effects for the different member countries of the EA, with clear different consequences for core and periphery. Our results support the existence of some room for monetary targeting. However, the decoupling between monetary aggregates and interest rates in many member countries and their increased exposure to financial global risk has narrowed the margin of manoeuver for monetary policy over time.

5 Conclusions

A prerequisite for a monetary-targeting strategy to work is a stable money demand function, which in turn requires the stability of velocity. In modern economies, neglecting the existence of possible changes in equilibrium money velocity leads to large relative errors in the estimation of inflation and output. Traditional theories of money demand identify income as the main determinant of velocity, so that economic growth would induce a secular downward trend in velocity with some fluctuations depending on the opportunity cost of money, driven by inflation and interest rates.

The success of the ECB in achieving and maintaining low inflation in the EA may also have contributed to blur the distinction between monetary and financial assets as credibility gains may have reduced inflation risk premia, giving rise to a decoupling between money aggregates and inflation ("inflation puzzle"). In such a context, the acceleration in the decline of M3 income velocity might be another illustration of the "paradox of credibility" that can be reinforced under the current "liquidity trap" environment.

To fully take the dynamics of income velocity and its determinants into account, we use a very flexible and comprehensive state-space framework. Our research extends the simple canonical model usually employed in the state-space literature into a panel data time-varying parameter framework, combining fixed (both common and country-specific) and varying components. Regarding the transition equation, our specification allows for the estimation of different autoregressive alternatives and includes control instruments, whose coefficients can be set up either common or idiosyncratic.

Our results confirm the role of permanent income per capita to explain the long-run decreasing trend in M3 velocity in the Eurozone, which would point to continued monitoring of the monetary aggregates in the long run. However, our results also show country heterogeneity across the EA and an asymmetric impact of the business cycle position that determines the evolution of velocity. These results are in line with recent evidence stressing the role of a declining potential output as the most relevant explanatory elements of the "inflation puzzle." Failing to properly adjust for underlying movements in natural growth or equilibrium velocity may obscure the fundamental link between money growth and inflation but does not in any sense reduce its significance and value for monitoring inflation. As we can derive from our results, over short horizons, the relationship between monetary aggregates and inflation can be in large part overshadowed by cyclical factors. Moreover, without the proper adjustment for changes in potential income growth and velocity, money growth and inflation may not appear to track each other even over medium-run horizons. Finally, another interesting contribution is the finding of a negative link between our measure of global risk aversion, the baaspread and money velocity. This relationship suggested by the literature as a negative link between confidence and money velocity would affect asymmetrically to peripheral and core EMU countries. Indeed, the market sentiment may slow down money velocity in peripheral countries of the EA, offsetting the effects of an expansionary monetary policy from the ECB.

To sum up, our results suggest that the quantitative targeting strategy is still valid for European monetary policy, as the relationship of money velocity with its determinants (mainly with $GNIpc^{p}$) is stable in the long run. Nevertheless, our findings point toward an asymmetric pattern for M3V, as a result of an exogenous global risk repricing, proxied by baaspread. For these countries, the reduction of money velocity due to the financial turmoil would offset, even totally, the intended impact of expansive monetary initiatives. Hence, although monetary aggregates can be used as a nominal anchor, it is essential, especially for the case of peripheral EMU countries, to complement these measures with other policies, not only unconventional monetary policies, but also the use of fiscal and structural initiatives.

From a policy stand, although our results point to the validity of the long-run monetary targeting strategy, they also show that monetary policy cannot be the engine of growth, as some delicate potential inter-temporal policy trade-offs may arise. Legacy debt and risk aversion, which are the consequence of the global crisis, as well as uncertainty related to a deep transformation of our economies may weight on both output and inflation in a persistent way. To ensure sustainable growth, a more balanced policy mix is in order with a more active role of structural reforms and sensible fiscal frameworks, using resources to boost sagging public investment.

Acknowledgements We acknowledge the financial support from the Agencia Estatal de Investigación-FEDER ECO2017-83255-C3-2-P project, from the Generalitat Valenciana, Prometeo/2018/102 project and European Commission project 611032-EPP-1-2019-1-ES-EPPJMO-CoE (C. Tamarit and M. Camarero) and the Betelgeux-Christeyns Chair (J. Sapena). The European Commission support does not constitute an endorsement of the contents which reflects the views only of the authors, and the Commission cannot be held responsible for any use which may be made of the information contained therein. All remaining errors are ours.

Appendix: The Kalman Filter

The objective of the state-space formulation is to define the state vector ξ_t in a way that guarantees the minimization of the number of elements and the comprehension of all the available information at time t. To estimate the model, we use the maximum likelihood technique.

Essentially, the Kalman[22] filter (hereafter KF) is, in fact, an algorithm composed by a set of equations, which, performed sequentially, allows to obtain in a forward prediction procedure (called "filtering"), the sequence of linear least squares forecasts of the state vector, $\hat{\xi}_{t|t-1} = \mathrm{E}[\xi_t|\zeta_{t-1}]$ (and hence for the vector of dependents, $\hat{y}_{t|t-1}$), on the basis of the information available at $t-1$, summarized by the vector $\zeta_{t-1} \equiv (y_{t-1}, y_{t-2}, \ldots, y_1, x_{t-1}, x_{t-2}, \ldots, x_1)$:

In general, the KF, described in Zadeh and Desoer (1963), Harvey (1989) or Hamilton (1994a),[23] is particularly suited for Bayesian analysis where, at each time t, there are prior distributions for state variables ξ_t and parameters produced at time $t-1$, from which posterior distributions at time t are calculated as observation y_t becomes available.

The iteration is started by assuming that the initial value of the state vector ξ_1 is drawn from a normal distribution with mean denoted $\hat{\xi}_{1|0}$ and unconditional variance $P_{1|0}$,

[22]The Kalman filter owes its name to the Hungarian Rudolf E. Kalman and his contributions in Kalman (1960), Kalman and Bucy (1961) and Kalman (1963), although similar algorithms had been developed earlier by Thiele (1880) and Swerling (1959), and also at Zadeh and Desoer (1963), constituting an efficient way to formulate the likelihood (usually Gaussian) for many complex econometric models for estimation and prediction purposes.

[23]This section, and the notation employed, draws heavily on the exposition in Chap. 13 of Hamilton (1994b).

$$\xi_1 \sim N\left(\hat{\xi}_{1|0}, P_{1|0}\right) \tag{32}$$

In this case, the distribution of ξ_t conditional on ζ_{t-1} turns out to be normal for $t = 2, 3, \ldots, T$. The mean of this conditional distribution is represented by the $(r \times 1)$ vector $\hat{\xi}_{t|t-1}$, and the variance of this conditional distribution is represented by the $(r \times r)$ matrix $P_{t|t-1} = \mathrm{E}\left(\xi_t - \hat{\xi}_{t|t-1}\right)\left(\xi_t - \hat{\xi}_{t|t-1}\right)'$.

When all elements of the state vector ξ_t defined by 11 are stationary, i.e., if the eigenvalues of F are all inside the unit circle, the initial means, variances and covariances of these initial state elements can be derived from the model parameters, the system is stable, and $\hat{\xi}_{1|0}$ would be the unconditional mean[24] of this stationary vector, while $P_{1|0}$ would be the unconditional variance that can be calculated from:

$$\mathrm{vec}\left(P_{1|0}\right) = [I_{r^2} - (F \otimes F)]^{-1} \cdot \mathrm{vec}\,(Q) \tag{33}$$

where vec $\left(P_{1|0}\right)$ is the $(r_2 \times 1)$ vector formed by stacking the columns of $P_{1|0}$, one on top of the other, ordered from left to right, I_{r^2} represents a r^2 dimension identity matrix and the $\otimes$ operator represents the Kronecker product.

On the contrary, if (at least some of) the elements of the state vector ξ_t are non-stationary, the process of starting up the series is said to be diffuse initialization of the filter, as at least some of the elements of $\hat{\xi}_{1|0}$ and $P_{1|0}$ are unknown. Hence, for time-variant or non-stationary systems, $\hat{\xi}_{1|0}$ represents a guess as to the value of ξ_1 based on prior information, while $P_{1|0}$ measures the uncertainty associated with this guess.[25] This prior cannot be based on the data, since it is assumed in the derivations to follow that ν_{t+1} and w_t are independent of ξ_1 for $t = 1, 2, \ldots, T$. Harvey and Phillips (1979) propose a simple approximate technique for the diffuse initialization of the filter that consists in to initialize non-stationary components of the state vector by any value (say, $\hat{\xi}_{1|0} = 0$), and an arbitrary large variance, $P_{1|0}$, relative to the magnitude of the series, and then use the standard Kalman filter. The larger the variance, the lesser informative the initialization is for the filter. Koopman (1997) and Koopman and Durbin (2003) propose a more transparent treatment of diffuse filtering and smoothing based on Ansley and Kohn (1985).

Departing from the initial conditions, the algorithm works sequentially in a two-step process. Having described the values of $\hat{\xi}_{t|t-1}$ and $P_{t|t-1}$ for $t = 1$, once the outcome of the next measurement (possibly corrupted with some amount of error, including random noise) is observed, these estimates are updated into the a posteriori estimate, as in (34) and (35):

$$\hat{\xi}_{t|t} = \hat{\xi}_{t|t-1} + \cdot P_{t|t-1} H (H' P_{t|t-1} H + R)^{-1} (y_t - A' x_t - H' \hat{\xi}_{t|t-1}) \tag{34}$$

[24] When $\hat{\xi}_{1|0}$ is covariance stationary, a candidate value for $\hat{\xi}_{1|0}$ is zero so that all state variables are initially in steady state.

[25] The greater our prior uncertainty, the larger the diagonal elements of $P_{1|0}$.

$$P_{t|t} = \cdot P_{t|t-1} - \cdot P_{t|t-1} H (H' P_{t|t-1} H + R)^{-1} H' P_{t|t-1} \quad (35)$$

The predictive phase uses the state estimate from the previous timestep to produce an estimate of the state at the current timestep. This predicted state estimate is also known as the a priori state estimate, along with their uncertainties:

As $\hat{\xi}_{t+1|t} = F\hat{\xi}_{t|t}$ and $P_{t+1|t} = F \cdot P_{t|t}F' + Q$, to calculate the sequence $\left\{\hat{\xi}_{t+1|t}\right\}_{t=1}^{T}$ and $\left\{P_{t+1|t}\right\}_{t=1}^{T}$, one has simply to iterate on Eqs. (36) and (37) for $t = 1, 2, \ldots, T$.

$$\hat{\xi}_{t+1|t} = F\hat{\xi}_{t|t-1} + F \cdot P_{t|t-1} H (H' P_{t|t-1} H + R)^{-1} (y_t - A' x_t - H' \hat{\xi}_{t|t-1}) \quad (36)$$

$$P_{t+1|t} = F \cdot P_{t|t-1} F' - F \cdot P_{t|t-1} H (H' P_{t|t-1} H + R)^{-1} H' P_{t|t-1} F' + Q \quad (37)$$

The output obtained for step t is used sequentially as the input for the step $t + 1$.

When the values of the matrices F, Q, B, A, H and R are unknown, we collect the unknown elements of these matrices in a hyper-parameter vector, θ, and obtain their maximum likelihood estimates. Casual starting values are assigned to the ξ_t vector and to the unknown elements of matrices included in the vector of parameters θ, and the estimation procedure maximizes the likelihood function. In general, Q and R are assumed to be positive semi-definite (which includes the possibility that some of the error terms may be zero).

In some cases, it is desirable to use information through the end of the sample (date T) to help improving the inference about the historical value that the state vector took on at any particular date t in the middle of the sample. Such an inference is known as a smoothed estimate, $\hat{\xi}_{t|T} = \mathrm{E}[\xi_t|\zeta_T]$. The mean squared error of this estimate is denoted $P_{t|T} = \mathrm{E}\left(\xi_t - \hat{\xi}_{t|T}\right)\left(\xi_t - \hat{\xi}_{t|T}\right)'$. The estimation of the sequence $\left\{\hat{\xi}_{t|T}\right\}_{t=1}^{T}$ can be calculated by iterating in reverse order for $t = T - 1, T - 2, \ldots, 1$ on:

$$\hat{\xi}_{t|T} = \hat{\xi}_{t|t} + J_t(\hat{\xi}_{t+1|T} - \hat{\xi}_{t+1|t}) \quad (38)$$

$$P_{t|T} = P_{t|t} + J_t(P_{t+1|T} - P_{t+1|t})J_t' \quad (39)$$

where

$$J_t = P_{t|t} F' P_{t+1|t}^{-1} \quad (40)$$

References

Andrews, D. W. K., & Ploberger, W. (1994). Optimal tests when a nuisance parameter is present only under the alternative. *Econometrica, 62*(6), 1383.

Ansley, C. F., & Kohn, R. (1985). Estimation, filtering, and smoothing in state space models with incompletely specified initial conditions. *The Annals of Statistics*, 1286–1316.

Bai, J., & Carrion-I-Silvestre, J. (2009). Structural changes, common stochastic trends, and unit roots in panel data. *The Review of Economic, 76*(2), 471–501.

Bai, J., & Ng, S. (2002). Determining the number of factors in approximate factor models. *Econometrica, 70*(1), 191–221.

Bai, J., & Ng, S. (2004). A panic attack on unit roots and cointegration. *Econometrica, 72*(4), 1127–1177.

Bai, J., & Perron, P. (1998). Estimating and testing linear models with multiple structural changes. *Econometrica, 66*(1), 47–78.

Bai, J., & Perron, P. (2003). Computation and analysis of multiple structural change models. *Journal of Applied Econometrics, 18*(1), 1–22.

Benati, L. (2009). Long run evidence on money growth and inflation. *ECB Working Paper Series* (1027).

Bernoth, K., & Erdogan, B. (2012). Sovereign bond yield spreads: A time-varying coefficient approach. *Journal of International Money and Finance, 31*(3), 639–656.

Benati, L. (2019). Money velocity and the natural rate of interest. *Journal of Monetary Economics*, 1–18.

Biefang-Frisancho Mariscal, I., & Howells, P. G. (2012). Income velocity and non-GDP transactions in the UK. *International Review of Applied Economics, 26*(1), 97–110.

Bobeica, E., & Jarociński, M. (2017). Missing disinflation and missing inflation: The puzzles that aren't. *ECB Working Paper Series* (2000).

Bordes, C., Clerc, L., & Marimoutou, V. (2007). Is there a structural break in equilibrium velocity in the euro area? To cite this version: HAL Id: hal-00308654. Bank de France. Notes d'Études et de Recherche.

Bordo, M. D., & Jonung, L. (1987). *The long-run behavior of the velocity of circulation: The international evidence*. Cambridge: Cambridge University Press.

Bordo, M. D., & Jonung, L. (1990). The long-run behavior of velocity: The institutional approach revisited. *Journal of Policy Modeling, 12*(2), 165–197.

Bordo, M. D., Jonung, L., & Siklos, P. L. (1997). Institutional change and the velocity of money: A century of evidence. *Economic Inquiry, 35*(4), 710–724.

Brand, C., Gerdesmeier, D., & Roffia, B. (2002). Estimating the trend of M3 income velocity underlying the reference value for monetary growth. *ECB Occasional Paper Series* (3).

Broto, C., & Perez-Quiros, G. (2015). Disentangling contagion among Sovereign CDS spreads during the European debt crisis. *Journal of Empirical Finance, 32*, 165–179.

Bruggeman, A., Camba-Méndez, G., Fischer, B., & Sousa, J. (2005). Structural filters for monetary analysis. The inflationary movements of money in the Euro area. *ECB Working Paper Series* 470.

Calza, A., & Zaghini, A. (2009). Nonlinearities in the dynamics of the euro area demand for M1. *Macroeconomic Dynamics, 13*(1), 1–19.

Camarero, M., Sapena, J., & Tamarit, C. (2019). Modeling time-varying parameters in panel data state-space frameworks: An application to the Feldstein-Horioka puzzle. *Computational Economics, 56*, 87–114.

Carrion-i Silvestre, J. L., Kim, D., & Perron, P. (2009). GLS-based Unit Root tests with multiple structural breaks under both the null and the alternative hypotheses. *Econometric Theory, 25*(Special Issue 06), 1754–1792.

Cochrane, J. H. (2017). The radical implications of stable quiet inflation at the zero bound. *mimeo*.

Crowder, W. (1998). The long-run relation between money growth and inflation. *Economic Inquiry, 36*(2), 229–243.

De Grauwe, P., & Polan, M. (2005). Is inflation always and everywhere a monetary phenomenon? *Scandinavian Journal of Economics*, *107*(2), 239–259.

Dreger, C., & Wolters, J. (2009). Money velocity and asset prices in the euro area. *Empirica*, *36*(1), 51–63.

El-Shagi, M., & Giesen, S. (2010). Money and inflation: The role of persistent velocity movements. *Mimeo*, 1–24.

El-Shagi, M., Giesen, S., & Kelly, L. (2011). The quantity theory revisited: A new structural approach. *mimeo*, 1–26.

European Central Bank. (2017). Base money, broad money and the APP. *ECB Economic Bulletin*, *6*, 62–65.

Faragher, R. (2012). Understanding the basis of the Kalman filter via a simple and intuitive derivation [lecture notes]. *IEEE Signal Processing Magazine*, *29*(5), 128–132.

Fitzgerald, T. J. (1999, August). Money growth and inflation: How long is the long-run? *Federal Reserve Bank of Cleveland Economic Commentary*, 1–5.

Friedman, M. (1963). *Inflation: Causes and consequences*. New York: Asia Publishing House.

Friedman, M., & Schwartz, A. J. (1963). *A monetary history of the United States, 1867–1960*. Princeton: Princeton University Press.

Gerlach, S., & Svensson, L. E. O. (2003). Money and inflation in the euro area: A case for monetary indicators? *Journal of Monetary Economics*, *50*(8), 1649–1672.

Gourieroux, C., & Monfort, A. (1997). *Time series and dynamic models*. Cambridge: Cambridge University Press.

Gravy, G. (1959). Structural aspects of money velocity. *The Quarterly Journal of Economics*, *73*(3), 429–447.

Hamilton, J. D. (1989). The long-run behavior of the velocity of circulation. A review essay. *Journal of Monetary Economics*, *23*(January 1987), 335–344.

Hamilton, J. D. (2018). Why you should never use the Hodrick-Prescott filter. *Review of Economics and Statistics*, *100*(5), 831–843.

Hamilton, J. D. (1994a). State-space models. In R. F. Engle & D. L. McFadden (Eds.), *Handbook of econometrics* (Vol. 4, Chap. 50, pp. 3039–3080). Amsterdam: Elsevier.

Hamilton, J. D. (1994b). *Time series analysis*. Princeton, NJ: Princeton University Press.

Harvey, A. (1989). *Forecasting, structural time series models and the Kalman filter*. Cambridge: Cambridge University Press.

Harvey, A. C. (1993). *Time series models* (2nd ed.). New York: Harvester Wheatsheaf.

Harvey, A. C., & Phillips, G. D. A. (1979). Maximum likelihood estimation of regression models with autoregressive-moving average disturbances. *Biometrika*, *66*(1), 49–58.

Haug, A. A., & Dewald, W. G. (2012). Money, output, and inflation in the longer term: Major industrial countries, 1880–2001. *Economic Inquiry*, *50*(3), 773–787.

Kim, C.-J., & Nelson, C. R. (1999). *State-space models with regime switching: Classical and Gibbs-sampling approaches with applications* (Vol. 1). Cambridge, MA: MIT Press Books.

Kalman, R. E. (1960). A new approach to linear filtering and prediction problems. *Journal of Fluids Engineering*, *82*(1), 35–45.

Kalman, R. E. (1963). New methods in Wiener filtering theory. In J. L. Bogdanoff, & F. Kozin (Eds.), *Proceedings of the First Symposium on Engineering Applications of Random Function Theory and Probability*. New York: Wiley.

Kalman, R. E., & Bucy, R. S. (1961). New results in linear filtering and prediction theory. *Journal of Basic Engineering*, *83*(1), 95–108.

Koopman, S. J. (1997). Exact initial Kalman filtering and smoothing for nonstationary time series models. *Journal of the American Statistical Association*, *92*(440), 1630–1638.

Koopman, S. J., & Durbin, J. (2003). Filtering and smoothing of state vector for diffuse state-space models. *Journal of Time Series Analysis*, *24*(1), 85–98.

Leão, P. (2005). Why does the velocity of money move pro-cyclically? *International Review of Applied Economics*, *19*(1), 119–135.

Lucas, R. E. (1980). Two illustrations of the quantity theory of money. *The American Economic Review*, *70*(5), 1005–1014.
McCallum, B. T. (1984). On low-frequency estimates of long-run relationships in macroeconomics. *Journal of Monetary Economics*, *14*(1), 3–14.
Mele, A., & Stefanski, R. (2019). Velocity in the long run: Money and structural transformation. *Review of Economic Dynamics*, *31*, 393–410.
Moon, H. R., & Perron, B. (2004). Testing for a unit root in panels with dynamic factors. *Journal of Econometrics*, *122*(1), 81–126.
Morgan, D. P. (1993). Asymmetric effects of monetary policy. *Federal Reserve Bank of Kansas City Economic Review*, second quarter, 21–33.
Orphanides, A., & Porter, R. D. (2000). P * revisited: Money-based inflation forecasts with a changing equilibrium velocity. *Journal of Economics and Business*, *52*, 87–100.
Orphanides, A., Porter, R. D. (2001). Money and inflation: The role of information regarding the determinants of M2 behavior. In *Monetary analysis: Tools and applications* (pp. 77–95). European Central Bank.
Perron, P. (1989). The great crash, the oil price shock, and the unit root hypothesis. *Econometrica*, *57*(6), 1361–1401.
Perron, P., & Yabu, T. (2009). Testing for shifts in trend with an integrated or stationary noise component. *Journal of Business & Economic Statistics*, *27*(3), 369–396.
Perron, P., & Zhu, X. (2005). Structural breaks with deterministic and stochastic trends. *Journal of Econometrics*, *129*(1–2), 65–119.
Raj, B. (1995). Institutional hypothesis of the long-run income velocity of money and parameter stability of the equilibrium relationship. *Journal of Applied Econometrics*, *10*(September 1994), 233–253.
Reynard, S. (2006). Money and the great disinflation. *Swiss National Bank Working Papers*, 06(7).
Sargan, J. D., & Bhargava, A. (1983). Testing residuals from least squares regression for being generated by the Gaussian random walk. *Econometrica*, *51*(1), 153–174.
Stock, J. H., & Watson, M. W. (1996). Evidence on structural instability in macroeconomic time series relations. *Journal of Business & Economic Statistics*, *14*(1), 11–30.
Stock, J. H., & Watson, M. W. (2005). Implications of dynamic factor models for VAR analysis. *NBER Working Paper Series*, 11467 (pp. 1–67).
Stock, J. H., & Watson, M. (1999). A comparison of linear and nonlinear models for forecasting macroeconomic time series. In R. F. Engle, & H. White (Eds.), *Cointegration, causality and forecasting: A Festchrift in honour of Clive W.J. Granger* (pp. 1–44). Oxford: Oxford University Press.
Svensson, L. E. (2010). Inflation targeting. In B. M. Friedman, & M. Woodford (Eds.), *Handbook of monetary economics* (Vol. 3B, Chap. 22, pp. 1237–1303). Oxford: Elsevier.
Swamy, P., & Tavlas, G. S. (2001). Random coefficient models. In B. H. Baltagi (Ed.), *A companion to theoretica econmetrics* (pp. 410–428). Malden, MA: Blackwell Publishing Ltd.
West, M., & Harrison, J. (1997). *Bayesian forecasting and dynamic models* (2nd ed.). New York, NY, USA: Springer.
Swerling, P. (1959). First-order error propagation in a stagewise smoothing procedure for satellite observations. *The Journal of the Astronautical Sciences*, *6*, 45–52.
Thiele, T. N. (1880). Om Anvendelse af mindste Kvadraters Methode i nogle Tilfælde, hvor en Komplikation af visse Slags uensartede tilfældige Fejlkilder giver Fejlene en 'systematisk' Karakter. *Det Kongelige Danske Videnskabernes Selskabs Skrifter-Naturvidenskabelig og Mathematisk Afdeling* (pp. 381–408).
Zadeh, L. A., & Desoer, C. A. (1963). *Linear system theory: The state space approach*. New York: McGraw-Hill.

Revisiting Wealth Effects in France: A Double-Nonlinearity Approach

Olivier Damette and Fredj Jawadi

1 Introduction

Household consumption is an important component that is central to several macroeconomic theories, but the specifications of its drivers and determinants are not unanimously agreed upon among economists. Indeed, while both income and wealth seem to be major drivers of consumption, several different consumption theories have been developed since the 1930s. They mainly differ according to the definitions of these drivers (Disposable Income (Keynes 1936); Permanent Income (Friedman 1957); Life-Cycle model; Housing Wealth and Financial Wealth (Case et al. 2005 etc.). Furthermore, the question of household consumption dynamics and wealth/income effects has recently been the focus of several debates and studies further to the recent global financial crisis (2008–2009) which severely affected wealth, income and consumption, raising various questions[1] and monetary policy implications for central banks. It is thus crucial to assess the effects variation in aggregate consumption when wealth changes (due to macro shocks, interest rates changes etc.). Increasing wealth could encourage consumption and increase the borrowing capacity of agents while decreasing wealth could decrease consumption and reduce the value of the available collateral.

As for the wealth–consumption relationship, it seems to be a priori complex for at least three reasons. First, this relationship depends on the type of wealth and on its importance, which varies over time and across countries according to the economic

[1]See Jawadi (2008), Jawadi and Léoni (2013) for a recent Literature survey on the income-consumption relationship.

O. Damette
University of Lorraine, Nancy, France

F. Jawadi (✉)
University of Lille, Lille, France
e-mail: fredj.jawadi@univ-lille.fr

G. Dufrénot and T. Matsuki (eds.), *Recent Econometric Techniques for Macroeconomic and Financial Data*, Dynamic Modeling and Econometrics in Economics and Finance 27, https://doi.org/10.1007/978-3-030-54252-8_6

environment. The wealth effect on consumption thus depends on households' expectations, which may change over time due to changes in savings, interest rates, demographics, social insurance, tax institutions, imperfect knowledge of their own wealth (Alexandre et al. 2007; Gabriel et al. 2008), etc. For example, the importance of Housing Wealth (note HW hereafter) might change over time after changes in savings and tax institutions. Additionally, when central banks keep the interest rate very low—as in the aftermath of the global financial crisis (2008–2009)—this can affect savings, credit conditions and financial wealth (noted FW hereafter) and therefore consumption. In such a context, Whelan (2008) and Slacalek (2009) pointed to the instability of the cointegration vector for income-wealth–consumption due to a change in the long-run interest rate. Gabriel et al. (2008) outlined that the volatility of financial markets and assets might lead to different impacts on consumption according the regime of the economy. Alexandre et al. (2007) also found that the volatility of asset wealth might lead to biased wealth effects estimates. Carroll et al. (2011a) developed an alternative method over linear cointegrating regressions to take further changes in financial market institutions, demographics or expected income growth into account.

Second, according to Kahneman and Tversky's Prospect Theory, households behave differently after a wealth increase than after a wealth decrease (Genesove and Mayer 2001), and changes in wealth could thus have different effects on consumption that vary across the sign and size of the wealth shock (asymmetric effects).[2]

Third, wealth effects on consumption may be lagged, due to smooth and heterogeneous changes in consumption, habit formation and sticky expectations (Jawadi 2008; Slacalek 2009; Carroll et al. 2011a; Jawadi and Léoni 2013).

In the literature, several empirical studies have studied the wealth–consumption relationship using different approaches and data. Campbell and Cocco or Arrondel et al. (2015) for the French case studied wealth effects using microeconomic data, but most existing studies refer to macroeconomic data and measured wealth effects through different time series approaches, in particular the cointegration approach (Byrne and Davis 2003; Catte et al. 2004) in line with the seminal studies from Lettau and Ludvigson (2001, 2004). Indeed, Lettau and Ludvigson (2001, 2004) showed that it is possible to obtain a super-consistent cointegrating estimation of the parameters linking consumption, labor income and asset wealth in a linear model.

All previous studies are however often concentrated on the US market and all in all their results are not unanimously conclusive. Overall, these studies propose to use disaggregated data while distinguishing HW from FW in order to better understand wealth effects.

Case et al. investigated the relationship between wealth effects and consumption in a panel of US states over the period 1982–2009, extending their previous studies of 2005. The authors showed a large HW effect on consumption which, interestingly, is stronger than the FW effect. Indeed, the HW effect is bounded between 2 and 6% while the FW effect varies between 4 and 6%. This result confirms their previous findings and remains true even when the volatility excess of wealth and consumption

[2]According to this theory, regret due to loss of wealth has different psychological effects to an increase in wealth, which frees up opportunities to use that wealth to finance consumption.

over the decade 1999–2009 is taken into account.[3] Furthermore, Case et al. estimated different specifications to better capture the wealth–consumption relationship and their findings are robust. Interestingly, these authors highlighted further evidence of asymmetry in the wealth effect, as consumption increases after a HW increase but declines only marginally after a HW decrease.[4] According to the authors, during the steady performance of the US economy over the period 1983–2000, the US housing market increased significantly and the price did not decline after 1975. The rise in house price since 1990, due to easy money and excessive optimism, accelerated in 2001, even as the stock market collapsed, and significantly stimulated household consumption. Kennickell and Lusardi reached the same conclusion while pointing to the usefulness of home equity for consumption. Steindel and Luvigston showed the unstable wealth-income relationship in the US market, while Desnoyers suggested that the effect of wealth on US consumption is temporary. Poterba and Samwick also pointed to a positive correlation between stock prices and consumption.

Studies looking at the wealth–consumption relationship outside the US market are rather scarce. In a panel of developed countries, Caroll et al. also found a significant HW effect on household consumption while proposing an alternative approach to measure wealth. Unlike previous studies, Ludwing and Slok found that the FW effect is larger than that of HW for 16 OECD countries, and also that the wealth effect increased over time. For Girouard and Blönd (2001), the wealth effect is significantly variable across countries, but the comparison of HW effect to FW effect is not conclusive. Recently, Jawadi and Sousa (2014) assessed the significant relationship between consumption, wealth and income in a context of financial crisis for the US, the UK and the Euro area and showed that the crisis induced a fall in permanent income, asset wealth and consumption. Jawadi et al. (2015) showed that the wealth–consumption relationship exhibits asymmetry, time variation and switching regime and that this relationship varies across regime and with the economic state. In particular, while the HW significantly affects US consumption and drives transition between consumption regimes, the effect is smaller and insignificant for the UK and the Euro area.

Only a few studies have investigated consumption–wealth behaviors for France, often using macroeconomic approaches.[5] Boone et al. (2001) pointed to a positive wealth effect on consumption but their specification did not take the potential endogeneity of the covariates into account. Recently, this bias was corrected using more robust specifications including the Dynamic Ordinary Least Squares (DOLS) method (Chauvin and Damette 2010, 2011), the Unrestricted Error Correction (EC) Models (Barrel and Davies 2007; Byrne and Davis 2003), and the VECM-Generalized Least Squares models (Chauvin and Damette 2010, 2011). The findings are still rather heterogeneous, however. While the wealth effect on French consumption exceeds

[3] According to Case et al. due to the housing boom over the period 2001–2005, $700 billion of equity was extracted each year by home equity loans, which obviously encouraged aggregate spending.

[4] While Case et al. found a positive correlation between HW and consumption, some other authors suggest little or no welfare gain after an increase in house value because a house is both an asset and a necessary portion of outlays.

[5] Except Arrondel et al. (2015) who recently used microeconomic data.

18% for Barrell and Davis (2007), a wealth increase of 100% induces only a marginal effect on consumption (2.5%) for Aviat et al. Also, in contrast with studies on the US market, it seems that FW effects are larger than HW effects for France (Catte et al. 2004; IMF 2004; Chauvin and Damette 2010, 2011). Slacalek, however, reported the opposite result in favor of HW.

All in all, first, related previous studies provide further evidence of a significant wealth effect on consumption, but these findings and their significance level vary not only across countries but also according to the sample and methodology under consideration, and findings are very heterogeneous. Second, even the breakdown of total wealth into HW and FW is statistically significant and improves the specification of the consumption–wealth relationship. The importance of the type of wealth for consumption is not sufficiently proven, which can be explained by the lack of consensus regarding the definition of wealth. Third, we have noted a significant focus on the US market in previous studies. Fourth, the wealth–consumption relationship seems to be rather unstable and this instability leads to asymmetry[6] and nonlinearity. But the previous literature often studied the wealth effect in a linear context, which is not appropriate for reproducing such asymmetry, nonlinearity and instability in the consumption–wealth relationship. Indeed, few works have studied the wealth effect using nonlinear techniques. Skinner found asymmetrical wealth effects on the youngest people. Stevans (2004) applied the M-TAR model of Enders and Siklos to investigate FW effects. Apergis and Miller (2006) and more recently Donihue and Avramenko (2007), Mac Donald et al., Marquez et al. (2013) and Jawadi et al. (2015) proposed nonlinear specifications for wealth effects using TAR, MTAR models and quantile regressions. Alexandre et al. (2007) and Gabriel et al. (2008) computed Markov switching models to test the instability of the wealth effect according to the state of the economy (interest rates, term structure, demographics, monetary policy changes). However, none of them focused on the French market and also all of them applied a particular type of nonlinear model and only in the short term.

This paper fills this gap and contributes differently. First, investigating the wealth effect on French consumption using different nonlinear approaches, we provide the first nonlinear study on wealth effects for France. Second, the breakdown of wealth into HW and FW and their time variation per regime according to the state of the economy improves the specification of the French consumption–wealth model in the short term. Third, within the long-run macroeconomic data available, not only do we test the wealth effect in both calm and crisis periods, but we also investigate these effects in the long term. Fourth, the combination of nonlinear and cointegration tests in adjustment and in the long-run - which is the main novelty of this paper - helps to better understand the complexity associated with wealth effects, while also testing different nonlinearity types.

[6]This asymmetry can be explained by time variation in wealth, habit formation utility functions and loss aversion. Also, wealth changes are often perceived as temporary, etc.

Our findings showed significant time-varying wealth effects on French consumption confirming an unstable relationship between wealth and consumption. In particular, this relationship exhibits asymmetry and nonlinearity. Interestingly, the introduction of nonlinearity in the long-term relationship showed a significant volatility excess in the cointegration relationship around the year 2000 and during the subprime crisis, suggesting an increase in the wealth effect during these periods.

The rest of the paper is organized into four sections. Section 2 briefly presents the econometric methodology. Data and main empirical results are discussed in Sect. 3. Section 4 concludes.

2 Econometric Methodology

In this section, we briefly discuss the econometric methodology used to investigate wealth effects on consumption. In particular, we discuss three different specifications: (i) A Linear Error Correction Model (LECM) based on a linear cointegration relationship, (ii) A threshold ECM (TCEM) based on a linear cointegration relationship and (iii) A nonlinear specification introducing nonlinearity in the cointegration relationship.

2.1 Linear Cointegration Specification for Wealth Effects

First, we specify the consumption–wealth relationship differently while assessing for the effect of total wealth (Eq. 1) and that of disaggregate wealth (Eq. 2) since households might react differently to shocks on financial assets or on property prices. Indeed, in line with the theoretical framework from Lettau and Ludvigson (2001, 2004), we can write the following log-linear model:

$$c_t = \alpha + \beta_1 TW_t + \beta_2 y_t + \varepsilon_t \tag{1}$$

$$c_t = \alpha + \beta_1 FW_t + \beta_2 HW_t + +\beta_3 y_t + \varepsilon_t \tag{2}$$

where: c_t, W_t, FW_t, HW_t and y_t refer to consumption, total wealth (TW), financial wealth (FW), housing wealth (HW) and disposable income respectively. All variables are in logarithm.

Considering Lettau and Ludvigson (2001, 2004) in line with the life-cycle approach of wealth effects, Eqs. (1) and (2) are estimated in a cointegration framework. Indeed, Lettau and Ludvigson (2001) used the Campbell and Mankiw (1989) micro-funded model of consumption to show that consumption tends to a stationary fraction of wealth. The so-called cointegration-based approach from Lettau and Ludvigson (2001, 2004) lead directly to the estimations of wealth effects elasticities.

The framework of linear cointegration following the two-step method of Engle and Granger was first retained to assess for wealth effects. We tested the presence of a linear cointegration relationship between wealth and consumption in a first step (long-run relationship), and estimated a LECM to capture further short-term wealth effects in the second step if the cointegration tests did not reject the cointegration hypothesis. To do so, we estimated the cointegration relationship using OLS and specifically using Dynamic Ordinary Least Squares (DOLS) following Saikkonen and Stock and Watson (1993) in order to take further endogeneity bias into account (Eqs. 3 and 4):

$$c_t = \alpha + \beta_1 TW_t + \beta_2 y_t + \sum_{i=-m}^{i=m} \phi_i \Delta TW_t + \sum_{i=-n}^{i=n} \rho_i \Delta y_t + \varepsilon_t \tag{3}$$

$$c_t = \alpha + \beta_1 TW_t + \beta_2 y_t + \sum_{i=-m}^{i=m} \phi_i \Delta FW_t + \sum_{i=-k}^{i=k} \psi_i \Delta HW_t + \sum_{i=-n}^{i=n} \rho_i \Delta y_t + \varepsilon_t \tag{4}$$

The DOLS consists in adding k leads and p lags of the first difference of the regressors. The DOLS estimator has recently become a very popular estimator (Forest and Turner, 2013): it is indeed super-consistent and lags and leads eliminate (asymptotically) the possible bias due to endogeneity or serial correlation (see Montalvo 1995 for a discussion[7]).

2.2 Threshold ECM Specification for Wealth Effects

In a second time, we consider potential thresholds into the ECM framework and we extend the LECM by introducing nonlinearity in the adjustment and consideration of a Nonlinear ECM. In particular, we propose two different regressions for the wealth–consumption short-term adjustment dynamics: TAR-ECM and M-TAR-ECM models in order to capture different forms of nonlinearities in the adjustment. This extension aims to capture further asymmetry and nonlinearity in the wealth–consumption adjustment relationship that may escape the LECM. It is thus a more flexible methodology to investigate cointegration. The difference between the TAR and MTAR models is in the definition of the Heaviside indicator function, which is based on the level value of the threshold of the indicator variable in the TAR model, while it is based on the difference of the indicator variable for the MTAR model.

In practice, TAR-ECM and M-TAR-ECM are estimated using the Enders and Siklos (2001) methodology which generalized the Engle and Granger procedure

[7]Using Monte Carlo simulations about small sample performances of time series estimators in small ($T = 50$) samples, he outlined that DOLS has smaller bias and RMSE than OLS and CCR estimators.

to allow for nonlinear adjustments. Indeed, the estimation of the TAR-ECM and M-TAR-ECM processes is based on the estimated residuals from the long-run relationship between the two variables, consumption and wealth [models (1) and (2)]. This also requires the existence of a single cointegration relationship (see for instance Marquez et al. 2013 for the UK case).

The M-TAR-ECM model can be written as:

$$\Delta\hat{u}_t = I_t\rho_1\hat{u}_{t-1} + (1 - I_t)\rho_2\hat{u}_{t-1} + \sum_{i=1}^{k}\gamma_i\Delta\hat{u}_{t-1} + \varepsilon_t \tag{5}$$

with the indicator function I_t defined as: $I_t = \begin{Bmatrix} 1 \text{ if } \Delta\hat{u}_{t-1} \geq 0 \\ 0 \text{ if } \Delta\hat{u}_{t-1} < 0 \end{Bmatrix}$.

A TAR-ECM corresponds to:

$$\Delta\hat{u}_t = I_t\rho_1\hat{u}_{t-1} + (1 - I_t)\rho_2\hat{u}_{t-1} + \sum_{i=1}^{k}\gamma_i\Delta\hat{u}_{t-1} + \varepsilon_t \tag{6}$$

with the indicator function I_t defined as: $I_t = \begin{Bmatrix} 1 \text{ if } \hat{u}_{t-1} \geq 0 \\ 0 \text{ if } \hat{u}_{t-1} < 0 \end{Bmatrix}$.

The testing strategy of Enders and Siklos (2001) is based on two types of tests. The first one refers to the Φ statistic that tests the null hypothesis of no cointegration ($H0 : \rho_1 = \rho_2 = 0$) using a Fisher distribution. However, this statistic does not follow a standard distribution. Enders and Siklos (2001) computed quite suitable critical values. If we reject the null hypothesis of no cointegration on the basis of the Φ statistic, the considered variables are cointegrated. The second statistic, called t-Max, is performed to test the hypothesis of symmetric adjustment $H0 : \rho_1 = \rho_2$ against its alternative of an asymmetrical adjustment. This statistic follows a standard t-Student distribution.

2.3 *Time-Varying VECM Specification for Wealth Effects*

This specification is more original as it enables both the long-run relationship (cointegration relationship) and the ECM to exhibit nonlinearity.[8] This novel specification allows a generalization of both first and second specifications and offers a more original econometric framework to investigate complex wealth effects.

Formally, following Bierens and Martins (2010), we first compute the multivariate time-varying cointegration. Bierens et Martins (2010) explained that long-run coefficients of the VAR(p) are allowed to change with time and can be approximated by a finite sum of Chebyshev polynomials. In this way, the Bierens and Martins

[8]To our knowledge, this is the first study that introduces nonlinearity in both adjustment and cointegration equations to assess for wealth effects, making our paper the first essay in this field.

methodology considers a multivariate VECM framework for which the Johansen (1991) model is a special case.

Thus we start with the following TV-VECM of order p:

$$\Delta Z_t = \mu + \alpha\beta_t' Z_{t-1} + \sum_{j=1}^{p-1} \Gamma_j \Delta Z_{t-j} + \varepsilon_t, \varepsilon_t \sim i.i.d.N_k(0, \Omega), t = 1, \ldots, T. \tag{7}$$

With $Z_t = (C_t, TW_t, \text{Income}_t)$ for the model with aggregate data or $Z_t = (C_t, FW_t, HW_t, \text{Income}_t)$ when considering the disaggregate data. μ, α and β are 3×1 fixed coefficients vectors.

Contrary to the standard VECM from Johansen (1991), the coefficients may be time-varying. Assuming that the function of discrete time β_t is smooth in line with Bierens and Martins (2010), we thus have the following: $\beta_t = \beta_m\left(\frac{t}{T}\right) = \sum_{i=0}^{m} \xi_{i,T} P_{i,T}(t)$ where the orthonormal Chebyshev time polynomials $P_{i,T}(t)$ are defined by $P_{0,T}(t) = 1, P_{i,T}(t) = \sqrt{2}\cos\left(\frac{i\pi(t-0.5)}{T}\right), t = 1, 2, \ldots, T, i = 1, 2, \ldots, m$ and $\xi_{i,T} = \frac{1}{T}\sum_{t=1}^{T} \beta_t P_{i,T}(t)$ are unknown $k \times r$ matrices with k the number of variables and r the rank.

Hence, note that when $m = 0$, the time-varying model becomes the usual time-invariant cointegration model as in Johansen (1996). Thus, testing the time-varying cointegration is equivalent to testing the null hypothesis $m = 0$ or equivalently $\beta_t = \xi_0$ against the alternative hypothesis: $m > 0$, suggesting a finite number of breaks. In practice, as in Bierens and Martins (2010), the estimation is carried out using the Maximum Likelihood method, while the likelihood ratio (LR) test is applied to test the null hypothesis.

3 Data and Empirical Analysis

3.1 *Data and Preliminary Analysis*

Data are quarterly and cover the period 1987Q1 to 2011Q4. They concern France and are obtained from financial and non-financial national accounts. Consumption is defined as the household's total expenditures, while Income corresponds to the flow of human wealth and is measured by disposable income net of property and imputed rents. Financial wealth consists in the household's financial assets net of debts, whereas Housing wealth consists in tangible assets (land and housing). Our study extended the one by Chauvin and Damette (2011), who used similar data over the period 1987–2008, by focusing on nonlinearity in the wealth–consumption relationship. It also extended their study through the use of more recent data to outline the effect of the subprime crisis on the Consumption/Wealth relationship. More details about the data are reported in Fig. 1.

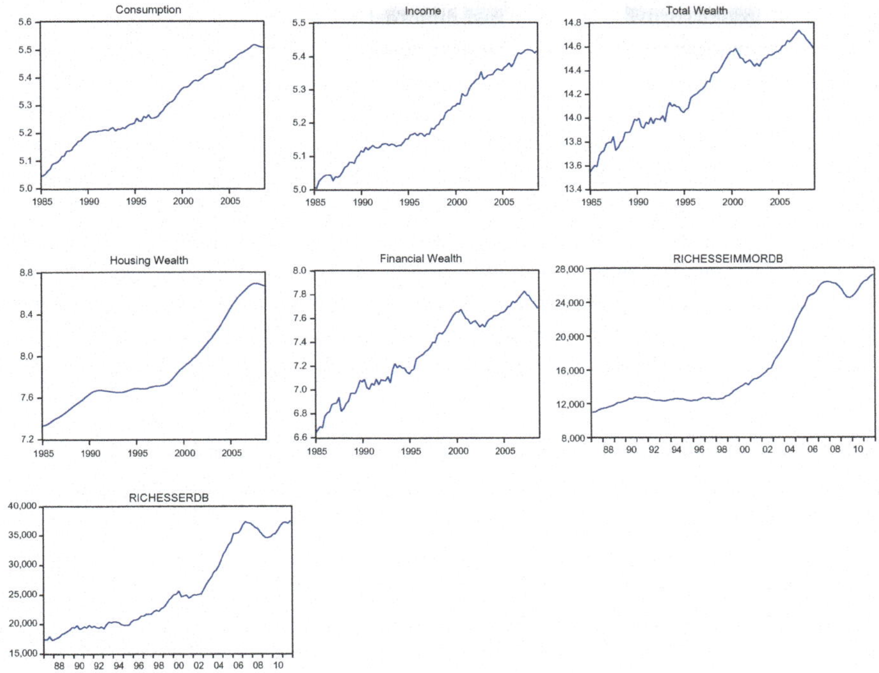

Fig. 1 Dynamics of the series

First, the analysis of Fig. 1—which reports consumption, income, total wealth, HW and FW in logarithms—shows that series are a priori non-stationary in level. Furthermore, consumption and HW indicate some smoothness and seem less volatile than income, FW and Total Wealth (TW). We also plot the dynamics of the FW/Income and HW/Income ratios, using the disposable income net of property and imputed rents. These ratios show some French stylized facts associated with the preference of French householders for real estate investments to financial investments. This fact is more marked after the 2000 dotcom bubble.

Second, we tested for the presence of a unit root in the data. To this end, we performed both the usual unit root tests—ADF of Dickey-Fuller (1979) and DF-GLS of Elliot, Rothenberg, Stock (1996)—and also a unit root test with structural breaks of Zivot and Andrews (1992) and Kapetanios et al. (2003) in the nonlinear STAR framework. Accordingly, all series are integrated of order one, noted I(1).[9] We focused thereafter on the variables in first difference.

[9]Results of Unit root tests are not reported to save space but are available upon request. Note that the previous study relating to the French case by Damette and Chauvin also performed the Lee and Strazicich test on the period 1987–2008 and derived similar conclusions. It is thus not surprising to find that consumption, wealth and income variables are I(1).

Next, in order to provide an initial insight into these relationships between income, wealth and consumption, we computed the unconditional correlation matrix reported in Table 1.

We noted a significant and positive correlation between total wealth and consumption. Such a correlation is stronger and significant when we consider housing wealth. This is not unexpected and is in line with our analysis in Fig. 1 suggesting the importance of housing wealth, in particular over the most recent period.

In Table 2 we also report the main descriptive statistics for consumption and wealth variations. We checked their distribution in order to better investigate the statistical properties of these data. In this way, we showed that the symmetry hypothesis is not accepted for the consumption, FW and total wealth series, while a leptokurtic excess characterizes the distribution of most series. Additionally, normality is not accepted

Table 1 Unconditional correlation matrix

Correlation *t*-statistic	D(conso)	D(housing)	D(income)	D(financial)	D(total wealth)
D(consumption)	1.000				
D(housing)	0.329[a] (3.452)	1.000			
D(income)	0.071 (0.709)	0.170[c] (1.704)	1.000		
D(financial)	0.042 (0.414)	0.030 (0.293	0.090 (0.895)	1.000	
D(total wealth)	0.207 (2.092)[b]	0.573[c] (6.923)	0.151 (1.510)	0.831[c] (14.775)	1.000

Note Values in brackets refer to the t-ratios. [a, b] and [c] denote significance at 1%, 5% and 10% respectively

Table 2 Descriptive statistics and normality test

	D(consumption)	D(housing)	D(income)	D(financial)	D(total)
Mean	0.004	0.013	0.004	0.009	0.012
Median	0.004	0.015	0.004	0.013	0.012
Maximum	0.017	0.038	0.032	0.11	0.04
Minimum	−0.012	−0.008	−0.024	−0.116	−0.029
Std. Dev.	0.005	0.012	0.008	0.033	0.014
Skewness	−0.232	0.081	0.063	−0.414	−0.437
Kurtosis	3.748	2.058	4.513	4.714	2.692
Jarque-Bera	3.237	3.802	9.607	15.103	3.588
Probability	0.198	0.149	0.008	0	0.166
Sum	0.432	1.346	0.411	0.909	1.206
Sum Sq. Dev.	0.003	0.015	0.006	0.108	0.021
Observations	100	100	100	100	100

for the income and financial wealth series. The rejection of normality implies that the analysis of unconditional correlation—which required the normality—should be carefully reconsidered. Also, the rejection of normality and symmetry and the negativity of the skewness coefficient suggest that distribution exhibits further asymmetry and potential nonlinearity which may escape linear specification. A nonlinear econometric methodology might be useful to take into account this kind of distribution.

Before moving on to the implementation of the nonlinear analysis, we tested the wealth–consumption relationship in a linear framework using the two-step method of Engle and Granger. We also applied Johansen cointegration tests to check the presence of several cointegration relationships between consumption, income, TW, HW and FW.

3.2 *The Linear Cointegration Analysis*

First, we estimated models (1) and (2) by the Ordinary Least Squares (OLS) method as a benchmark since OLS is a super-consistent estimator in a cointegration framework. We first applied multivariate Johansen (1991) cointegration tests (see Appendix 1) and found evidence of one cointegrating relationship, suggesting further evidence of wealth effects on French consumption in the long term.[10] In order to take a potential endogeneity bias into account—induced by a reverse causality of the explanatory variables and the poor properties of the OLS estimator with 100 observations with cointegrating series—we re-estimated these models using the Dynamic Ordinary Least Squares (DOLS) method of Stock and Watson (1993) in line with models (3) and (4). We report the estimated results in Table 3.

Accordingly, our results showed that the wealth effect is still positive but only about 8%, rather than 13% in Chauvin and Damette (2010, 2011), suggesting that the wealth effect slightly decreased after the subprime crisis. To take better account of the global financial crisis (2008) and control for this potential significant break in the data, we introduced a dummy variable for the 2008 crisis. We report the results in Table 4. As Table 4 shows, the dummy variable is significant and negative, but the estimation of the total wealth effect is almost unchanged. Barrell et al. (2015), using rolling DOLS regressions found evidence that MPC's (Marginal Propensity to Consume) out of financial wealth were decreasing during the recent financial crisis in UK and slightly increasing in Italy. So, the French consumption and wealth effect dynamics are closed to UK (more negative impacts of credit constraints in France and UK than Italy, different habit formation behaviors etc.).

As for the estimation of disaggregate wealth effects, we noted that the FW effect has slightly increased from 0.08 estimated in Chauvin and Damette (2010, 2011) to 0.09 in our new estimates (Table 2), while the HW effect has decreased from 0.08 to 0.05. Furthermore, the wealth effects are statistically significant and pointed to the

[10]We identify the only one cointegrating relationship estimating a VAR model.

Table 3 DOLS estimates of long-run elasticity of total consumption

Variables	Coefficient	Std. Error
Total wealth		
Total wealth	0.078[a]	0.03
Income	0.782[a]	0.08
Constant	0.035	0.05
Disaggregated wealth		
Financial wealth	0.093[a]	0.01
Housing wealth	0.051[a]	0.01
Income	0.632[a]	0.05
Constant	−0.063	0.05

Note We use 2 leads and 0 lags considering the SIC criterion concerning total wealth estimation
[a] Denotes the significance at 1% statistical level

Table 4 DOLS estimates of long-run elasticity of total consumption with crisis dummy

Variables	Coefficient	Std. Error
Total wealth		
Wealth	0.077[a]	0.00
Income	0.805[a]	0.00
Constant	−0.076	0.18
Dummy 2008 Crisis	−0.017[a]	0.00
Disaggregated wealth		
Financial	0.083[a]	0.00
Housing	0.051[a]	0.00
Income	0.0661[a]	0.00
Constant	−0.087[c]	0.07
Dummy 2008 crisis	−0.008[b]	0.04

Note We use 2 leads and 0 lags considering SIC criterion concerning total wealth estimation. [a], [b] and [c] Denote the significance at 1%, 5% and 10% statistical level respectively

superiority of the FW effect over that of HW. The opposite effect was found for the US market and French wealth effects are lower in magnitude than those in the US market. This is in line with Jawadi and Sousa (2014) and Jawadi et al. (2015) who found that wealth effects in the Euro zone are lower than those in the USA.

After analyzing the long-run wealth effect, using the residuals computed from previous Dynamic Ordinary Least Squares estimates of models (3) and (4), we estimated by OLS linear ECM models for the cointegration relationship between wealth and consumption. Our main results are reported in Table 5. We showed that in the linear ECM for total wealth, there is a significant long-run wealth effect (see the ECM term). We also noted a significant linear mean reversion indicating that in the

Table 5 Linear ECM estimates

Variables	Coefficient	Std. Error
Total wealth		
D(Wealth)	0.036	0.04
D(Income)	0.066	0.07
D(Wealth)(−1)	0.120[a]	0.04
D(Income)(−1)	0.099	0.07
Constant	0.002[b]	0.00
Error correction term	−0.183[a]	0.01
Disaggregated wealth		
D(Financial)	0.005	0.01
D(Housing)	0.242[c]	0.13
D(Income)	0.041	0.06
D(Financial)(−1)	0.039[a]	0.01
D(Housing)(−1)	−0.117	0.13
D(Income)(−1)	0.095	0.06
Constant	0.002[b]	0.00
Error correction term	−0.380[a]	0.00

Note We use 1 lag considering the BIC criterion concerning total wealth and disaggregated wealth estimations. [a], [b] and [c] Denote the significance at 1%, 5% and 10% statistical levels respectively

long-term wealth constitutes a driver for households. When considering disaggregate wealth, we noted a significant current HW effect. Interestingly, the mean reversion in consumption still exists and its intensity is stronger than in the ECM with total wealth.

To sum up, our first estimates pointed to significant wealth effects on consumption, as in previous studies. However, findings vary according to the specifications under consideration, the sample (including the recent global financial crisis event or not) and the horizon (short versus long term). As discussed in the first section, this may be associated with the fact that wealth is naturally time-varying and the wealth–consumption relationship is probably unstable. Furthermore, several previous studies including Case et al. and Marquez et al. (2013) have suggested further evidence of asymmetry in the wealth–consumption relationship: positive and negative changes in wealth have a different impact on household's consumption expenditure. Our preliminary analysis (Table 2) also pointed to further asymmetry in our data and to a rejection of normality. Distributions are leptokurtic and are characterized by a significant and negative skewness, which may be an indication of a nonlinear effect in the data.

In order to take these properties into account, we extended the analysis so that the adjustment of consumption dynamics after a shock on wealth was time-varying, asymmetrical and nonlinear. Formally, this is made possible using the class of nonlinear ECMs that provide a better understanding of wealth effects in a nonlinear context in both the short and long term.

3.3 Nonlinear Cointegration with Asymmetric Adjustment

The reaction of consumption to a wealth shock depends on its sign and on its size and also varies according to the market and the economic state. Case et al. suggest that this type of reaction might be a source of asymmetry and nonlinearity. Asymmetrical adjustment TAR and M-TAR models present the advantage of capturing this asymmetry. In particular, this modeling gives different specifications of consumption dynamics, and therefore of the transmission of wealth effects to consumption, according to the business cycle phase.

In practice, we re-parameterize these nonlinear adjustment processes so that models (3) and (4) are still used to define the long-term relationship between consumption and wealth while the short-term adjustment will be captured using a two-regime TAR-ECM and a two-regime M-TAR-ECM.

To this end, in a first step we tested the asymmetry in the adjustment using Hansen and Seo's (2002) threshold cointegration test. Second, if the hypothesis of threshold effect was not rejected, we estimated a two-regime TAR-ECM and a two-regime M-TAR-ECM. As in Case et al. the estimation of different regressions (nonlinear regressions in our case) helps to check the robustness of our results.

3.3.1 TAR-ECM and M-TAR-ECM Estimates

These models are specified and estimated in line with Enders and Siklos (2001) for total wealth and for the disaggregate wealth data.

Total Wealth Specification

Table 6 reports the TAR-ECM and M-TAR-ECM estimate results bearing in mind the total wealth specification. In model (5), the threshold is constrained to be zero while in model (6) the threshold value is estimated. Our results showed that the null hypothesis of no cointegration is rejected at a robust confidence level when the threshold value is estimated [specification (6)]. The magnitude of the adjustment coefficient is lower when the residuals exceed the estimated threshold. For instance, in the TAR (1) model, the adjustment coefficient is -0.276 when the residuals are below the estimated threshold (assumed to be zero) but it turns out to be -0.356 when the residuals exceed their estimated threshold. The same results are derived when the threshold is endogenously estimated (see TAR(2) and MTAR(2) columns). This finding suggests that French households seem to react more strongly to a negative

Table 6 TAR-ECM and M-TAR-ECM model estimations

	TAR (1)-ECM	M-TAR (1)-ECM	TAR (2)-ECM	M-TAR (2)-ECM
$\rho_1 = I\hat{u}_{t-1}$	-0.276^{a} (0.11)	-0.277^{a} (0.11)	-0.224^{b} (0.10)	0.109 (0.18)
$\rho_2 = (1 - I)\hat{u}_{t-1}$	-0.356^{a} (0.13)	-0.349^{a} (0.12)	-0.485^{a} (0.14)	-0.394^{b} (0.09)
Lags (BIC)	1	1	1	1
Threshold value	Constrained to 0	Constrained to 0	−0.009	0.008
t-max	-2.476^{a}	-2.427^{b}	-2.178^{b}	6.840^{a}
Φ	6.331^{*}	6.305^{*}	7.561^{b}	10.057^{a}

Note Standard errors are in parentheses. [a] and [b] denote significance at 1% and 5% levels respectively. The simulated critical values at 5% significant level are 7.08 and 6.86 considering a model with one lag (using the BIC criterion) for the **Φ** TAR and **Φ** MTAR statistics respectively; −1.90 and −1.94 for t-Max TAR and t-Max MTAR respectively

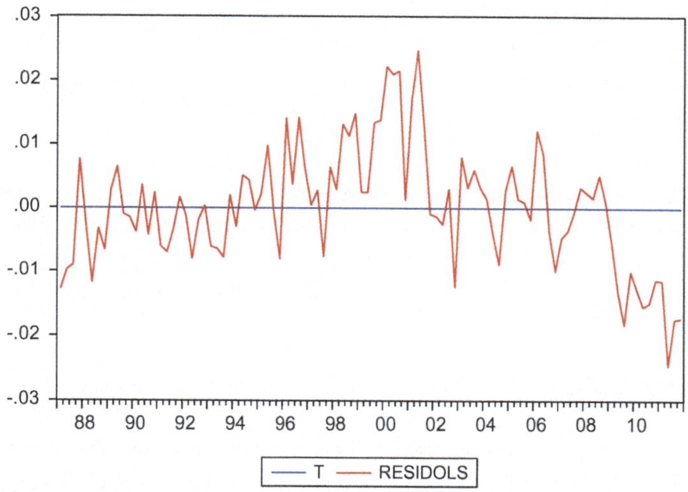

Fig. 2 Residuals of the long-run relationship versus threshold value: regimes of high and low adjustments (total wealth). *Note RESIDOLS* denotes the residuals from model (1) estimated by DOLS and T refers to the threshold of the TAR model

than a positive wealth shock and then more rapidly adjust their consumption to income and wealth targets.

Overall, this points to a nonlinear mean reversion in consumption and suggests further evidence of nonlinear wealth effects. Interestingly, with regard of the LECM results, we noted that the speed of adjustment is stronger when nonlinearity is considered and when the hypothesis is switched in the adjustment. In order to grasp the contribution of nonlinearity, Fig. 2 plots the residuals and the threshold value and highlights the low and high adjustment regimes over time. When the red curve is above the threshold value (which is in blue) as in the 1994–2001 period, then the residuals are positive and thus the adjustment of consumption to wealth is relatively

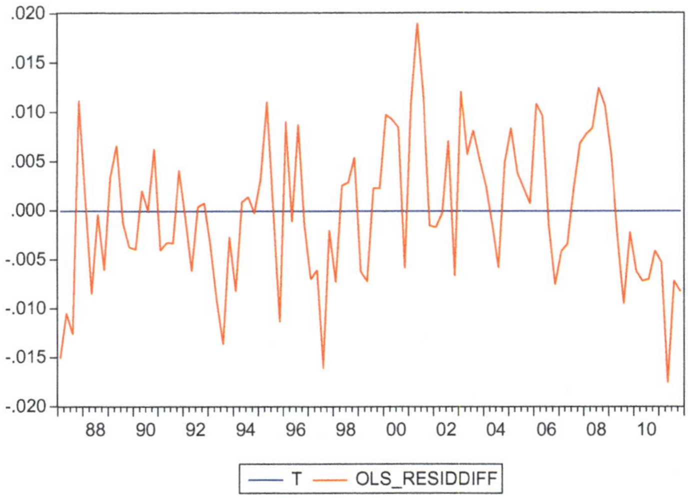

Fig. 3 Residuals versus threshold value: regimes of high and low adjustments (disag. wealth)

slow. Historically, this period ranges from the post–1993 French economic crisis and the beginning of the housing bubble in USA in the early 2000s. In contrast, when the residuals curve is above the blue line as in the 2008–2011 period, the shocks are negative and the adjustment of consumption to wealth is achieved faster, suggesting that the adjustment occurs faster in crisis periods when wealth and income are decreasing.

Disaggregated Wealth

When both TF and TH are considered (Fig. 3), we confirmed the results associated with the TW specification as the adjustment of the residuals is faster below the threshold value. However, historically, the length of the high adjustment regime (1998–2008) is greater with the specification for disaggregate data than in the specification with total wealth (ten years versus seven years). This confirms the interest of using disaggregate data (Table 7).

3.4 NECMs with Nonlinearity in the Long Run

This specification also aims to consider nonlinearity in the adjustment between wealth and consumption, while enabling the long-run relationship to be nonlinear too. To our knowledge, this specification has never been done in previous studies and constitutes an original contribution by our current paper. Why this hypothesis? Recently, Carroll et al. (2011a, p. 56) emphasize that "*basic consumption theory does not imply the existence of a stable cointegrating vector; in particular a change in the*

Table 7 TAR-ECM and M-TAR-ECM model estimations

	TAR (1)	M-TAR (1)	TAR (2)	M-TAR (2)
$\rho_1 = I\hat{u}_{t-1}$	-0.593^a (0.16)	-0.392^a (0.17)	-0.514^a (0.15)	-0.392^a (0.16)
$\rho_2 = (1 - I)\hat{u}_{t-1}$	-0.76^a (0.17)	-0.899^a (0.15)	-0.898^a (0.17)	-0.933^a (0.15)
Lags (BIC)	2	2	2	2
Threshold value	Constrained to 0	Constrained to 0	−0.006	−0.001
t-max value	-3.698^a	-2.374^b	-3.362^a	-2.454^a
Φ	13.354^a	17.513^a	15.402^a	18.293^a

Note Standard-errors are in parentheses. [a] and [b] denote significance at 1% and 5% level respectively. The simulated critical values at 5% significant level are 7.08 and 6.86 considering a model with one lag (on the basis of the BIC criterion) for the TAR and MTAR statistics respectively; −1.90 and −1.94 for *t*-Max TAR and *t*-Max MTAR respectively

long-run growth rate or the long-run interest rate should change the relationship". As previously explained, the main approach computed to estimate wealth effects in the literature is the Lettau and Ludvigson cointegration approach. It is however possible that this relationship could have been subject to regime shifts (Alexandre et al. 2007) due to the existence of different regimes in financial markets linked with asset price volatility and the role played by asset wealth volatility; as a consequence the relationship between consumption and wealth might lead to an indeterminate solution. In addition, from Slacalek (2009) and Carroll et al. (2011a), the cointegration vector might be unstable due to changes in features of the economy relevant for the consumption/savings decision and the basic consumption theory does not imply a stable vector if the long-run growth rate, demographics or long-run interest rate are changing in the long term. Carroll et al. (2011a) developed a new method of estimation based on three steps: in the first one, the consumption growth is estimated by taking into account habits and stickiness; in the second one, short run effects are estimated and finally the previous steps derive the eventual long-run marginal propensity to consume (see also Barrell et al. (2015) for a recent application for Italy and UK).

This means that the attractor may also exhibit nonlinearity and time variation. In order to take further long-run instability into account, we propose to test the linearity of the cointegration relationship by applying the recent Bierens-Martins Time-Varying Cointegration (TVC).

The likelihood ratio (LR) test in Bierens and Martins (2010) is distributed as a Chi2(r m k) and empirical values are close to asymptotical critical values for $T = 100$, as in our paper. In practice, we compute the LR test for $m = 1$ to $m = 10$, i.e., when allowing for the first Chebyshev polynomial as the only source of time variation.[11] The estimates of the time-varying cointegration relationship are based on $r = 1$ (that is one cointegrating relationship since r is the rank of the matrix) and $p = 1$ or $p = 4$ lags (the choice of the VAR order is based on HQ and SC criteria).

[11] The choice of the maximum value of m that is the dimension of the Chebyshev time polynomials is based on the rule defined by Martins: $m_{\max} = T/10$ with T the number of time series observations.

Table 8 Time-varying cointegration test: total wealth specification

p	m	$r = 1$	Log likelihood	HQ	BIC	r = 2	Log likelihood	HQ	BIC
1	1	9.18 (0.027)	1010.29	−19.97	−19.64	19.19 (0.004)	1010.29	−20.55	**−20.08**
	2	23.42 (0.000)	1017.41	−20.02	**−19.64**	39.55 (0.000)	1017.41	−20.57	−20.00
	3	32.23 (0.000)	1021.82	−20.01	−19.59	71.13 (0.000)	1021.82	**−20.70**	−20.04
	4	44.04 (0.000)	1027.72	−20.04	−19.57	88.05 (0.000)	1027.72	−20.68	−19.94
	10	109.46 (0.000)	1060.43	**−20.15**	−19.40	184.09 (0.000)	1060.43	−20.55	−19.23

Note Minimizing the HQ criterion results in $m = 10$ and $m = 3$ with $r = 1$ and $r = 2$ being retained respectively. Numbers in (.) are p-values for null hypothesis of standard time-invariant cointegration against the alternative hypothesis of TVC. Values in bold are the m Chebyshev polynomials according to the HQ criteria. Computed with Gauss software

Table 9 Time-varying cointegration test: disaggregated wealth specification

p	m	$r = 1$	Log likelihood	HQ	BIC
1	1	10.60 (0.000)	1342.90	−26.41	**−25.91**
	2	24.05 (0.000)	1349.63	−26.42	−25.86
	3	44.46 (0.000)	1359.83	−26.51	−25.88
	4	62.78 (0.000)	1368.99	−26.57	−25.88
	10	168.86 (0.000)	1422.04	**−26.91**	−25.84

Note Minimizing the HQ criterion results in $m = 10$ and $m = 3$ with $r = 1$ being retained. Computations with $r = 2$ lead to problems of . Numbers in (.) are p-values for null hypothesis of standard time-invariant cointegration against the alternative hypothesis of TVC. Values in bold are the m Chebyshev polynomials according to the HQ criteria. Computed with Gauss software

In Tables 8 and 9, our results point to the rejection of the null hypothesis of time-invariant cointegrating vector in favor of time-varying cointegration for $m = 1$ to $m = 10$.

Figures 4 and 5 plot the normalized parameter estimates from the TVC cointegration model with one cointegrating relationship ($r = 1$), with $m = 10$ (choice based on HQ criterion) and for $p = 1$[12] for total wealth and disaggregated wealth models respectively. The parameters are quite unstable over the 1987–2011 period.[13] The volatility of the total wealth parameter has increased since the beginning of 2000s: the coefficient increased drastically from 2000 to 2005 and then decreased from 2005 to 2008. Historically, this period from 2000 to 2008 corresponds to the French housing

[12]We also performed results for $p = 4$ to check robustness.

[13]Results over the 1977–2011 period lead to very similar results and are available upon request.

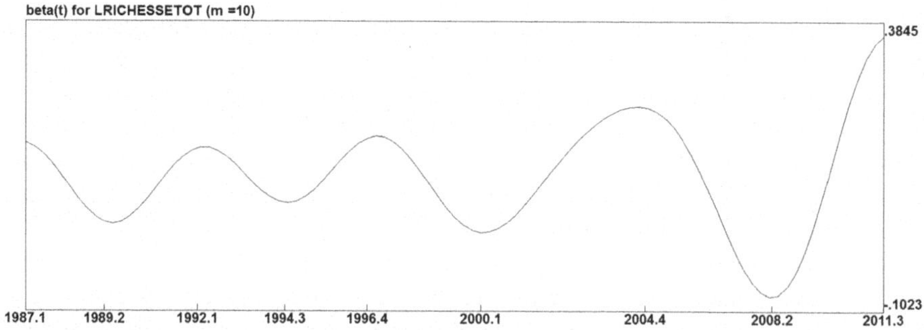

Fig. 4 Time-varying coefficients of the VECM ($m = 10, p = 1$): total wealth specification

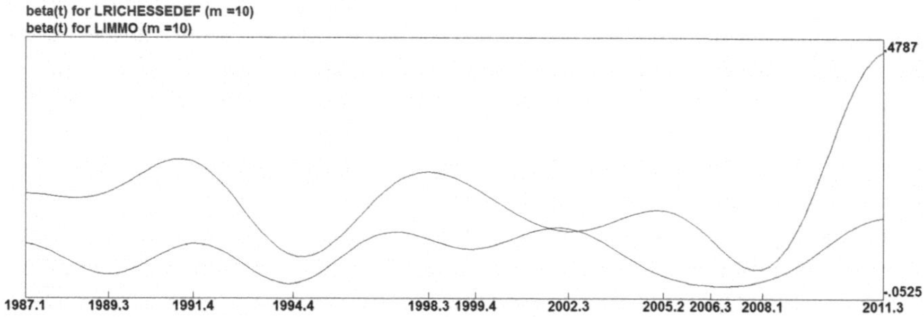

Fig. 5 Time-varying coefficients of the VECM ($m = 10, p = 1$): disaggregated wealth specification

bubble, leading to a peak in 2007–2008 at the same time as the subprime crisis. In Fig. 5, we show that the volatility of the housing effect was slightly greater than the volatility of the wealth effect over the 1987–2008 period. We can conclude that the variations in the wealth effect in France since the financial liberalization of 1987 are due to financial wealth, but also to housing wealth. This is not unexpected as the volatility due to the housing effect stems from the fact that 63% of French households owned their homes while home ownership only stands at 53% in Germany and 50% on average for the EU.

4 Conclusion

Several prior studies have investigated the wealth effect on household consumption, although those with a focus on the US market are more numerous. Overall, these studies have suggested two interesting results. First, they showed that wealth effects are greater for Anglo-Saxon countries including the US and the UK than for other European countries. Second, they highlighted further time variation in the wealth–consumption relationship, suggesting evidence of time-varying wealth effects.

To our knowledge, while some studies including Salacalek (2009), Jawadi and Sousa (2014, 2015) have studied wealth effects for the Euro Area, only a few—except Boone et al. (2001) and more recently Damette and Chauvin—have focused on the wealth effect in France. Yet French wealth has increased and gone through various episodes over the three last decades. Furthermore, the study by Damette and Chauvin only carried out a linear investigation that was unable to capture further instability in the consumption–wealth relationship.

Our paper fills this gap by investigating this relationship in a nonlinear context and offering, as far as possible, a flexibility for the modeling to capture different forms of wealth effects that might be further computed for other countries. Interestingly, while the estimation of TAR-ECM and M-TAR-ECM reproduces further asymmetry, discontinuities and nonlinearity in the consumption mean reversion toward wealth, the implementation of the nonlinear long-run relationships by Bierens and Martins (2010) also allows the attractor to be time-variable in order to capture further instability in the long-run equilibrium between wealth and consumption. Furthermore, the application of these different models and specifications, as in Case et al. for the US data, enables us to check the robustness of our findings. Last but not least, the estimation of these models for both aggregated and disaggregated data is helpful in checking the conclusion with regard to further error measurements.

Our findings showed time-varying wealth effects on French consumption, confirming an unstable relationship between wealth and consumption. In particular, this relationship exhibits asymmetry and nonlinearity. Interestingly, the introduction of nonlinearity in the long-term relationship showed a significant volatility excess in the cointegration relationship around the year 2000 and during the subprime crisis, suggesting an increase in the wealth effect during these periods. Overall, this paper has shown that the focus on nonlinearity improves the specification of the wealth–consumption relationship, while the application of these tests to disaggregated data helps to identify the driver of household consumption for each regime of the business cycle and identifies the relative importance of HW and FW.

Appendix 1 Unit Root and Cointegration Tests in the Linear Model

Multivariate linear Johansen cointegration test: total wealth
Using the VAR order selection criteria (AIC, HQ and BIC), we set a VAR with $p = 1$ lag and find that $r = 1$, that is, there is one only cointegrating relationship between Consumption, Income and Total Wealth.

H0	Lambda max	5% critical value
$r = 0$	24.4	20.8
$r = 1$	7.3	14.0

(continued)

(continued)

H0	Lambda max	5% critical value
$r = 2$	1.4	4.0
HO	Trace	5% critical value
$r = 0$	33.1	29.5
$r = 1$	8.7	15.2
$r = 2$	1.4	4.0

Multivariate linear Johansen cointegration test: disag. Wealth
Using the VAR order selection criteria (AIC, HQ and BIC), we set a VAR with $p = 2$ lags and find that $r = 2$, that is, there is one only two cointegrating relationships between Consumption, Income, Financial Wealth and Housing Wealth.

H0	Lambda max	5% critical value
$r = 0$	31.2	27.6
$r = 1$	21.2	21.1
$r = 2$	4.8	14.3
$r = 3$	4.1	3.8
HO	Trace	5% critical value
$r = 0$	66.4	27.6
$r = 1$	30.2	21.1
$r = 2$	9.0	14.2
$r = 3$	4.1	3.8

References

Alexandre, F., Baçao, P., & Gabriel, V. (2007). The consumption-wealth ratio under asymmetric adjustments. *Studies in Nonlinear Dynamics and Econometrics, 12*(4), 1–32.

Altissimo, F., Georgiou, E., Sastre, T., Valderrama, M. T., Sterne, G., Stocker, M., Weth, M., Whelan, K., Willam, A. (2005). Wealth and asset price effects on economic activity. *ECB occasional paper series, 29.*

Apergis, N., & Miller, S. (2006). Consumption asymmetry and the stock market: Empirical evidence. *Economics Letters, 93,* 337–342.

Arrondel, L., Lamarche, P., & Savignac, F. (2015). Wealth effects on consumption across the wealth distribution: Empirical evidence. *Working Papers, 552,* Banque de France.

Barrell, R., & Davis, E. P. (2007). Financial liberalisation, consumption and wealth effects in seven OECD countries. *Scottish Journal of Political Economy, 54*(2), 254–267.

Barell, R., Costantini, M., & Meco, I. (2015). Housing wealth, financial wealth, and consumption: New evidence for UK and Italy. *International Review of Financial Analysis, 42,* 316–323.

Bierens, H., & Martins, L. (2010). Time varying cointegration. *Econometric Theory, 26,* 1453–1490.

Boone, L., Girouard, N., & Wanner, I. (2001). Financial market liberalisation, wealth and consumption. *OECD Economics Department Working Papers, 308.*
Buiter, W. H. (2008). Housing wealth isn't wealth. *CEPR Discussion paper, 6920.*
Byrne, J. P., & Davis, E. P. (2003). Disaggregate wealth and aggregate consumption: An investigation of empirical relationships for the G7. *Oxford Bulletin of Economics and Statistics, 65*(2), 197–220.
Campbell, J. Y., & Mankiw, N. G. (1989). Consumption income and interest rates: Reinterpreting the time series evidence. *NBER Macroeconomics Annual, 25*(2), 185–216.
Carroll, C., Ostuka, M., & Slacalek, J. (2011a). How large are housing and financial wealth effects? A new approach. *Journal of Money Credit and Banking, 43*(1), 55–79.
Carroll, C. D., Slacalek, J., Sommer, M. (2011b). International evidence on sticky consumption growth. *The Review of Economics and Statistics, MIT Press, 93*(4), 1135–1145, November, 2011.
Carroll, C. D., Slacalek, J., & Tokuoka, K. (2014). The distribution of wealth and the MPC: implications of new European data, *American Economic Review, American Economic Association, 104*, 5, 107–11, May, 2014.
Case, K. E., Quigley, J. M., & Shiller, R. J. (2005). Comparing wealth effects: The stock market versus the housing market. *Advances in Macroeconomics, 5*(1), 1–32.
Catte, P., Girouard, N., Price, R., & Andre, C. (2004). Housing markets, wealth and the business cycle. *OECD Economics Department Working Papers, 394.*
Chauvin V., & Damette, O. (2010). Wealth effects on private consumption: the french case. In De Bandt et al. (Ed.) *Housing Markets in Europe. A Macroeconomic Perspective* (pp. 263–282), Springer.
Chauvin, V., & Damette, O. (2011). Effets de richesse: le cas français. *Economie et Statistique, 438–440,* 111–141.
Cochrane, J. H. (1994). Permanent and transitory components of GDP and stock prices. *Quarterly Journal of Economics, 109*(1), 241–265.
Donihue M., & Avramenko, A. (2007). Decomposing consumer wealth effects: Evidence of the role of real estate assets following the wealth cycle of 1990–2002. *The BE Journal of Macroeconomics, 7*, article 25.
Enders, W., & Siklos, P. L. (2001). Cointegration and threshold adjustment. *Journal of Business & Economic Statistics, 19,* 166–176.
European Commission. (2008). Wealth household consumption, what are the risks attached to falling house prices and high debt? *European Commission, Quarterly Report on the Euro Area, III.*
Friedman M. (1957). A theory of the consumption function, Milton Friedman. Princeton: Princeton University Press.
Forest, J. J., & Tuner, P. (2013). Alternative estimators of cointegrating parameters in models with nonstationary data: An application to US export demand. *Applied Economics, 45,* 629–636.
Gabriel, V., Alexandre, F., & Baçao, P. (2008). The consumption-wealth ratio under asymmetric adjustment. *Studies in Nonlinear Dynamics and Econometrics, 12*(4), 1–32.
Gonzalo, J., & Ng, S. (2001). A systematic framework for analyzing the dynamic effects of permanent and transitory shocks. *Journal of Economic Dynamics and Control, 25*(10), 1527–1546.
IMF. (2004). Modelling consumption behavior. *IMF Country Report, Selected Issues,* 04/346, 6–18.
Jawadi, F. (2008). Does Nonlinear Econometrics Confirm the Macroeconomic Models of Consumption? *Economics Bulletin, 5*(17), 1–11.
Jawadi, F., & Léoni, P. (2013). Nonlinearity, cyclicity and persistence in consumption and income relationships: Research in honor of Melvin J. Hinich. *Macroeconomic Dynamics, 16*(S3), 376–393.
Jawadi, F., & Sousa, R. (2014). The relationship between consumption and wealth: A quantile regression approach. *Revue d'Economie Politique, 124,* 639–652.
Jawadi, F., Soparnot, R., & Sousa, R. (2015). Assessing financial and housing wealth effects through the lens of a nonlinear framework. *Research in International Business and Finance.*
Johansen, S. (1991). Estimation and hypothesis testing of cointegration vectors in gaussian vector autoregressive models. *Econometrica, 59*(6), 1551–1580.

Johansen, S. (1995). *Likelihood-based inference in cointegrating vector autoregressive models.* Oxford: Oxford University Press.

Kapetanios, G., Shin, Y., & Snell, A. (2003). Testing for a unit root in the nonlinear STAR framework. *Journal of Econometrics, 112,* 359–379.

Keynes, J. M. (1936). Théorie générale de l'emploi, de l'intérêt et de la monnaie (livres IV à VI).

Lettau, M., & Ludvigson, S. (2001). Consumption, aggregate wealth, and expected stock returns. *Journal of Finance, 56,* 815–849.

Lettau, M., & Ludvigson, S. (2004). Understanding trend and cycle in asset values: Reevaluating the wealth effect on consumption. *American Economic Review, 94*(1), 276–299.

Marquez, E., Martinez-Canete, A. R., & Pérez-Soba, I. (2013). Wealth shocks, credit conditions and asymmetric consumption response: empirical evidence for the UK. *Economic Modelling, 33,* 357–366.

Mastrogiacomo, M. (2010). Testing consumers' asymmetric perception of changes in household financial situation. *Review of Income and Wealth, 56*(2), 327–350.

Montalvo, J. G. (1995). Comparing cointegration regression estimators: Some additional Monte Carlo results. *Economics Letters, 48,* 229–234.

Muellbauer, J. (2008). Housing, credit and consumption expenditure. *CEPR Discussion paper, 6782.*

Muellbauer, J., & Lattimore, R. (1995). The consumption function: A theoretical and empirical overview. In M. H. Pesaran & M. Wickens (Eds.), *Handbook of applied econometrics: macroeconomics* (pp. 221–311). Oxford: Blackwell.

Nan-Kuang, C. A., Shiu-sheng Chen, A., & Yu-Hsi, C. (2010). House prices, collateral constraint, and the asymmetric effect on consumption. *Journal of Housing Economics, 19,* 26–37.

Rudd, J., & Whelan, K. (2006). Empirical proxies for the consumption wealth ratio. *Review of Economic Dynamics, 9,* 34–51.

Skudelny, F. (2009). Euro area private consumption: is there a role for housing wealth effects? *Oxford Bulletin of Economic and Statistics, 54,* 257–287.

Slacalek, J. (2006a). What drives personal consumption? The role of housing and financial wealth. *DIW working paper, 647.*

Slacalek, J. (2006b). International wealth effects. Discussion Papers of DIW Berlin 596, DIW Berlin, German Institute for Economic Research.

Slacalek, J. (2009). What drives personal consumption? The role of housing and financial wealth. *ECB working paper, 1117.*

Sousa, R. M. (2009). Wealth effects on consumption. *ECB working paper, 1050.*

Stock, J. H., & Watson, M. W. (1993). A simple estimator of cointegrating vectors in higher order integrated systems. *Econometrica, 61,* 783–820.

Shirvani, H., & Wilbratte, B. (2000). Does consumption respond more strongly to stock market declines than to increases? *International Economic Journal, 14*(3), 41–49.

Stevans, L. (2004). Aggregate consumption spending, the stock market, and asymmetric error correction. *Quantitative Finance, 4,* 191–198.

Whelan, K. (2008). Consumption and expected asset returns without assumptions about unobservables. *Journal of Monetary Economics, 55,* 1209–1221.

Zivot, E., & Andrews, D. (1992). Further evidence on the great crash, the oil-price shock, and the unit-root hypothesis. *Journal of Business and Economic Statistics, 10*(3), 251–270.

Productivity Spillovers in the Global Market

Nazmus Sadat Khan and Jun Nagayasu

JEL Classification: C32 · O47

1 Introduction

Globalization has accelerated especially after World War II due to the initiatives of the Bretton Woods institutions, such as the International Monetary Fund and the World Bank. Economic globalization has been considered to be particularly successful at regional level. For example, the European Union (EU) and the Association of South-East Asian Nations (ASEAN), among others, have removed virtually all trade barriers in order to promote intra-regional trade. Furthermore, the creation of single currency areas, like the Eurozone, represents the ultimate form of regional liberalization. Such economic liberalization is often related to the free movement of goods, services, capital, and labor and has been advanced on the basis of the standard economic theory, which predicts that countries will benefit from removing cross-border trade barriers.

However, advanced countries have recently shifted toward protectionism. The US president, Donald Trump, expressed strong interest in protecting domestic industries, i.e., the 'America First' policy, to correct prolonged and large US trade deficits. Conservative movements have also occurred in Europe. Following its 2016 referendum,

N. S. Khan
The World Bank (Macro, Trade and Investment Global Practise) and University of Muenster, Plot E-32, Agargaon, Sher-E-bangla Nagar, Dhaka, Bangladesh
e-mail: nkhan12@worldbank.org

J. Nagayasu (✉)
Graduate School of Economics and Management, Tohoku University, 27-1 Kawauchi, Aoba-ku, Sendai, Miyagi 980-8576, Japan
e-mail: jun.nagayasu.d8@tohoku.ac.jp

G. Dufrénot and T. Matsuki (eds.), *Recent Econometric Techniques for Macroeconomic and Financial Data*, Dynamic Modeling and Econometrics in Economics and Finance 27, https://doi.org/10.1007/978-3-030-54252-8_7

the UK has decided to leave the EU to gain more independent control of its immigration policies. In 2017, Germany imposed a cap on immigration, particularly that from the Middle East. In short, many people nowadays doubt the outcome of economic liberalization. There is a significant gap between the standard economic theory and actual economic policies implemented by many advanced countries.

Against this background, our study attempts to investigate the role of spillover effects of productivity shocks in the context of economic growth. More specifically, we analyze the following spillover effects within the last three decades. First, we quantify the magnitude of US spillover effects on the rest of the world because the USA possesses many advanced industries such as information technology. Second, in order to investigate the significance of spillovers originating in the US, we compare them with productivity shocks originating in other parts of the world. In short, we confirm that there are asymmetric effects in external spillovers and that US spillovers exert more influence over global markets.

This paper contributes to the literature in the following ways. First, at least in theory, productivity improvements, along with capital and labor, are known to be an essential factor for economic growth. However, despite efforts to remove cross-country barriers, productivity spillovers in the international context have rarely been studied, in particular by using historic macroeconomic data. In the past, researchers often investigated international trade as an engine of economic growth, but paid little attention to cross-boarder movements of productivity (see the literature survey section).[1] We primarily study the extent to which US productivity has affected its own economic growth and that of other country groups. Whereas country-specific outcomes are important, we are interested in a more general outcome that is relevant to formulating future international economic policies in many countries.

Second, we use a Global Vector autoregressive (GVAR) model (Pesaran et al. 2004) in order to investigate cross-country spillover effects of productivity shocks. The empirical literature on productivity spillover remains limited largely due to technical challenges involving the handling of multi-country models with many variables. Three broad approaches that have been used recently to handle this problem include the GVAR, factor-augmented VAR (FAVAR) and spatial models. To the best of our knowledge, the first two methods have not been used before in the context of productivity shocks.

The GVAR is unique in several respects. First, compared to the FAVAR, the GVAR approach can combine different macroeconomic variables of a large number of countries in a much more efficient manner. GVAR models tackle the ‘curse of dimensionality’ problem by imposing restrictions directly on the parameters of the model. In particular, all foreign economies are typically approximated by one representative economy, constructed as a (trade-) weighted average of foreign economies. Here, individual country models are added by means of a consistent econometric approach to create a global model, where cointegration is allowed for variables within and

[1]The inexplicit treatment of spillover effects in empirical studies is attributable to the standard statistical approach, such as panel data estimation methods, that often use common time dummies to capture cross-sectional dependence across countries.

across countries. On the other hand, FAVAR and spatial models capture country-specific dynamics only through idiosyncratic components (i.e., the residuals) with the assumption of homogeneous impacts of international spillovers across countries and regions.[2]

Also, in GVAR models, data can be used in levels, and therefore, long-run information in the data is retained. The FAVAR and spatial models often assume the stationarity of the data and no cointegration; therefore, economic trends that are ignored in these models can be specified in the GVAR.

Third, this paper contributes to the GVAR literature. Instead of generalized impulse-response functions (GIRFs) that are commonly used in the GVAR literature, structural generalized impulse response functions (SGIRFs) are used for dynamic analysis, in order to better identify the shocks. This goes beyond most previous studies to identify shocks and at least partially overcomes one of the main criticisms regarding the identification and use of GIRFs in GVAR models. Also, instead of commonly used fixed trade weights, time-varying trade weights are used to construct foreign-specific variables of individual country models. This is more realistic, as it controls for the changing relationship between the participating countries for different parts of the business cycle (e.g., rapid growth in the US economy throughout most of the 1990s, the dot-com bubble in 2001, and the financial crisis of 2008).

The subsequent parts of the paper are organized as follows: Sect. 2 reviews the literature on productivity spillover. Section 3 explains the GVAR model, and Sect. 4 discusses the data and some important statistical properties of the GVAR model. In Sect. 5, our results are presented and discussed. Finally, Sect. 6 presents our conclusions.

2 Literature Survey on Economic Growth and Productivity Spillovers

Economic growth has been recognized as an important research and policy topic. Therefore, many researchers have developed economic theories and attempted to discover what factors contribute to economic growth using time-series and/or panel data. Here, we provide only a limited number of studies in order to illustrate the relationship between economic growth and productivity spillovers, which is of our interest.

The classic economic theory (Solow 1956) points to labor and capital as essential ingredients for economic growth. Therefore, as summarized in Eq. (1), the standard economic growth model can be expressed in terms of the Cobb-Douglas production function:

$$Y_t = K_t^{\alpha}(A_t L_t)^{1-\alpha} \tag{1}$$

[2]Spatial models are designed to estimate homogeneous effects across countries while often making a priori assumption about the importance of neighbors (i.e., spatial weights).

where Y_t is aggregate production at time t ($t = 1, \ldots, T$), K_t is capital, and L_t is labor. A weight α represents the significance of capital in the production ($\alpha \in [0, 1]$) and a higher value of α shows that the country produces more capital-intensive goods. In the classic economic growth theory, these production factors (K and L) are assumed to be exogenous, whereby increases in these factors contribute to economic growth.

However, previous studies have found that capital and labor alone fail to explain the historical experiences of many countries.[3] In this regard, more recent literature considers productivity, human capital, education, and research and development (R&D) as additional factors contributing to economic growth.[4] For example, Lucas (1988) discussed the importance of education and diversification in skills, and Mankiw et al. (1992) confirmed human capital proxied by education as an important factor of economic growth using a comprehensive set of countries. The importance of R&D is empirically confirmed in India (Raut 1995) and in a panel of countries [Coe and Helpman (1995), Coe et al. (1997)]. These extensions are equivalent to the recognition of investment in humans. Efforts to improve human capital are now widely included as part of economic policies.

In terms of Eq. (1), A may capture these extra elements missing in the classic economic theory and is known as the Solow residual or as the total factor productivity (TFP). A can be thought of as factors, such as innovation and R&D, to improve productivity. When A that is generated endogenously by R&D or human capital is included, the model can explain sustainable economic growth. Therefore, economic growth has traditionally been analyzed without considering the explicit role of cross-border spillovers, which is not surprising when the market has only limited access to global markets.

Alternatively, education effects (S) and productivity can be treated separately [Hall and Charles (1999), Frankel and Romer (1999)]. Education is expected to improve labor efficiency, closely following Mincer's wage equation (Mincer 1974).

$$Y_t = K_t^{\alpha} (e^{\vartheta(S_t)} A_t L_t)^{1-\alpha} \tag{2}$$

A theoretical link between international trade and economic growth has been developed over the last two decades. For example, Grossman and Helpman (1991) explained the role of international trade in economic growth, and Ventura (1997) provided theoretical justification for quick economic growth experienced by export-oriented countries. Lucas (1993) argued that knowledge spillovers are closely associated with international trade, so is productivity growth. Ertur and Koch (2007) proposed the economic theory to link physical capital externality and economic growth, consistent with spatial model specification, and concluded that a model without international spillovers suffers from a misspecification problem.[5] More broadly,

---

[3]See Jones (2016) for a comprehensive survey on empirical results.

[4]These factors may not be mutually exclusive.

[5]See these studies about economic theories to link between international spillovers and economic growth.

productivity spillovers contributed more to economic growth than did international trade of goods (Weil 2009), and similarly, knowledge spillovers contributed more to economic growth than did capital developments (Lucas 1993).

Previous studies looked at spillover effects of productivity improvements in a particular country, such as Spain (Barrios and Eric 2002), China (Lin and Kwan 2016), England (Girma 2005), and India (Raut 1995). Gorg and Strobl (2001) reviewed earlier literature and found that spillover effects are more clearly observed in microdata analysis. Similarly, they observed large spillover effects in open and developing markets. By identifying trade, financial, and indirect linkages, Arora and Athanasios (2005) showed that international trade as an engine of economic growth generally holds true for both advanced and developing countries. A similar result was also reported by Ho et al. (2013) using spatial models. The indirect linkage is a channel through which business and consumer confidence in advanced countries affects confidence in developing countries. Keller (2002) documented that productivity at home is influenced by R&D originating from both home and foreign countries. The 20% of home productivity improvements are found to be caused by R&D in foreign countries. Keller and Yeaple (2013) pointed out that knowledge transfer may occur through exchanging intermediaries, as well as direct communication. They found evidence that direct communication becomes more difficult as the distance between companies increases.

We investigate spillover effects using macroeconomic data that have not been analyzed as extensively as microdata, but are nonetheless useful in order to draw a general conclusion. Furthermore, our focus is not on cross-boarder spillovers arising from international trade, but productivity itself. Here, an economic growth equation is expressed in terms of per capita, which is the most standard measure for aggregate economic activities when comparing economic development across countries. This measure differs from another popular specification based on GDP per labor. While GDP per labor may be more consistent with the concept of productivity, we study GDP per capita because of its prevalence in economics and its significant implications for economic policies. Furthermore, only a few countries under our investigation release labor statistics. This is an important factor in our decision, since this study covers a wide range of countries. When multiple countries become research targets and population (N) replaces labor, Eq. (1) can be re-stated for countries i as:

$$\frac{Y_{it}}{N_{it}} = \left(\frac{K_{it}}{N_{it}}\right)^{\alpha} A_{it}^{1-\alpha} \tag{3}$$

Panel data allow us to utilize cross-sectional information resulting in statistically more reliable outcomes, in addition to being expected to offer a general conclusion relevant to most investigated countries. In natural logarithmic form, we can derive the standard growth equation that has been investigated by many researchers.

$$y_{it} = \alpha k_{it} + (1 - \alpha) a_{it} \tag{4}$$

where $y_{it} = ln(Y_{it}/N_{it})$, $k_{it} = ln(K_{it}/N_{it})$, and $a_{it} = lnA_{it}$. Similarly, Eq. (2) can be expressed in terms of per capita as:

$$\frac{Y_{it}}{N_{it}} = \left(\frac{K_{it}}{N_{it}}\right)^{\alpha} (A_{it} e^{\vartheta(S_{it})})^{1-\alpha} \quad (5)$$

or

$$y_{it} = \alpha k_{it} + (1-\alpha)a_{it} + (1-\alpha)\vartheta(S_{it}) \quad (6)$$

Importantly, these equations are the classic specification for panel data analyses, which assume cross-sectional independence and thus cannot capture spillover effects from overseas. In other words, they are designed for closed economies and do not contain foreign variables that directly influence home economic growth. This assumption has become increasingly unrealistic in the modern world. To mitigate it, we use the GVAR model to extend these specifications to include spillover effects. This extension has been widely recognized among researchers and is indeed in line with Lucas (1988), who introduced human capital spillovers in order to explain different patterns in economic growth. Given that people (or labor) and capital are less mobile than knowledge across countries, it is more natural to consider the effect of cross-border technological transfers on per capita income. In connection to this, Comin and Bart (2010), using data from 166 countries, showed that the speed of technology diffusion has been accelerated in modern times. Over 25% of differences in per capita income can be explained by cross-country variation in technological adoption.

Whenever economic growth at home can also be explained by productivity improvements in other countries (a_{it}^*), Eq. (6) will be stated in a more general form by including varieties of external shocks originating from other variables. For example,

$$y_{it} = \alpha k_{it} + (1-\alpha)a_{it} + \beta a_{jt}^* + (1-\alpha)\vartheta(S_{it}) \quad (7)$$

where $i \neq j$, and $\beta > 0$ when there are positive productivity spillovers from overseas. However, β can be negative when productivity improvements in other countries lead to a deterioration in external competitiveness of a home country.

Furthermore, we extend this specification using the GVAR model and analyze the heterogeneous effects of productivity shocks on economic growth, rather than the homogeneous effects that are assumed in Eq. (7). It is more natural to assume that impacts of international spillovers on economic growth differ among countries or groups of countries, and estimating heterogeneous effects on economic growth distinguishes this study from previous studies using the standard panel data and spatial models.

3 The GVAR Model

The GVAR approach introduced by Pesaran et al. (2004) gives a relatively simple, yet effective way to model present global economy, where each country, and different macroeconomic factors within countries, are related to each other. The methodology of GVAR modeling consists of two different stages. First, a model known as the VARX is estimated for each country separately, where the letter X indicates the presence of an exogenous component in the VAR. If any of the variables have unit roots and are cointegrated, the model is estimated in their error-correcting form. In these individual VARX (or the vector error-correction model with exogenous variables, VECMX, in the presence of cointegration) models, each country has two different types of variables: domestic and foreign. Domestic variables are endogenous in the model, while foreign variables are exogenous and have their corresponding foreign variables. These foreign variables are constructed using a weight matrix, so that the relative importance of different countries is reflected properly in the analysis. They provide a connection between the evolution of the domestic economy and that of the rest of the world. The foreign variables must be weakly exogenous, an assumption that needs to be tested. In the second step, the individual VARX (or VECMX) models are combined together in a consistent manner with the help of a link matrix to build a global model.

- **Individual country model** Let there be $N + 1$ countries in the model, indexed by $i = 0, 1, 2, \ldots, N$, where country 0 is the reference country. Each country i then follows the VARX (p, q) model, which can be defined as:

$$y_{i,t} = a_{i,0} + a_{i,1}t + \sum_{j=1}^{p} \alpha_{i,j} y_{i,t-j} + \sum_{j=1}^{q} \beta_{i,j} y^*_{i,t-j} + u_{i,t} \tag{8}$$

for $t = 1, 2, \ldots, T$. Here, $k_i \times 1$ matrix $y_{i,t}$ represents the endogenous domestic variables and $k_i^* \times 1$ matrix $y^*_{i,t}$ represents the corresponding (weakly) exogenous foreign variables. k and k^* are the numbers of domestic and foreign variables, respectively, $a_{i,0}$ is a $k_i^* \times 1$ vector of fixed intercepts and $a_{i,1}$ is a $k_i^* \times 1$ vector of coefficients on the deterministic time trends. p and q are the lag lengths of domestic and foreign variables, respectively. They are selected according to the Schwartz Bayesian (SB) information criterion. Finally, $u_{i,t} \sim iid(0, \Sigma_{u_i})$.

The vector of foreign country-specific variables, $y^*_{i,t}$, is obtained from weighted averages of each variable across all other countries in the sample. More specifically, for any $i, j = 0, 1, \ldots, N$,

$$y^*_{i,t} = \sum_{j=0}^{N} w_{i,j,t} y_{j,t}, \tag{9}$$

where $w_{i,j,t}$ is a weighting factor that captures the importance of a country j for a country i, with $\sum_{j=0}^{N} w_{i,j,t} = 1$ and $w_{i,i,t} = 0$. Most of the GVAR literature uses fixed trade weights based on bilateral trade volumes. However, these may be subject to temporal changes, and as a result, a fixed weight might confuse the results. In order to account for the changes that took place throughout the sample period, this paper uses time-varying weights to construct foreign variables in country-specific models. These are constructed as three-year moving averages to smooth out short-run business cycle effects in the bilateral trade flows. More compactly, setting $p_i = \max(p, q)$, Eq. (8) can be written as:

$$A_{i,0}z_{i,t} = a_{i,0} + a_{i,1}t + \sum_{j=1}^{p_i} A_{i,j}z_{i,t-j} + u_{i,t} \tag{10}$$

where vector $z_{i,t} = (x_{i,t}{}', x^{*}_{i,t}{}')'$ represents both domestic and foreign variables and coefficient matrices are $A_{i,0} = (I_{k,i}, -\beta_{i,0})$ and $A_{i,j} = (\alpha_{i,j}, \beta_{i,j})$.

Because of the characteristics of the macroeconomic variables and to allow for the cointegrating relationship within and between countries, the country-specific VARX models are estimated in the following error-correction form (VECMX):

$$\Delta y_{i,t} = c_{i,0} - \alpha_i \beta_i{}'(z_{i,t-1} - a_{1,t}(t-1)) + \beta_{i,0}\Delta y^{*}_{i,t} + \sum_{j=1}^{p_i-1} \phi_{i,j}\Delta z_{i,t-j} + u_{i,t} \tag{11}$$

Here, α_i is a $k_i \times r_i$ matrix of rank r_i and β_i is a $(k_i + k_i^{*}) \times r_i$ matrix of rank r_i. Country-specific VECMX models are estimated using reduced rank regressions conditional on weakly exogenous foreign variables. This takes into account the possibility of cointegration within domestic variables and across domestic and foreign variables. This way, estimates for r_i, β_i, and α_i are obtained. Other parameters are estimated by OLS from this equation:

$$\Delta y_{i,t} = c_{i,0} + \delta ECM_{i,t-1} + \beta_{i,0}\Delta y^{*}_{i,t} + \phi_i \Delta z_{i,t-1} + u_{i,t} \tag{12}$$

where $ECM_{i,t-1}$ are the error-correction terms referring to the r_i cointegrating relations of the ith country model.

- **Global model**

The next step is to combine individual country-specific parameter estimates into a single global model. All country-specific variables are considered as a single $k \times 1$ global vector $y_t = (y'_{0t}, y'_{01}, \ldots, y'_{Nt})'$ where $k = \sum_{i=0}^{N} k_i$, so that all the variables are endogenous in the system as a whole. For each country, the corresponding VARX model is obtained from the VECMX model that was estimated. The link matrix W_i, which is the $(k_i + k_i^{*}) \times k$ matrix collecting the trade weights w_{ij}, $\forall i, j = 0, 1, 2, \ldots, N$, is used to obtain the identity $z_{i,t} = W_i y_t$. From Eq. (10), it follows that:

$$A_{i,0}W_iy_t = a_{i,0} + a_{i,1}t + \sum_{j=1}^{p_i} A_{i,j}W_iy_{t-j} + u_{i,t} \tag{13}$$

for $i = 0, 1, \ldots, N$. Then, the $N + 1$ systems in Eq. (13) are combined to obtain the global model in levels:

$$G_0y_t = a_0 + a_1t + \sum_{i=1}^{p} G_iy_{t-i} + u_t \tag{14}$$

Here, $G_0 = (A_{00}W_0, A_{10}W_1, \ldots, A_{N0}W_N)'$ is a known nonsingular $k \times k$ matrix that depends on the trade weights and parameter estimates $G_i = (A_{0i}W_0, A_{1i}W_0, \ldots, A_{Ni}W_N)'$ for $i = 1, 2, \ldots, p$, $a_0 = (a_{00}, a_{10}, \ldots, a_{N0})'$, $a_1 = (a_{01}, a_{11}, \ldots, a_{N1})'$, $u_t = (u_{0t}, u_{1t}, \ldots, u_{Nt})$ and $p = \max(p_i)$ across all i. Pre-multiplying Eq. (14) by G_0^{-1}, the GVAR (p) model is obtained as

$$y_t = b_0 + b_1t + \sum_{i=1}^{p} F_iy_{t-i} + \varepsilon_t \tag{15}$$

where, $b_0 = G_0^{-1}a_0$, $b_1 = G_0^{-1}a_1$, $F_i = G_0^{-1}G_i$ for $i = 1, 2, \ldots, p$ and $\varepsilon_t = G_0^{-1}u_t$.

The dynamic properties of the GVAR model in Eq. (15) can then be examined using structural generalized impulse-response functions (SGIRFs).

4 Data and Relevant Tests

Prior to the formal analysis, this section describes the data, the model specification, and the results of the weak exogeneity test.

4.1 Data and Model Specification

We use quarterly data from 1991 to 2016 for advanced and developing countries. There are 18 countries in the sample, divided into four groups. (1) The EU group consists of seven countries including Belgium, France, Germany, Italy, Netherlands, Switzerland, and the UK.[6] (2) The non-OECD group consists of all the countries in the sample that are not part of OECD. It includes Brazil, China, Hong Kong, India, Mexico, Russia, and Taiwan. (3) Countries in the sample that do not fall into the above two categories are included in the 'Others' group. These are Canada, Japan,

[6]Though Switzerland is not part of the EU and the UK will soon exit from the EU, they are included in this group for geographical proximity and historical links with the EU.

and Korea. (4) The USA is considered separately, as we are primarily interested in investigating spillover effects from the USA to other countries.

Regional variables for the groups 'EU,' 'Non-OECD,' and 'Others' were constructed by aggregating country-specific variables over N countries.

$$y_t = \sum_{i=1}^{N} \omega_i y_{i,t}$$

where y_t denotes a regional variable, $y_{i,t}$ is the value of that variable for country i, and ω_i represents the relative importance of a country i within the region. Following Dees et al. (2007), ω_i is computed by dividing the PPP-GDP figure of each country by the total sum across the N countries of the region, such that their weights add up to unity.

Spillover effects may be related to the level of trade with the rest of the world. Therefore, a country's openness to other countries is calculated as (Import + Export)/GDP. Table 1 shows the openness of each country vis-a-vis the USA. This table shows the ratio being often greater than one. It follows that the majority of countries are indeed more open than the USA. Only two countries exhibit evidence of lesser openness than the USA, but even in these cases, the ratios are close to the US level.

Table 1 Country's openness

Country group	Country	Country's openness relative to the USA
EU	Belgium	5.508
	France	2.047
	Germany	2.188
	Italy	1.923
	Netherlands	4.945
	Switzerland	3.960
	UK	2.188
Non-OECD	Brazil	0.917
	China	1.910
	HK	13.551
	India	1.417
	Mexico	2.132
	Russia	2.306
	Taiwan	4.331
Others	Canada	2.712
	Japan	0.986
	Korea	2.929

Notes The openness is calculated as (Export + Import)/GDP. The statistics show the ratio of country's openness vis-a-vis that of the USA

The model will be estimated with data on real GDP growth, expenditure on R&D, capital stock, and the TFP. The TFP is used as a proxy for productivity. The source of real GDP and population data is *Oxford Economics*. Production data are expressed in US dollars to make international comparisons later in our analysis. The TFP and capital stock data are collected from *Penn World Table Ver.9*. R&D data are collected from the OECD and are only available for the countries in the sample that are part of the OECD. For non-OECD countries, we assume no significant R&D activities. Annual data are converted to quarterly data, using a quadratic function of the Eviews.

The TFP of selected countries is shown in Fig. 1. Since this statistic indicates TFP of a country relative to the USA, it shows a marked difference in TFP among countries. Germany exhibits steady improvements in TFP and convergence with that of the USA. Brazil and Japan fail to improve TFP after 1990 and show divergence from the USA. Chinese TFP shows rather different trajectories compared to other countries. It is highly volatile and has shown steady increases since 2000, which is consistent with its rapid economic growth. However, TFP of all four countries is less than one and thus is below the level of the USA, thereby confirming the significant size of the US TFP compared to all other countries.

Figure 2 shows the ratio of R&D to GDP for selected countries and regions. The OECD data suggest that most countries spend about 1–3% of GDP on R&D.

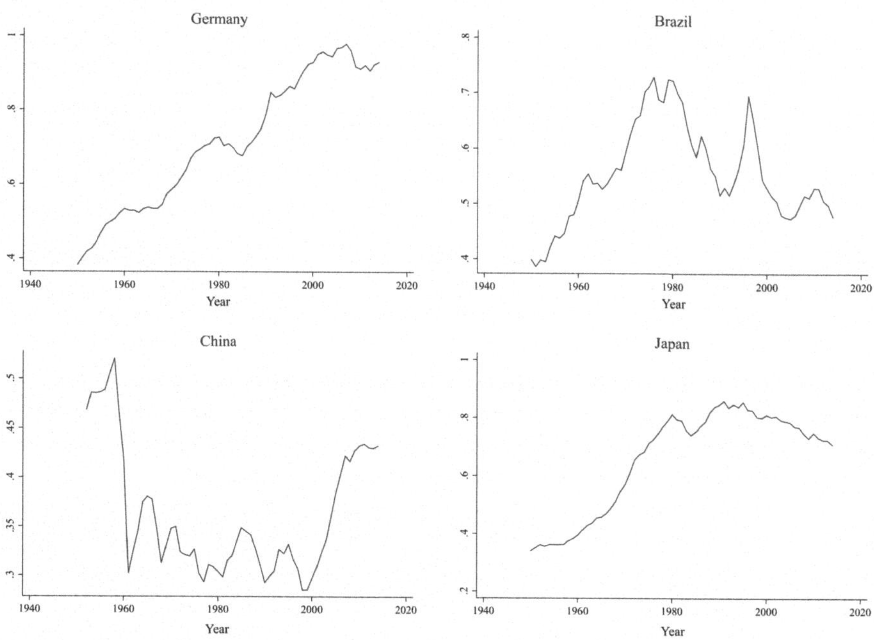

Notes: The data are from the Penn World Table Ver. 9. The US TFP is equal to one.

Fig. 1 Total factor productivity of selected countries. *Notes* The data are from the Penn World Table Ver. 9. The US TFP is equal to one

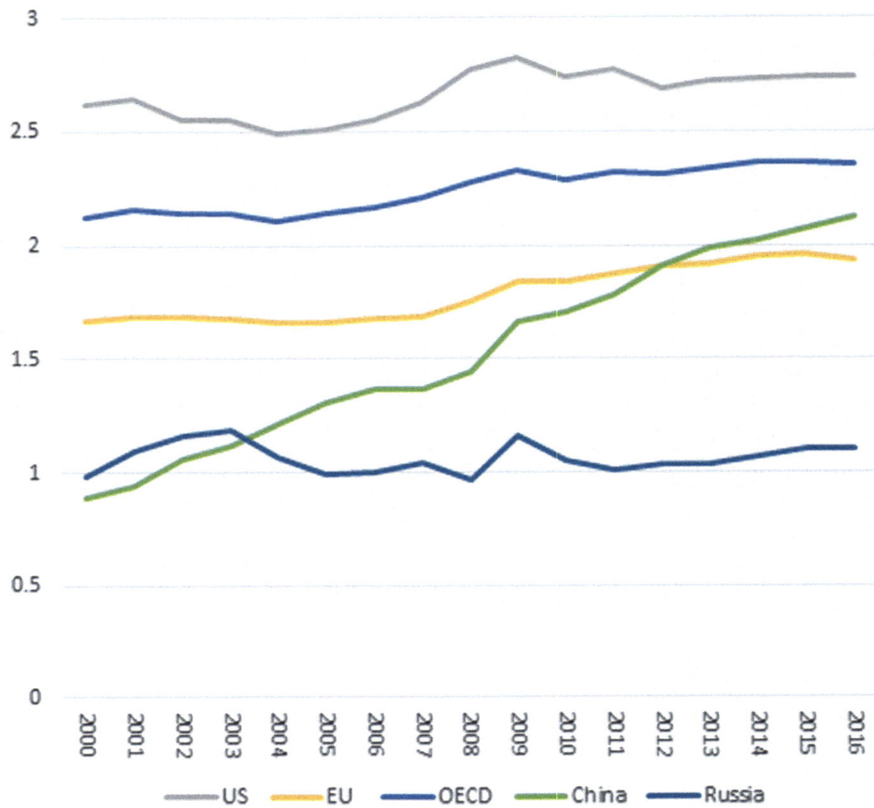

Notes: Data source: `http://stats.oecd.org/Index.aspx?DataSetCode=MSTI_PUB`

Fig. 2 R&D (% of GDP). *Notes* Data source: http://stats.oecd.org/Index.aspx?DataSetCode=MSTI_PUB

However, there is a variation in the trend of this variable among countries. While this ratio is stable over time for most countries and regions, China has increased R&D expenditure remarkably in recent times. Nonetheless, this ratio remains relatively high in the USA over this century, while remaining low in Russia.

The correlation among key variables is summarized for each country group in Table 2. Generally, the variables are positively correlated with each other. Notably, the TFP is positively correlated with GDP growth. On the other hand, contrary to expectations, capital has a negative relationship with other economic variables.

Next, we check the stationarity and cointegration of data. First, the stationarity of data is examined for all the country-specific domestic variables and their corresponding foreign variables. In addition to the commonly used Augmented Dickey Fuller (ADF) test, the Weighted-Symmetric Dickey Fuller (WS) test suggested by Park and

Table 2 The correlation of variables within country groups

	TFP	R&D	GDP	Capital
USA				
TFP	1			
R&D	0.366	1		
GDP	0.816	0.374	1	
Capital	−0.589	−0.692	−0.446	1
EU				
TFP	1			
R&D	0.542	1		
GDP	0.329	0.686	1	
Capital	−0.686	−0.925	−0.643	1
Non-OECD				
TFP	1			
R&D	0.361	1		
GDP	0.436	0.936	1	
Capital	−0.685	−0.822	−0.820	1
Others				
TFP	1			
R&D	0.338	1		
GDP	0.388	0.990	1	
Capital	−0.572	−0.896	−0.916	1

Notes Full sample

Fuller (1995) is also used for this purpose.[7] The lag length was selected according to the SB criterion. The unit root tests are conducted with variables in levels with an intercept and time trend. The results for the level data for domestic variables are shown in Table 3. The t-values are shown with the 5% critical values. As expected, there is a mix of stationary and non-stationary data across country groups. However, the majority of data seem to follow a non-stationary process, as we often fail to reject the null hypothesis of the unit root process.

The next step is to define the country-specific VARX models. The GVAR model has the flexibility of handling different specifications for different countries (i.e., the number of domestic and foreign variables that goes into each country-specific model). However, since the countries in the sample have trading relations and are expected to affect each other, all the country models initially include all four variables as domestic and foreign variables (constructed using the weight matrix mentioned in

[7]The WS test exploits the time reversibility of a stationary autoregressive process in order to increase their power performance. Many authors like Leybourne et al. (2005) and Pantula et al. (1994) show evidence of superior performance of the WS test as compared to the ADF test.

Table 3 Unit root tests for the domestic variables

Domestic variable	Statistic	Critical value	EU	Non-OECD	Others	USA
TFP	ADF	−3.45	−2.51	−1.80	−4.19	−1.88
	WS	−3.24	−2.70	−1.01	−4.09	−2.20
R&D	ADF	−3.45	−2.67	−1.63	−0.49	−5.12
	WS	−3.24	−2.18	−1.15	−0.73	−4.88
GDP	ADF	−3.45	−1.35	−2.29	−2.32	−1.50
	WS	−3.24	−1.16	−1.87	−2.30	−1.27
Capital	WS	−3.24	−0.93	−2.03	−3.13	−1.54
	ADF	−2.89	1.06	0.20	0.22	−0.71

Notes Unit root tests examine the null hypothesis of the unit root. The 5% critical values are reported

Table 4 Order of the VARX model and the number of cointegrating relations

	p_i	q_i	# Cointegrating relations
EU	2	1	3
Non-OECD	2	1	4
Others	2	1	4
USA	2	1	3

Notes The lag length is determined by the Schwartz Bayesian information criterion. The number of cointegrating relationships is based on trace statistics of Johansen tests

the previous section).[8] Next, the order of the country-specific VARX (p_i, q_i) model is selected using the SB criterion. While selecting the lag order, p_i and q_i were not allowed to go over 2, because of the small sample size compared to the large number of parameters to be estimated. VARX model is shown in Table 4.

For all country groups, the VARX (2, 1) model is selected. Given that some of the variables are non-stationary, Johansen's cointegration test is conducted in order to determine the number of cointegrating relations for each country group. Here, the specifications consider Case IV according to Pesaran et al. (2000), where a linear deterministic trend is implicitly allowed for the cointegration space, but can be eliminated in the dynamic part of VEC models. The number of cointegration relations for each country group based on the trace statistics is also shown in Table 4. All country groups seem to have either three or four cointegrating relations. Next, individual country-specific VECMX models were estimated, subject to the reduced rank restrictions. Corresponding error-correcting terms were then derived. These ECMs were subsequently used to conduct the weak exogeneity test.

[8]Several robustness checks were also conducted by leaving out some variables as foreign variables for countries that are less integrated with other countries in terms of trade. However, the main findings of the paper are not affected.

4.2 Weak Exogeneity Test

As mentioned earlier, one of the main assumptions of the GVAR model is the weak exogeneity of the country-specific foreign variables $y^*_{i,t}$. In general, a variable in a VARX model is considered weakly exogenous if it is not dependent on contemporaneous values of endogenous variables, but is likely to depend on the lagged values of these endogenous variables. More formally, $y^*_{i,t}$ is considered weakly exogenous if $y_{i,t}$ does not affect $y^*_{i,t}$ in the long run, but $y^*_{i,t}$ is said to be 'long-run forcing' for $y_{i,t}$. As shown in Johansen (1992), this assumption allows proper identification of the cointegration relations. In the formal test, the joint significance of the estimated error-correction terms in auxiliary equations for the country-specific foreign variables $y^*_{i,t}$ is tested. Specifically, for each lth element of $y^*_{i,t}$, a regression of the following form is conducted:

$$\Delta y^*_{i,t,l} = a_{i,l} + \sum_{j=1}^{r_i} \delta_{i,j,l} \widehat{ECM}_{i,j,t-1} + \sum_{s=1}^{p^*_i} \phi'_{i,s,l} \Delta y_{i,t-s} + \sum_{s=1}^{q^*_i} \Psi_{i,s,l} \Delta \widetilde{y}^*_{i,t-s} + \eta_{i,t,l} \tag{16}$$

where $\widehat{ECM}_{i,j,t-1}$, for $j = 1, 2, \ldots, r_i$ are the estimated error-correction terms corresponding to the r_i cointegrating relations found for the i th country, and p^*_i and q^*_i are the orders of the lagged changes for the domestic and foreign variables, respectively. The test for the weak exogeneity is an F-test of the joint hypothesis that $\delta_{i,j,l} = 0$ for $j = 1, 2, \ldots, r_i$ in the above equation. It is not necessary that lag orders of p^*_i and q^*_i are the same for the underlying country-specific model. They are selected using the SB criterion.

The results are shown in Table 5. As can be seen, all the variables pass the weak exogeneity test, as the assumption of exogeneity cannot be rejected at the 5% level. This is a very desirable result, as it confirms the suitability of the GVAR model for this region. Based on the eigenvalues, the model was also found to be stable.

Table 5 Weak exogeneity test

Country	F-test	Critical value	TFP	R&D	GDP	Capital
EU	$F(2,71)$	3.12	1.27	2.48	2.29	0.28
Non-OECD	$F(5,68)$	2.34	1.89	0.86	2.24	1.09
Others	$F(3,70)$	2.73	0.08	1.27	1.73	1.62
USA	$F(3,70)$	2.73	1.95	1.78	0.23	1.17

Notes The 5% critical values are reported

5 Results

By means of the SGIRF analysis, this section investigates the effects of US TFP shocks on the home economy (i.e., real GDP, R&D spending, and capital stock). Similarly, we study the effects of TFP shocks of the major trading partners of the USA (grouped into the EU, non-OECD, and other groups).

5.1 *Structural Generalized Impulse-Response Function (SGIRF) Analysis of US Productivity Shocks*

The identification of shocks has been a major issue in GVAR models. In order to conduct dynamic analysis, the vast majority of research papers using GVAR models rely on the GIRF proposed by Koop et al. (1996) and further developed by Pesaran and Shin (1998). The identification of shocks in a GVAR model is complicated due to the cross-country interactions and high dimensionality of the model.

The identification in a traditional VAR analysis is usually achieved by using the orthogonalized impulse-response functions (OIRFs) that require a certain ordering of variables. This approach is often not suitable for GVAR models, as it requires ordering not only of the variables, but also countries. As a result, when a large number of variables and countries are included in the model, it becomes difficult to justify such ordering based on economic theory and empirical findings. The advantage of GIRFs is that they are invariant to the ordering of countries and variables. This is very convenient for models like GVAR that involve many countries and variables. However, it comes at a cost. Critics often argue that in GIRFs, the error terms are not orthogonal and it allows correlation among them. This, in turn, makes economic interpretation of shocks difficult.

We take this into account by using SGIRFs instead of GIRFs. The SGIRF allows the most dominant economy in the model to be ordered first and also its variables to have certain ordering. Since the main aim of this paper is to investigate spillover effects of productivity shocks arising in the USA, the largest economy in the model, the USA and its variables are ordered first. This means that the identifying scheme for the model of the USA is based on a lower-triangular Cholesky decomposition and has the following ordering: [R&D, TFP, capital, GDP]'. Thus, for the USA, R&D is ordered first, followed by TFP because greater expenditure on R&D could increase TFP. This assumes that R&D spending affects TFP contemporaneously, but not vice versa. TFP is then followed by capital and GDP. This ordering system assumes that GDP is the most endogenous variable, which is a realistic assumption to make. Other countries and their variables are kept unrestricted. More about the GIRF and SGIRF is discussed in the appendix.

The main aim of this paper is to evaluate the effects of US productivity shocks on the USA itself as well as the spillover effects on other countries. This result is depicted in Fig. 3. It shows the SGIRF for the USA, the EU, non-OECD, and the rest

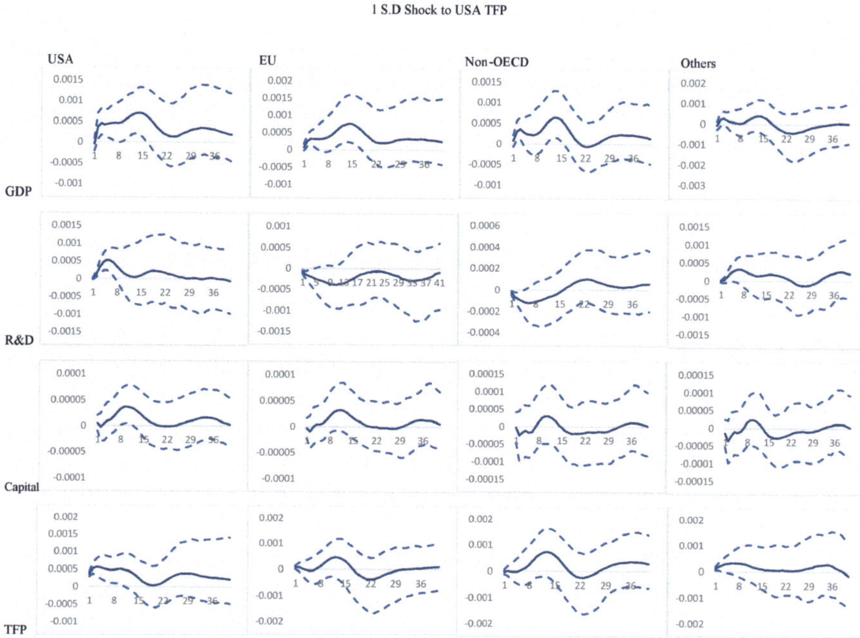

Fig. 3 Impulse-response functions for an expansionary 1 SD shock to US TFP. *Notes* The figure reports structural generalized impulse-response functions (SGIRFs) for the GDP, R&D, capital and TFP. The graphs show bootstrap median estimates with the associated 90% bootstrap confidence bands computed on the basis of 1000 replications of the SGIRFs, where the forecast horizon extends up to 40 quarters and is recorded along the horizontal axis

of the countries (i.e., Others) to one standard deviation (SD) positive TFP shock to the USA. For each country, reactions of GDP, R&D, capital, and TFP are depicted for up to 40 quarters. The associated 90% bootstrap confidence bands, computed on the basis of 1000 replications, are also displayed by dotted lines.

The results show that a TFP shock in the USA not only has a positive effect on real GDP of the USA, but also has positive spillover effects on the real GDP of its main trading partner countries. This effect is statistically significant for the USA, the EU, and non-OECD country groups, but not highly significant for others. For all countries, the effect is much smaller on impact. It increases over time and stays significant for up to 16 quarters (i.e., four years) for most country groups. This implies quick cross-country transmission of innovation, in line with Ho et al. (2013) that showed a short-lived effect of international trade on economic growth.

After a US TFP shock, R&D spending shows a statistically significant positive reaction in the USA and some negative reaction in the EU and non-OECD countries (though significant only for the first few quarters). This result indicates that US productivity shocks are positively associated with its own R&D spending, but it is not strongly correlated with other countries' R&D spending. In fact, the initial reactions of the EU and the non-OECD countries' R&D are negatively associated with US productivity shocks.[9] Similarly, the reaction of capital after a US TFP shock is mostly insignificant for all country groups. The TFP instead shows some positive reaction to shocks in the USA, but does not show any significant results for other countries.

5.2 Productivity Shocks in the EU, Non-OECD, and Others

Figures 4, 5 and 6 show the response of the same variables to one standard deviation (SD) positive shock in 'the EU,' 'non-OECD,' and 'Others' country groups, respectively. The first row of each figure shows the response of GDP to a TFP shock. The EU's TFP shock does have some positive and significant effects on GDP of the EU and the other country groups. Compared to the US TFP shock, these reactions are smaller in magnitude. EU's TFP shock also increases US R&D and has a further positive effect on its own TFP. Such reactions are, however, significant for a very short period of time. A shock to the non-OECD group's TFP has some positive significant effect on its own GDP, but spillover effects are not highly significant. A shock to the final group 'Others' has no significant effect on the GDP of any country groups either. In terms of the effect on other variables, results are not very significant.

Interestingly, a positive TFP shock is associated with increases in R&D spending for the USA and other country groups. This might be due to the size of the US R&D and the fact that US multinational firms are much more global in terms of investing in other country groups. According to the Forbes Global 2000 that lists top 2000 companies in the world, the USA was ranked first in terms of the number of firms included in this list. An increase in productivity in the rest of the world creates greater incentives for them to expand their business by spending more on R&D. Put it differently, productivity improvements in other country groups are dependent upon productivity advances in the USA. On the contrary, R&D spending in the EU decreases when there is a positive TFP shock in 'Non-OECD' and 'Others' country groups. These results might explain why the EU is still lagging behind the USA in terms of research and innovation and are in line with findings of Miller and Atkinson (2014). While the USA is able to make best use of productivity improvements in other countries, the rest of the world fails to do so.

5.3 The Asymmetric Effect of Productivity Shocks

In order to compare the magnitude of US shocks with that of other country groups, we calculate the ratio of impulse responses (Table 6). For country groups X and

[9] R&D shocks to the EU and non-OECD countries (results not showed here, but available on request) actually also have no significant effect on the GDP of their own countries.

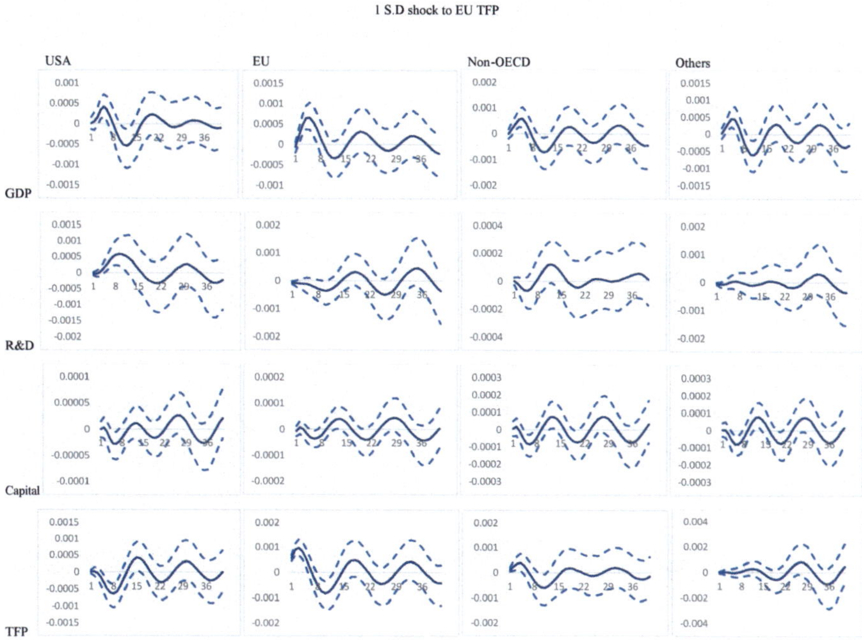

Fig. 4 Impulse-response functions for an expansionary 1 SD shock to EU TFP. *Notes* see Fig.3

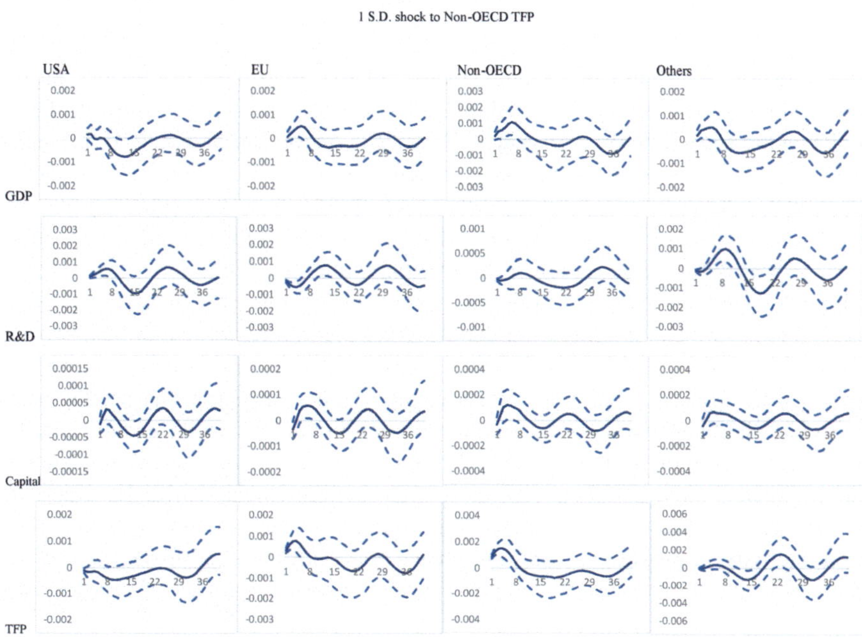

Fig. 5 Impulse-response functions for an expansionary 1 SD shock to non-OECD TFP. *Notes* see Fig. 3

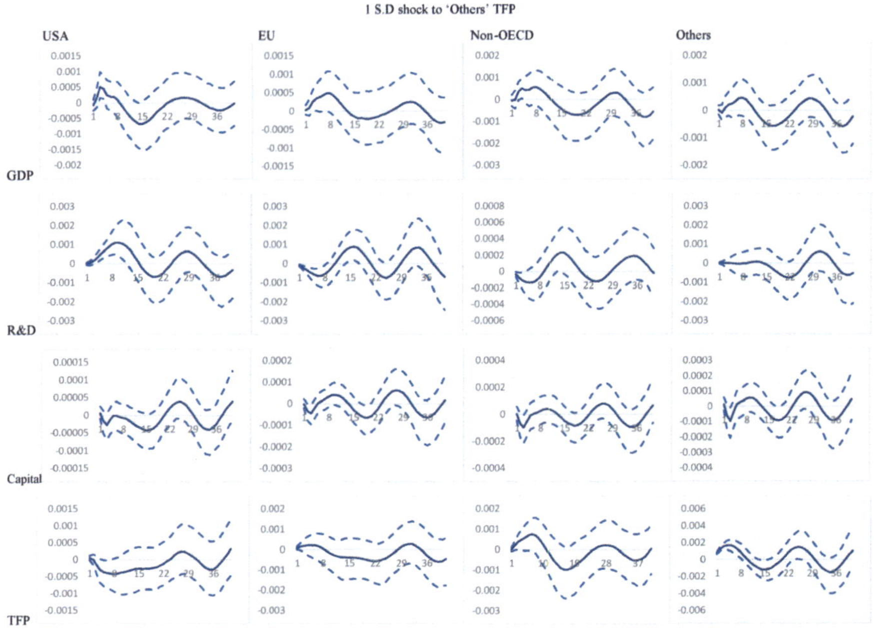

Fig. 6 Impulse-response functions for an expansionary 1 SD shock to OTHERS TFP. *Notes* see Fig. 3

Y, the ratio of (X→X)/(Y→Y) indicates the relative importance of shocks in the home economy. Therefore, (EU→EU)/(USA→USA) compares the impact of EU shocks on the EU economy with that of US shocks on the US economy. If this ratio equals unity, it follows that their impact on their own economy is identical. Similarly, (X→Y)/(Y→X) gives us information about which direction of productivity shocks are relatively more important between X and Y. When this ratio is higher than unity, it means that economic shocks from X are greater than that from Y.

Table 6 shows the average of this ratio over 40 time horizons. Since this ratio is less than unity, we can interpret that the US shocks are more influential than shocks from other country groups over its own economy as well as that of other country groups. Moreover, we sometimes find a negative ratio, implying that productivity shocks have had negative impacts on the economy in one of country group pairs. In short, there is asymmetry in the effect of productivity improvements across country groups. Given that the USA is not the most open economy in the world, the relatively more significant effect of US shocks is attributable to the amount of expenditure on productivity improvements and the presence of multinational firms that utilize advanced technology. Thus, we confirm heterogeneous responses of productivity

Table 6 The impact of productivity shocks

	(EU→EU)/ (USA→USA)	(EU→USA)/ (USA→EU)	(NonOECD→NonOECD)/ (USA→USA)	(NonOECD→USA)/ (USA→NonOECD)	(Others→Others)/ (USA→USA)	(Others→USA)/ (USA→Others)
Average	0.244	0.060	−0.423	−0.255	−0.192	−0.099
Std error	0.111	0.080	0.258	0.109	0.181	0.105
Medium	0.290	0.061	−0.385	−0.302	−0.231	−0.248
Mini	−1.172	−0.834	−5.168	−1.276	−2.335	−0.944
Max	1.594	1.324	2.384	1.866	1.717	1.571

Notes The ratio of the magnitude of productivity shocks is reported shocks on the economy. Our results provide the reasons why previous macroeconomic analyses often fail to detect the role of productivity spillovers in economic growth; that is, productivity spillovers are not long-lived and are insignificant for some countries.

6 Concluding Remarks

This paper empirically investigates the impact of productivity spillovers in global markets, where productivity shocks affect not only home economies, but other countries too. In order to capture such spillover effects, we have used the GVAR to model global economies, which allows us to estimate heterogeneous reactions to spillovers across country groups. By doing so, we have arrived at noteworthy results.

Generally, we find that while productivity shocks are asymmetric in global markets, productivity improvements have generally contributed to economic growth, leading us to conclude that liberalizing international markets is beneficial to many (but not all) countries. Productivity shocks originating from the USA are more influential over the rest of the world than those from other countries. Our country group analyses suggest that US shocks are positively correlated with economic growth of other countries, and thus, global markets benefit from the spillover of productivity improvements. In contrast, such a clear positive relationship can hardly be observed in the analysis of productivity shocks originating from other country groups. Our results of heterogeneous and short-lived effects of productivity spillovers provide explanations why previous macrodata analyses have failed to obtain a significant relationship between economic growth and international spillovers than microdata studies. Moreover, our results seem to illustrate and provide insight into the concerns of the USA behind the recent China-USA trade battle. That is, the root of this trade war is attributable to the transmission of more US technologies to the rest of the world (say China) than those in other countries.

Because our analysis is based on macroeconomic data, we cannot provide a detailed and insightful explanation regarding the asymmetric effect. However, our results are consistent with the extent of the US innovation. The USA spends more on innovation than any other country. In this regard, the finding that global markets benefit from US productivity shocks will be valuably supplemented by multi-country analyses using firm- or industry-level data in future studies.

Acknowledgements We would like to thank Gabriel Cordoba for the research assistance. This research was initiated when Khan was visiting Tohoku University. A financial support for travel expenses was provided by the Japan Investment Advisers Association.

Declarations of interest None.

Submission Declaration and Verification This paper is no under consideration for publication elsewhere.

Appendix

GIRF and SGIRF

The generalized impulse-response functions (GIRFs) was introduced by Koop et al. (1996) and further developed by Pesaran and Shin (1998). Let us consider the model obtained during the solution of the GVAR, expressed in terms of the country-specific errors given by Eq. (14). The GIRFs are based on the definition:

$$\text{GIRF}(y_t; u_{i,l,t}; h) = E(y_{t+h}|u_{i,l,t} = \sqrt{\sigma_{ii,ll}}, I_{t-1}) - E(y_{t+h}|I_{t-1}) \tag{17}$$

where I_{t-1} is the information set at time $t-1$, $\sigma_{ii,ll}$ is the diagonal element of the variance covariance matrix Σ_u corresponding to the lth equation in the ith country, and h is the horizon. GIRFs are invariant to the ordering of variables, and they allow for correlation of the error terms (the error terms are not orthogonal).

The structural generalized impulse-response functions (SGIRFs) used in this paper allows ordering of variables to one country. As the USA is the largest and the most dominant economy in this model, it is ordered first and its variables are ordered in the way mentioned in Sect. 4.1. The SGIRFs are invariant to the ordering of other countries and their variables. Let us consider the $VARX^*(p_1, q_1)$ model for the USA.

$$y_{1,t} = a_{1,0} + a_{1,1}t + \sum_{j=1}^{p} \alpha_{1,j} y_{1,t-j} + \sum_{j=1}^{q} \beta_{1,j} y^*_{1,t-j} + u_{1,t} \tag{18}$$

Let $V_{1,t}$ be structural shocks given by $V_{1,t} = P_1 u_{1,t}$, where P_1 is the $k_1 \times k_1$ matrix of coefficients to be identified. The identification conditions using the triangular approach of Sims (1980) require $\Sigma_{v,1} = \text{Cov}(v_{1,t})$ to be diagonal and P_1 to be lower triangular. Let Q_1 be the upper Cholesky factor of $\text{Cov}(u_{1,t}) = \Sigma_{u,1} = Q_1' Q_1$ so that $\Sigma_{v,1} = P_1 \Sigma_{u,1} P_1'$ with $P_1 = (Q_1')^{-1}$ Under this orthogonalization scheme, $\text{Cov}(v_{i,t}) = I_{k_0}$

Pre-multiplying the GVAR model in Eq. (14) by

$$P_{G_1}^1 = \begin{bmatrix} P_0 & 0 & 0 & 0 \\ 0 & I_{k_1} & 0 & 0 \\ 0 & 0 & \ddots & 0 \\ 0 & 0 & 0 & I_{k_n} \end{bmatrix}$$

It follows that

$$P_{G_1}^1 G_1 y_t = P_{G_1}^1 G_1 y_{t-1} + \cdots + P_{G_1}^1 G_p y_{t-p} + v_t \tag{19}$$

where $v_t = (v'_{1t}, u'_{1t}, \ldots, u'_{Nt})$ and

$$\Sigma_v = \mathrm{Cov}(v_t) = \begin{bmatrix} V(v_{1t}) & \mathrm{Cov}(v_{1t}, u_{1t}) & \ldots & \mathrm{Cov}(v_{1t}, u_{Nt}) \\ \mathrm{Cov}(u_{1t}, v_{1t}) & V(u_{1t}) & \ldots & \mathrm{Cov}(u_{1t}, u_{Nt}) \\ \vdots & \vdots & & \vdots \\ \mathrm{Cov}(u_{Nt}, v_{1t}) & \mathrm{Cov}(u_{Nt}, u_{1t}) & \ldots & V(u_{Nt}) \end{bmatrix}$$

with

$$V(v_{1t}) = \Sigma_{v,11} = P_1 \Sigma_{u,11} P_1^1 \text{ and } \mathrm{Cov}(v_{1t}, u_{jt}) = \mathrm{Cov}(P_1 u_{1t}, u_{jt}) = P_0 \Sigma_{u_{1j}}$$

By using the definition of the generalized impulse responses with respect to structural shocks given by

$$\mathrm{SGIRF}(y_t; v_{l,t}; h) = E(y_{t+h} | I_{t-1} \varrho'_l v_t = \sqrt{\varrho'_l \Sigma_v \varrho_l,}) - E(y_{t+h} | I_{t-1}) \quad (20)$$

it follows that for a structurally identified shock, v_{lt} such as a US trade shock the SGIRF is given by

$$\mathrm{SGIRF}(y_t; v_{l,t}; h) = \frac{\varrho_j A_n (P_{G_1} G_1)^{-1} \Sigma_v \varrho_l}{\sqrt{\varrho_l \Sigma_v \varrho_l}}, \; h = 0, 1, 2, ; j = 1, 2, \ldots, k$$

where $\varrho_l = (0, 0, \ldots, 0, 1, 0, \ldots 0)'$ is a selection vector with unity as the lth element in the case of a country-specific shock, Σ_v is the covariance matrix of structural shocks, and $P'_{G_1} G_1$ is defined by the identification scheme used to identify the shocks.

References

Arora, V., & Athanasios, V. (2005). How much do trading partners matter for economic growth. *IMF Staff Papers*, *52*(1), 24–40.

Barrios, S., & Eric, S. (2002). Foreign direct investment and productivity spillovers: Evidence from the spanish experience. *Weltwirtschaftliches Archiv*, *138*, 459–481.

Coe, D. T., & Helpman, E. (1995). International R&D spillovers. *European Economic Review*, *39*, 859–887.

Coe, D. T., Helpman, E., & Hoffmaister, A. (1997). North-south R&D spillovers. *Economic Journal*, *107*, 134–149.

Comin, D., & Bart, H. (2010). An exploration of technology diffusion. *American Economic Review*, *100*, 2031–2059.

Dees, S., Holly, S., Pesaran, M. H., & Smith, L. (2007). Long run macroeconomic relations in the global economy. *Economics—The Open-Access, Open-Assessment E-Journal*, *1*(3), 1–20.

Ertur, C., & Koch, W. (2007). Growth, technological interdependence and spatial externalities: Theory and evidence. *Journal of Applied Econometrics*, *22*(6), 1033–1062.

Frankel, J. A., & Romer, D. (1999). Does trade cause growth? *American Economic Review*, *89*(3), 373–399.

Girma, S. (2005). Absorptive capacity and productive spillovers from FDI: A threshold regression analysis. *Oxford Bulletin of Economics and Statistics*, *67*, 281–306.
Gorg, H., & Strobl, E. (2001). Multinational companies and productivity spillovers: A meta-analysis. *Economic Journal*, *111*, 723–739.
Grossman, G. M., & Helpman, E. (1991). *Innovation and growth in the global economy*. Cambridge: MIT Press.
Hall, R., & Charles, I. (1999). Why do some countries produce so much more output per worker than others? *Quarterly Journal of Economics*, *114*, 83–116.
Ho, C.-Y., Wang, W., & Yu, J. (2013). Growth spillover through trade: A spatial dynamic panel data approach. *Economics Letters*, *120*(3), 450–453.
Johansen, S. (1992). Cointegration in partial systems and the efficiency of single-equation analysis. *Journal of Econometrics*, *52*(3), 389–402.
Jones, C. (2016). The facts of economic growth. In *Handbook of macroeconomics* (Vol. 2, pp. 3–69). Elsevier.
Keller, W. (2002). Trade and the transmission of technology. *Journal of Economic Growth*, *7*, 5–24.
Keller, W., & Yeaple, S. R. (2013). The gravity of knowledge. *American Economic Review*, *103*, 1414–1444.
Koop, G., Perasan, H., & Potter, S. (1996). Impulse response analysis in nonlinear multivariate models. *Journal of Econometrics*, *74*(1), 119–147.
Leybourne, S., Kim, T., & Newbold, P. (2005). Examination of some more powerful modifications of the Dickey-Fuller test. *Journal of Time Series Analysis*, *26*, 355–369.
Lin, M., & Kwan, Y. K. (2016). FDI technology spillovers, geography, and spatial diffusion. *International Review of Economics & Finance*, *43*, 257–274.
Lucas, R. E, Jr. (1993). Making a miracle. *Econometrica*, *61*(2), 251–272.
Lucas, R. (1988). On the mechanics of economic development. *Journal of Monetary Economics*, *22*, 3–42.
Mankiw, N., David, R., & David, N. (1992). A contribution to the empirics of economic growth. *Quarterly Journal of Economics*, *107*, 407–437.
Miller, B., & Atkinson, R. (2014). *Raising European productivity growth through ICT*. Washington, D.C.: Information Technology and Innovation Foundation.
Mincer, J. A. (1974). *Schooling, experience, and earnings*. Number minc74-1 in NBER Books. National Bureau of Economic Research, Inc.
Pantula, S. G., Gonzalez-Farias, G., & Fuller, W. A. (1994). A comparison of unit-root test criteria. *Journal of Business & Economic Statistics*, *12*, 449–459.
Park, H., & Fuller, W. (1995). Alternative estimators and unit root tests for the autoregressive process. *Journal of Time Series Analysis*, *16*, 415–429.
Pesaran, M., Shin, Y., & Smith, R. J. (2000). Structural analysis of vector error correction models with exogenous I(1) variables. *Journal of Econometrics*, *97*(2), 293–343.
Pesaran, M. H., Schuermann, T., & Weiner, M. (2004). Modeling regional interdependencies using a global error-correcting macroeconomic model. *Journal of Business and Economic Statistics*, *22*, 129–162.
Pesaran, M. H., & Shin, Y. (1998). Generalized impulse response analysis in linear multivariate models. *Economics Letters*, *58*, 17–29.
Raut, L. (1995). R&d spillover and productivity growth: Evidence from Indian private firms. *Journal of Development Economics*, *48*, 1–23.
Sims, C. (1980). Macroeconomics and reality. *Econometrica*, *48*(1), 1–48. https://doi.org/10.2307/1912017
Solow, R. M. (1956). A contribution to the theory of economic growth. *Quarterly Journal of Economics*, *70*, 65–94.
Ventura, J. (1997). Growth and interdependence. *Quarterly Journal of Economics*, *112*, 57–84.
Weil, D. (2009). *Economic growth* (2nd ed.). Boston: Person Education.

Financial Econometrics

Commodity Prices in Empirical Research

Jean-François Carpantier

1 Introduction

Commodity prices are ingredients involved in many fields in economics. They are associated to natural resource curse in development economics, to trade policies in international economics, to diversification benefits in portfolio management, to commodity currencies in international finance, to civil wars and international conflicts in political economics, to the price puzzle in monetary economics. The empirical treatment of commodity prices has naturally followed a logic specific to each discipline, with empirical assumptions that have neither been harmonized, nor compared with each other. This monograph is designed to trigger a dialog and tentatively bridge some gaps by highlighting the empirical concerns associated to the commodity prices in their diversity and by illustrating the various solutions proposed to deal with potential endogeneity, with non-stationarity, and with data measurement.

We pick three emblematic topics involving the commodity prices, and we detail for each of them what empirical challenges do the commodity prices raise and how they are handled. The first topic consists in the existence of a long-run declining trend in commodity prices, the so-called Prebisch–Singer hypothesis. The second topic refers to the dependence of exchange rates of some commodity-exporting countries to the international commodity prices, that is, commodity currencies. The last topic builds on the rising weight of financial investors on the commodity markets and on their impact on commodity prices, what has been called the financialization of commodity markets.

We focus on these three topics for the following reasons. These topics allow us to make a comparison of treatments from three different fields, namely development economics, international economics, and finance. Further, these perspectives provide

J.-F. Carpantier (✉)
Aix-Marseille School of Economics, Marseille, France
e-mail: jf.carpantier@gmail.com

G. Dufrénot and T. Matsuki (eds.), *Recent Econometric Techniques for Macroeconomic and Financial Data*, Dynamic Modeling and Econometrics in Economics and Finance 27, https://doi.org/10.1007/978-3-030-54252-8_8

three different approaches on commodity prices. In the first topic, the Prebisch–Singer hypothesis, commodity prices are studied as such, in isolation from any other influence in a univariate setting. Inversely, the two other topics take the commodity prices as one variable among others in a multivariate setting. Commodity currency papers view commodity prices as a right-hand side exogenous variable, while the financialization literature sees commodity prices as a left-hand side endogenous variable.

2 Prebisch–Singer Hypothesis

2.1 Introduction

Terms of trade, defined as the ratio of export prices to import prices, have an impact on the macroeconomic performance of large exporters. A rising terms of trade signals a gain from international trade for that country, as it can pay for more imports per export units, which can ultimately translate into a more vivid economic growth (Collier and Goderis 2012).

For developing countries with commodities as main exports (energy, metals, agriculturals), the terms of trade consist in a ratio of country-specific commodity prices to import prices, which mainly consist in manufactured goods. Consequently, for these countries, the terms of trade are sometimes proxied by a commodity terms of trade, where the numerator reflects the evolution of the prices of the sole commodity exports.

Prebisch (1950) and Singer (1950) suggested, in what is known as the Prebisch–Singer hypothesis (PSH), that the commodity terms of trade should decline in the long run, as the income elasticity of demand for commodities is less than that of manufactured goods. Economic policies should thus consider protectionist policies and help commodity-dependent countries in diversifying their economies out of commodity exports and thereby to alleviate the negative long-run contribution of declining commodity prices.

The PSH was in conflict with the view of classical economists such as Ricardo, Malthus, Mill, and Jevons in the nineteenth century, who emphasized that commodities are characterized by physical finitude: Arable land is limited by earth surface, and minerals, such as oil and metals, have a limited available stock in the ground. Due to this inexorable limitedness, commodity prices, that is, terms of trade of developing countries, should rise in the long run, not decline (Hallam 2018).

Ultimately, whether the long-term trend of relative commodity prices is ascending or declining is an empirical question. As discussed below, and summarized in Table 1, a remarkable amount of empirical papers has been devoted to this question. We first show the different approaches followed in the literature to measure the (commodity) terms of trade and then discuss the incremental sophistications of the econometric strategies.

Table 1 PSH—review of empirical papers

	Methodology	Results
Grilli and Yang (1988)	TS approach on commodity indices. Breaks dating based on visual analysis.	Index of commodity prices decline by 0.5% per year, and non-fuel commodity prices by 0.6% per year. Breaks in 1921, 1932, 1945
Cuddington and Urzua (1989)	TS and DS approaches on commodity indices. Breaks tested with dummies.	Neither specification indicates evidence of secular deterioration in commodity prices, but only a permanent one-time drop in prices after 1920
Cuddington (1992)	TS and DS approaches on individual commodity price series. Breaks tested with dummies.	Half the series are TS, other half DS. 5 of the 26 individual commodity price series have a significantly negative trend. No break in 1921 for individual series, while break is found in index (aggregation concern)
Leon and Soto (1997)	TS and DS approaches on individual commodity price series. Recursive unit-root test of Zivot–Andrews (1992) to account for endogenous breaks.	5 of the 24 series are DS. 19 series are TS (with or without breaks). After consideration of endogenous breaks, 15 of the 24 series have a significantly negative trend
Kim et al. (2003)	Tests accounting for the uncertainty in choice of TS vs DS processes. Estimates on individual commodity price series. No consideration of breaks.	They find "*at least modest evidence of trend in only 8 of the series, and in just 6 of these was that of downward trend*"
Kellard and Wohar (2006)	Unit-root test of Lumsdaine and Papell (1997) that allows for two shifts in the mean and trend under the alternative hypothesis to individual commodity price series.	14 series are trend stationary after allowance for (up to) two breaks. Furthermore, for the majority of commodities, the trend is not well represented by a single downward slope, but instead by a shifting trend that often changes sign over the sample period

(continued)

Table 1 (continued)

	Methodology	Results
Ghoshray (2011)	Unit-root test of Lee and Strazicich (2003) who developed a test which incorporates structural change in the null hypothesis. On individual commodity price series	11 out of 24 commodity prices are found to be DS. The remaining 13 prices are found to exhibit TS behavior with either one or two structural breaks. Most of the commodities that do not exhibit DS behavior seem to contain no significant trends
Yamada and Yoon (2014)	Estimation of piecewise linear trends on individual commodity price series. Number of breaks is not fixed a priori.	Only one commodity has a negatively sloped trend function throughout the whole sample period. Most commodities exhibit negatively sloped trends during some of the sample periods (PSH only holds locally). Strength of PSH is weaker during the last decade
Arezki et al. (2014)	Use of the extended dataset of Harvey et al. (2010) starting in 1650. Panel data stationarity tests that allow for endogenous multiple structural breaks. Piecewise linear regressions on 25 commodity series.	4 commodities have 1 break, 13 have 2 breaks, 7 register 3 breaks, and 1 (gold) has 4 breaks. Out of the total of 80 slope estimates, 41 are negative and significant, 11 are negative but insignificant, 21 are positive and significant finally, and 7 are positive and insignificant
Di Iorio and Fachin (2018)	Panel cointegration bootstrap test on 3 commodity sub-indices. Restricted sample from 1955 onwards.	PSH rejected over the 1950–1980 period. Rejection not confirmed over the entire 1950–2011 sample for agricultural commodities

Note TS and DS stand for trend/difference stationary, respectively

2.2 How to Measure the Terms of Trade

2.2.1 Proxies

The empirical papers have progressively changed the focus variable of the PSH, going from the terms of trade, which is a country-specific measure, to a ratio of commodity to manufacture prices, that is, no longer a country-specific measure.

Different variants of this ratio of commodity to manufacture prices are found in the literature. Some studies focus on commodity indices, some on sub-indices (metals, energy, agriculturals), and others on specific commodities (oil, copper). The research question should drive the choice of the right measure.

The single-commodity approach is relevant for considering specific countries highly dependent on one commodity, as is the case of Chile for copper or of Saudi Arabia for oil. This approach is sometimes required by the econometric strategy, when we test the PSH with panel approach for example, in which case the single commodities are used as panel units for example.

Alternatively, many papers rely on commodity price indices, as set out in Eq. (1).

$$\mathrm{COMP}_t = \sum_{h=1}^{H} w_h P_{ht} \tag{1}$$

where COMP_t is the commodity price index, where P_{ht} is the logarithm of the price of commodity h, where H is the number of commodities considered in the index ($H = 24$ in the Grilli–Yang non-fuel commodity price index) and where w_h is the time-invariant weight associated to commodity h. The weights can be based on average shares of each commodity in world exports as in Grilli–Yang index or be equal to $1/H$ if we take equal weights.

Aggregation provides benefits and pitfalls. It has the benefit of offering a summary view of commodities in general, cleaning from commodity-specific noise but also removing their large heterogeneity. However, if the PSH is validated on an aggregate measure of commodity prices but not on single commodities, can we really neglect this inconsistency, given that many countries depend on one single commodity (typically crude oil)? Moreover, aggregation issues might arise. As an illustration, Cuddington (1992) tested the PSH separately on 24 commodities and found no break in any country in 1921, while a break was found for the aggregate commodity index (built as the trade-weighted average of the 24 series). Keeping in mind the context of the PSH, we would recommend to not discard too fast individual commodity measures, as there is certainly no developing country exporting a commodity basket similar to the aggregated ones.

2.2.2 Deflators

The PSH refers to a relative price where the denominator is expected to capture the evolution of import prices in developing countries. Three proxies are usually considered: the manufacturing unit values index, the U.S. manufacturing price index, and the consumer price indices.

The manufacturing unit values, or MUV, is a trade-weighted index of developed countries' (24 countries in the current version) exports of manufactured goods to developing countries. It is built from UNCTAD's Handbook of Statistics database and from IMF's World Economic Outlook database (Spatafora and Tytell 2009). It covers the period after 1900 with some gaps for 1914–1920 and 1939–1947 that have been filled by interpolation by Grilli and Yang (1988) and updated notably by

Pfaffenzeller et al. (2007). This proxy is by far the most frequently used (see for example, Aslam et al. 2016).

The US manufacturing price index was derived by Grilli and Yang (1988) as an index of domestic prices of manufactured products in the U.S. netting out energy, timber, and metal prices from the U.S. wholesale price index of industrial commodities. The key weakness of this indicator is that it only considers the mix of manufactured goods exports of one sole industrial country and also relies on the, strong, assumption of the law of one price whereby manufactured products prices in the US reflect the international prices.

The consumer price indices from major economies is a widely available alternative deflator. However, this proxy includes non-tradables, which unduly distort the terms of trade flavor that we want to capture.

As an alternative, commodity terms of trade could be measured against a price index of service sector outputs rather than manufacture, given the growing economic importance of the service sector. This suggestion certainly gains relevance as the weight of services reaches 23% of total world trade, growing by more than 7% in 2018, with the US and EU accounting together for 44% of world service trade.

2.2.3 Datasets

The most widely–used dataset is the one developed by Grilli and Yang (1988). Their indices are mainly based on World Bank commodity price data. Grilli and Yang considered 24 non-fuel individual commodities: aluminum, bananas, beef, cocoa, coffee, copper, cotton, hides, jute, lamb, lead, maize, palm oil, rice, rubber, silver, sugar, tea, timber, tin, tobacco, wheat, wool, and zinc. They construct an all commodities price index and three sub–indices for agricultural food commodities, non–food agricultural commodities, and metals. The indices are base period trade–weighted arithmetic averages of the commodity prices concerned. They also provide an alternative geometric average version of the same (sub-) indices. Grilli and Yang's original series ran from 1900 to 1986 but has been updated by a number of researchers (Pfaffenzeller et al. 2007).

For studying long–run price trends, the longer the series, the better. *The Economist* commodity price index goes back further to 1845, as used by Cashin and McDermott (2002). Harvey et al. (2010) assemble some data series back as far as 1650. Ensuring consistency and continuity over such a long period remains of course inexorably subject to cautious interpretations. Given the long-term perspective of the PSH, most researchers deal with annual frequency. Relying on monthly commodity price series of the World Bank, available as from 1960, would not bring additional relevant information.

2.3 *Review of Econometric Strategies*

We review the PSH empirical papers by presenting successively the trend-stationary and difference-stationary models. We then detail the different approaches designed to addressing the potential structural breaks and finally present the extensions based on panel methodologies. This chronological review is summarized in Table 1.

2.3.1 Deterministic Trends

First, let us assume that the relative commodity price series is generated by a trend-stationary (TS) data process as follows

$$\mathrm{COMP_t} = \alpha + \beta t + \varepsilon_t \tag{2}$$

where COMP_t is the logarithm of the commodity price indice, where t is an annual deterministic trend, where ε_t is a stationary process with mean equal to zero, an ARMA for example, and where the sign and significance of β lead to conclusions on the PSH. Most studies based on this methodology (Sapsford 1985; Grilli and Yang 1988) found support to the PSH, in other words β was found to be significantly negative.

Of course, these conclusions are subject to the validity of the stationarity assumption. Non-stationarity of the error terms could lead to spurious rejection of the null $\beta = 0$ and to spurious support of the PSH. Cuddington and Urzua (1989) were the first to carry out unit-root tests on the Grilli–Yang commodity price dataset. Similarly, Kim et al. (2003) showed that the 24 commodity price series contained in the standard Grilli–Yang commodity price index are characterized by unit-root behaviors (18 commodities) or quasi-unit roots (6 commodities). Similar results were reported by Cuddington (1992) and Newbold et al. (2005).

2.3.2 Stochastic Trends

Consequently, we can assume that the relative commodity price series are generated by a difference-stationary (DS) process as follows

$$\Delta \mathrm{COMP}_t = \beta + u_t \tag{3}$$

where ΔP_t is the differenced logarithm of the commodity price index, where u_t is a stationary process, an ARMA for example, with mean equal to zero, and where the sign and significance of β leads to conclusions on the PSH. Kim et al. (2003) accounted for non-stationarity and find much less support to the PSH. Indeed, using the same 24 commodity prices of the Grilli–Yang database, they observe that the null hypothesis of $\beta = 0$ is much less frequently rejected with a non-stationary process specification than in stationary models.

The finding that most commodity price series largely behave like random walks is not anodine. A shock to the price of, say, copper today would thus be permanent. Copper price would no longer revert to any stable, long-run values/trends. As a consequence, stabilization mechanism as the one implemented by Chile, whereby asset accumulation is conditioned on copper prices being above a long-term level, would theoretically no longer be sustainable as it relies on the concept of a stable level/trend.

Detection of unit roots remains subject to some caution. It might be spuriously derived from a bad specification of the data-generating process or due to the well-known lack of power of standard non-stationary tests (Schwert 1987). We now consider extensions related to these two possibilities and see that the conclusions supporting the PSH lose their strength.

2.3.3 Breaks

Structural breaks contribute to hiding the mean reversion behavior of a series. They can lead the econometrician to erroneously conclude to non-stationarity. Modelling breaks explicitly is therefore a key part of the empirical strategy. Let us now consider a general specification allowing for breaks before discussing the different choices open to the econometrician:

$$\Delta \mathrm{COMP}_t = \beta + \sum_{k=1}^{K} \delta_k D_{kt} + \sum_{k=1}^{K} \gamma_k DU_{kt} + u_t \tag{4}$$

where $\Delta\mathrm{COMP}_t$ is the differenced logarithm of the commodity price index, where K is the number of breaks, where D_{kt} is an indicator variable designed to capture breaks in the level and taking the value 1 if t is equal to the break date τ_k, the value 0 otherwise, where DU_{kt} is an indicator variable designed to capture breaks in the trend and taking the value 1 if t is equal to, or larger than, the break date τ_k, the value 0 otherwise, and where u_t is a stationary process with mean equal to zero, and where the sign and significance of β again leads to conclusions on the PSH.

We must first decide whether break dates are set exogenously or endogenously. The determination of break dates can rely on graphical or other external bases. As an illustration, Grilli and Yang (1988) tested the robustness of their results by imposing three breaks in 1921, 1932, and 1945, based on visual inspection. Cuddington and Urzua (1989) found breaks in 1921 and (“*to a lesser extent*”) in 1974. Alternatively, we can let the data speak and determine endogenously the break dates (or confirm the ones visually detected).

Leon and Sotto (1997) relied on the Zivot and Andrews (1992) tests by allowing the position of the break to be endogenously determined at the point at which the null hypothesis of a unit root is more easily rejected against the competing alternative

TS representation with shifts in either the level or the trend of the series. They showed that the unit roots found for 12 commodity price series by Cuddington (1992) were potentially related to misspecification problems, as they rejected the null of non-stationarity for 8 of their 12 series.

While the Zivot and Andrews (1992) test allows for one break ($K = 1$), Lumsdaine and Papell (1997) extended their methodology from one to two endogenously chosen break dates ($K = 2$). Allowing for up to two breaks, Kellard and Wohar (2006) found that 14 commodity price series, out of 24, were characterized by trend-stationarity.

These tests are questionable. There is a potential size distortion in the Zivot–Andrews and Lumsdaine–Papell approaches that can lead to spurious rejection of the null hypothesis of a unit root when the actual time series process contains a unit root with a structural break. Indeed, these approaches only allow for structural breaks under the alternative hypothesis and assume a liner non-stationary series under the null. Lee and Strazicich (2003) therefore proposed a minimum two-break LM test allowing for structural breaks under the null hypothesis. This test does not suffer from the size distortion discussed above. Using this methodology, Ghoshray (2011) found that 11 of the commodity price series were difference stationary, while the 13 other series were stationary with either one or two structural breaks but with no significant trend, thereby further weakening the support to the PSH.

Finally, and more critically, we should not lose sight of the underlying research question on the long-run trend of relative commodity prices. Identifying breaks is relevant and helps in building well-specified models, but neutralizing them, say sudden collapses, leads to underestimating the long-term trend.

2.3.4 Panels

Low power of non-stationary tests can also be addressed by considering a panel approach. Testing jointly the non-stationary of a set of commodity prices increases the power of the test relatively to testing individually each time series. This is the approach followed recently by Arezki et al. (2014), who considered panel stationary tests allowing for multiple endogenous breaks, with mixed support to the PSH. Di Iorio and Fachin (2018) similarly relied on panel cointegration methods but restricted their analysis to the "*post-colonial era*" and did not find support to the PSH, except for agricultural goods over the 1950–2011 period.

These approaches have the merit to partially account for heterogeneity of the commodity price series (via fixed effects) and in this sense dominate analyses based on simple commodity price aggregation (such as Grilli–Yang indices).

2.3.5 Other Extensions

By considering breaks, these empirical studies identify different periods (or regimes) that might be characterized by different dynamics. This can justify finer estimation methodologies such as piecewise linear regression approaches (Arezki et al. 2014) or quantile regressions incorporating unknown numbers and forms of breaks through a Fourier function (Bahmani et al. 2018).

Kellard and Wohar (2006) note that "*for the majority of commodities, the trend is not well represented by a single downward slope, but instead by a shifting trend that often changes sign over the sample period.*" They propose a measure of the prevalence of a negative trend (deterioration exists for more than 70% of the sample period) and find that eight commodity price series have a prevalent negative trend, a quite small support to the PSH. Instead of isolating a dominant trend, Yamada and Yoon (2014) estimate the piecewise linear trends of the relative primary commodity prices series and note that the PSH has become substantially weaker over the last decade.

2.4 Discussion

First, the abundance of studies on the PSH, that we summarize chronologically in Table 1, has probably brought more confusion than clarity. The interpretation of breaks, incorporated in most studies, is rarely straightforward. Is a sharp downward break just noise to be neutralized? or a shift revealing a secular declining trend? The econometric approach should inevitably be complemented by well-justified assumptions on which breaks are exogenous (due for example to wars or climatic events) and which ones are sharp adjustment contributing to the long-run trend (as the 1973 OPEC embargo for example). In principle, only the former should be neutralized. Few papers discuss the nature of the breaks (Di Iorio and Fachin 2018).

Second, the comparison benchmark of manufacture price is far from perfect, as: (a) it imperfectly proxies the manufacturing import of commodity-exporting countries; (b) it does not correct for manufacturing quality, which likely biases downward the estimates of the trend coefficient β; and (c) it does not include the price of the services that are a growing share of the global imports.

Finally, super cycles of commodity prices have also found some support in a parallel literature on commodity price booms and busts. Empirical studies investigating the PSH might take on board explicitly these developments.

Coming back to the PSH, and according to the review Table 1, we find that the more sophisticated the methodologies, the weaker the support to the PSH.

3 Commodity Currencies

3.1 Introduction

The real exchange rate of some commodity-exporting countries is driven by the fluctuating relative prices of the commodities they export, in other words their terms of trade. We call commodity currencies the currencies of such countries. Typically, it refers to currencies of Australia, Norway, Chile, or Ivory Coast. Before discussing empirical evidences on such dependence between real exchange rate and commodity prices, we first present the relevant definitions of the real exchange rate, then discuss the measurement issues of both commodity prices and real exchange rates and finally present the economic strategies designed for measuring the role of the terms of trade on the real exchange rate. We see that the arbitrary choices on measurement and methodologies have substantial consequences on the analysis.

3.2 RER Models

A common definition of the real exchange rate[1] is the nominal exchange rate adjusted by price levels:

$$\mathrm{RER}_t \equiv s_t - p_t^* + p_t \tag{5}$$

where RER is the real exchange rate, where s is the log exchange rate defined in units of foreign currency per unit of home, that is, s rising is an appreciation of the home currency, and p and p^* are log price levels, an asterisk denoting foreign or international prices. Now suppose that the price index is a geometric average of traded and nontraded good prices, then:

$$p_t = \alpha\, p_t^N + (1-\alpha) p_t^T$$
$$p_t^* = \alpha^* p_t^{N^*} + (1-\alpha^*) p_t^{T^*}$$

Substituting the price indices decomposition into the real exchange rate formula and re-arranging yields:

$$\mathrm{RER}_t \equiv \left(s_t - p_t^{T^*} + p_t^T\right) + \alpha\left(p_t^N - p_t^T\right) - \alpha^*\left(p_t^{N^*} - p_t^{T^*}\right)$$

This decomposition indicates that the real exchange rate can be expressed as the sum of three components: (i) the relative price of tradables, (ii) the relative price of nontradables in terms of tradables in the home country, and (iii) the corresponding relative price in the foreign country.

For the simplifying case where the weights of non-tradables in the aggregate price indices are identical, the second and third terms can be collapsed into an intercountry relative price of non-tradables:

$$\mathrm{RER}_t \equiv \left(s_t - p_t^{T^*} + p_t^T\right) + \alpha\left(\hat{p}_t^N - \hat{p}_t^T\right) \tag{6}$$

where the circumflex denotes the intercountry log difference.

[1] This section borrows from the primer on real effective exchange rates of Chinn (2006).

Firstly, if one assumes the law of one price holds for all goods, and consumption baskets are identical, then both terms on the right-hand side of Eq. (6) are zero, and PPP holds (since there are no non-tradables by definition). That is, the real exchange rate is a constant, far from empirical evidence.

Secondly, if instead PPP holds only for tradable goods, then the first term is zero, and the relative tradables–non-tradables price is the determining factor in the value of the real exchange rate. Fundamentals or mechanisms affecting the relative tradables–non-tradables price include productivity differentials (Balassa 1964; Samuelson 1964), differentials in production factor endowments (Baghwati 1984; Bodart and Carpantier 2016), government consumption (Ostry (1994)), fertility (Rose, Supaat and Braude 2009), etc.

Thirdly, another possibility is that all goods are tradable, but not perfectly substitutable; then one has an imperfect substitutes model. Both terms on the right-hand side of the above equation can take nonzero values. The imperfect substitutability means that the dynamics of export prices and import prices can be different. This imperfect substitutability can have different sources, such as imperfect mobility of capital or differences in natural resource endowments [specialization assumption in the models developed in Choudhri and Khan (2005)]. Papers relying on, or extending, these specific models relating terms of trade to real exchange rates include De Gregorio and Wolf (1994), Chen and Rogoff (2003), Cashin et al. (2004), Choudhri and Khan (2005), Ricci et al. (2013), as well as Bodart et al. (2012, 2015).

Commodity currencies models are derived from this imperfect substitutability definition of the real exchange rate.

3.3 How to Measure the Real Exchange Rate

3.3.1 Bilateral Versus Effective

Similar to nominal exchange rates, real exchange rates can be computed on a bilateral or on an effective basis.

In their seminal contribution, Chen and Rogoff (2003) studied commodity currencies with exchange rates expressed in USD. One obvious limit of bilateral exchange rates is that it does not isolate from factors that are specific to the reference currency area, namely the dollar zone. Checking the robustness of the results on alternative currencies is a necessity when working with bilateral exchange rates. Chen and Rogoff therefore compared their results with those obtained with exchange rates expressed in GBP.

Most studies on commodity currencies (Cashin et al. 2004; Bodart et al. 2012, 2015; Coudert et al. 2011) rely on the effective version of the exchange rate, defined as trade-weighted multilateral real exchange rate, where the weights are specific to each country trade network, as set out in Eq. (7)

$$\text{RER}_{it} = \sum_{j=1}^{J} w_{ij}\text{RER}_{ijt} \tag{7}$$

where RER_{it} is the real effective exchange rate of country i, RER_{ijt} is the real bilateral exchange of country i with country j, where w_{ij} is the weight associated to RER_{ijt}, and where J is the number of countries considered in the real effective exchange rate formula of country i. Such series are available from IMF-IFS database, or alternatively from Darvas (2012).

The strength of effective rates is that real exchange rates are measured in terms of a basket of currencies, thereby diluting the fluctuations due to country j shocks. The weakness is that the basket is country specific, that is, w_{ij} depends on i. Using country-specific trade weights is mainly justified for studies focusing on competitiveness. An alternative is to use a fixed basket of currencies (in the vein of special drawing rights or of the Libra), set identically for all investigated countries, that is, $w_{ij} = w_j$ in Eq. (7). Chen and Rogoff (2003) took this option by replicating their analysis on Canada, Australia, and New Zealand by looking at the exchange rate with the so-called broad index, a composite of over 30 non-US-dollar currencies. Surprisingly, this interesting practice has not been much followed.

3.3.2 Real Versus Nominal

As discussed by Chinn (2006), we often face a trade-off between using the most appropriate real exchange rate measure conceptually, and the most readily available data.

In practice, one only has a choice of a few price deflators. At the monthly frequency, they include the consumer price index (CPI), producer price index (PPI), wholesale price index (WPI), and export price index. At lower frequencies, such as quarterly, the set of deflators increases somewhat, to include the GDP deflator, unit labor costs, and price indices for the components of GDP, such as the personal consumption expenditure (PCE) deflator.

Typically, the CPI is thought of as weighting fairly heavily non-traded goods such as consumer services. Similarly, the GDP deflator and the CPI weigh expenditures on non-tradables in proportion to their importance in the aggregate economy. In contrast, the PPI and WPI exclude retail sales services that are likely to be non-traded.

Due to availability constraints for long periods and the need of a large enough set of developing countries, most studies use CPI-real exchange rates (Chen and Rogoff 2003; Cashin et al. 2004; Bodart et al. 2012, 2015), as provided by the IMF-IFS database for a wide set of countries and years.

Clearly, for purposes of calculating the relative price of goods and services that are tradable, the preferred measure would have been the exchange rate deflated by PPIs or WPIs were the data available. It is worth noting that a recent empirical paper of Ahn, Mano, and Zhou (2017),compared CPI, GDP, and ULC deflators in the context of the expenditure switching mechanism studies. It supports Chinn (2006) statement that the choice of the deflator may have considerable effects on the empirical conclusions.

3.4 How to Measure Commodity Dependence

3.4.1 Commodity Dependence Definitions

Commodity dependence is typically measured by the share of commodity-export earnings in total exports (IMF), in total merchandise exports (UNCTAD), and in GDP. Alternatively, commodity dependence can be measured by the percentage of people engaged in the production of commodities or by the share of government revenues due to commodity production and exports.

Part of the commodity currency literature picks some specific countries, without requiring any criterion, thereby presuming their commodity dependence, such as Russia, Saudi Arabia, and Norway in Habib and Kalamova (2007), or Australia, Canada, New Zealand, and South Africa in Chen and Rogoff (2003), Algeria in Koranchelian (2005) or Peru in Tashu (2015).

Another part of the literature casts the net wider by considering more systematically groups of countries specialized in commodities (Cashin et al. 2004; Bodart et al. 2015), or sub-groups such as energy (Coudert et al. 2011; Dauvin 2014). The main criterion used in this category of studies is the one originally set by the IMF where a country is classified as a commodity exporter when its primary commodity exports (categories SITC4 0, 1, 2, 3, 4, and 68 of the Standard International Trade Classification) account for at least 50% of the value of total exports of goods and services on average over a given time window. The list of countries accordingly established in Cashin et al. (2004) has been used in many subsequent studies.

A new IMF definition from Aslam et al. (2016) sets a dual criterion that commodity exporters are emerging market and developing economies for which gross exports of commodities constitute at least 35% of total exports and net exports of commodities constitute at least 5% of exports-plus-imports on average. This new criterion of net export is certainly relevant as it helps in excluding the commodity-exporting countries that are large commodity importers. We expect indeed the export commodity price fluctuations to be offset by the import ones when the net export of commodities is not positive and large enough.

3.4.2 Commodity Basket

The literature is divided on the way to build the relevant commodity price series. The least relevant approach in our context would be to relate the real exchange rates to a world commodity price index. Such approach would be appropriate if countries were dependent on the same commodities (which is not the case) or if the commodity prices were highly correlated (which is not the case). The commodity currency literature then only considers country-specific commodity price indices based on the country-specific commodity exports.

The first approach, followed by Chen and Rogoff (2003), Cashin et al. (2004) and Coudert et al. (2011), considers a country-specific commodity price index, CToT_{it}, constructed as follows

$$\text{CToT}_{it} = \sum_{h=1}^{H} w_{ih_i} P_{h_i t} \tag{8}$$

where $P_{h_i t}$ is the logarithm of the price of the commodity that ranks at the h-th position in commodity exports of country i, where w_{ih_i} is the weight of that commodity in commodity exports (Cashin et al. 2004) or home production (Chen and Rogoff 2003) of country i, normalized such that the weights sum to one, where H is the number of most exported commodities considered for the formula [$H = 3$ in Cashin et al. (2004) and $H = 5$ in Coudert et al. (2011)].

The second approach, that is used in Habib and Kalamova (2007) and Bodart et al. (2012, 2015), assesses whether the price of the top commodity exported by a country ($P_{1_i t}$) is an economically and statistically significant determinant of the long-run variations in their real exchange rate, therefore focusing on a single price rather than a constructed price index.

The relevance of these two approaches depends on the context. On the one hand, weighted averages provide finer measures of the commodity dependence, as they incorporate a richer set of information. On the other hand, the top commodity approach closely reflects the policy focus on a single commodity that we have in many oil exporting countries, in Chile for copper, or that we had in Columbia for coffee (Edwards 1986) or in South Africa for gold (MacDonald and Ricci 2005).

3.4.3 Deflators

In order to capture properly the relationship between real exchange rates and commodity prices, the latter have to be expressed in real terms (similarly to terms of trade). Following the practice of the PSH literature, commodity prices, or indices, are deflated by the IMF's index (of the unit value) of manufactured exports (MUV) expressed in US dollars. The use of the MUV index as a deflator is common in the commodity price literature and considered as a proxy of the price of developing country imports of manufactures (see for example Deaton and Miller 1996; Cashin et al. 2004).

3.5 Review of Econometric Strategies

3.5.1 Introduction

The choice of the empirical strategy best suited for assessing the elasticity of the real exchange rate to the commodity price index is generally preceded by a careful analysis of the stationarity of the series. Indeed, conclusions on the stationarity of the real exchange rate and of commodity prices are rarely clear-cut. Often, the papers devote a paragraph on diverging outcomes from alternative non-stationarity tests (Chen and Rogoff 2003; Ricci et al. 2013; Habib and Kalamova 2007; Bodart et al. 2012). Ultimately, most papers rely dominantly, or exclusively, on methodologies designed for non-stationary series. Those papers integrating an analysis designed for stationary series mainly rely on simple OLS regressions, with deterministic trend (Chen and Rogoff 2003) or with lags of commodity price index (Habib and Kalamova 2007). We now present the non-stationarity approaches that dominate, by far, the commodity currency empirical literature.

3.5.2 Cointegration Models

Most empirical papers on commodity currencies mainly rely on cointegration methods, which are characterized by three standard steps: first, documenting the non-stationarity of the series via standard unit-root tests; second, testing the cointegration of the real exchange rate with the commodity price index via Engle and Granger cointegration tests (Cashin et al. 2004), or Johansen cointegration tests (Habib and Kalamova 2007); and finally, estimating the cointegration coefficient through dynamic OLS (DOLS) or fully modified OLS (FMOLS).

Country-specific cointegration analyses rely on the following equation:

$$\mathrm{RER}_t = \alpha + \beta_0 \mathrm{CToT}_t + \sum_{l=1}^{L} \beta_l X_{lt} + \varepsilon_t \tag{9}$$

where RER_t is the logarithm of the real exchange rate, where CToT_t is the logarithm of the real price of commodity exports, that is, the country-specific commodity terms of trade, where the $X_{lt}s$ are a set of control variables, such as net foreign assets, productivity differentials, government consumption, and trade restriction index (see the review of Ricci et al. 2013), where ε_t is a stationary process with mean equal to zero and where RER_t and CToT_t are both I(1). The significance of β, the coefficient of interest, indicates whether the country is characterized by a commodity currency profile or not.

Such analysis has been carried out systematically by Cashin et al. (2004) on a set of 58 countries based on country-specific regressions. They found evidence in support of the long-run comovement of real exchange rate and real commodity-export price series for about one-third of the commodity-exporting countries. The median value of the elasticity β_0 is 0.42, indicating that a 10% rise in real commodity prices is typically associated with a 4.2% appreciation of the real exchange rate.

These results on non-stationarity, cointegration, and on elasticity estimation are subject to limitations due to low power of the tests, to omission of breaks, or to endogeneity issues. We now discuss the different solutions proposed in the literature to tackle these limits.

3.5.3 Panels

The panel approach expands the pool of observations and helps in reaching better power for the tests. It is also a convenient framework when we aim at interacting elasticity with country-specific variables such as currency regime or financial openness in a single-step estimation (Bodart et al. 2015).

This is not the aim of this contribution to review panel cointegration methodologies but to highlight commodity currency specificities. A key one refers to the degree of cross-sectional dependence of both real exchange rates and commodity prices. Indeed, we can suspect real exchange rates to be cross-sectionally dependent by construction. For example, we might expect from close trading partners, such as Canada and the USA, to have effective real exchange rates $RER_{USA,t}$ and $RER_{CAN,t}$ negatively correlated. Similarly, we can expect commodity prices, say $CToT_{gas,t}$ and $CToT_{oil,t}$, to be positively correlated. Cross-sectional dependence tests, such as the one of Pesaran (2004), confirm these presumptions (Bodart et al. 2012). Since original panel cointegration tools were assuming cross-sectional independence ($RER_{i,t} \perp RER_{j,t}$ and $CToT_{i,t} \perp CToT_{j,t}\ \forall i, j$), these were, for commodity currency empirical analyses, not supported by the data. A second generation of panel cointegration methods was then developed (see Breitung and Pesaran 2005; Hurlin and Mignon 2007; Hadri et al. 2015 for reviews), robust to cross-sectional dependence.

As discussed in the PSH section, commodity prices are subject to breaks over long periods. A third generation of non-stationary panel tests has emerged, which allows for breaks (Westerlund and Edgerton 2008; Banerjee and Carrion-i-Silvestre 2015). The same caveat applies here as to the interpretation and treatment of breaks. In addition, the difficulties expand in a bivariate and panel context. First, we might have two non-synchronous breaks, one affecting the real exchange rate, the other the commodity price index. Whether such breaks should be neutralized or not, depends on whether we allow for potential delayed response of the real exchange rate to large commodity price shocks. Second, whether breaks should be country specific or global is another arbitrary choice that affects the tests and elasticity estimates. Needless to say, that choices made by the econometrician need a careful justification.

3.5.4 Endogeneity Concerns

The point of working with the commodity component of the terms of trade rather than with the standard terms of trade measure, the point of taking average export weights, instead of time-varying weight, in the construction of the country-specific commodity price indices, the preference for selecting small economies and finally the choice of the econometric strategies all result to some extent from the need to address endogeneity concerns. We now present the four sources of endogeneity that must be addressed in these analyses.

Commodity Terms of Trade

The identification of terms of trade shocks in explaining real exchange rate fluctuations is not an easy task, as terms of trade are generally not exogenous to the domestic economy. However, there is a component of the terms of trade that is largely considered to be exogenous to small economics. Indeed, commodity prices are set on international markets and commodity terms of trade can reasonably be considered as exogenous, that is $\mathrm{CToT}_t \perp \varepsilon_t$. In their review of natural experiments in macroeconomics Fuchs–Schuendeln and Hassan (2016), note that commodity price variations are viewed as quasi-natural experiments for small economies. This assumption of commodity prices' exogeneity explains the development of the commodity currency literature and the construction of commodity terms of trade databases as the one of Gruss and Kebhaj (2019).

Terms of Trade Endogeneity

Although commodity prices can be viewed as exogenous, this is generally not the case of country-specific commodity terms of trade. Indeed, commodity terms of trade reflect fluctuations in commodity prices but also variations in the mix/weight of commodity exports. This mix is partially endogenous and can reflect new export and development policies, which are determined at the domestic level.

As a consequence, and to ensure that endogenous supply responses to price changes do not affect the analysis, it is a common practice to build the commodity price indices by taking commodity fixed weights, computed as average commodity weights in total commodity weights over a few years.

What is a "*few years*" is not consensual (three years, the full sample, mid-window average, etc.), but this choice has potentially substantial consequences. Indeed, the mix of commodity exports is not stable over time, and this variability can sometimes be extreme. For example, the share of aluminum in Mozambique exports was below 10% before 2001 and more than 50 in 2001. By taking the average weight of aluminum over say 20 years, we would largely overestimate the weight of aluminum in commodity exports before 2001 and largely underestimate it as from 2001. An alternative solution to the weight endogeneity problem consists in selecting the first main exported commodity (say oil in Saudi Arabia and copper in Chile), but this approach remains subject to large changes as the one discussed above for Mozambique and aluminum.

A visualization of the commodity-export basket over time helps in detecting such jumps. Gruss and Kebhaj (2019) propose to complement terms of trade measures based on fixed weights, with an alternative measure where weights are time-varying and based on three years rolling averages.

Market Power

One underlying assumption of the commodity currency literature is that fluctuations of the commodity price indices are exogenously affecting the real exchange rates, in other words, CToT_t is independent of ε_t. This is reasonable as long countries be small enough, compared to the globalized commodity markets. Gruss and Kebhaj (2019) showed that "*the world export market share of individual countries is larger than 40% in only a few food commodities: palm oil (Malaysia), soybeans (US), corn (US), olive oil (Spain), and soybeans oil (Argentina).*" For minerals, the world market shares of Chile, Niger, and Australia for copper, uranium, and coal, exceed 20%, 30%, and 20%, respectively.

To test the market power of these countries, Gruss and Kebhaj (2019) tested the hypothesis that GDP Granger causes the commodity terms of trade for different group of countries over 1970–2014. They do not reject the null of Granger non-causality and confirm that commodity terms of trade can be taken as exogenous from the perspective of individual countries. This might come as a surprise as the market shares of 20% and more, mentioned above, are quite substantial but two caveat apply here. First, the market shares focus on total exports and not on total production. Second, substitution across similar commodity products mitigate the market power that large commodity exporters have, "*even within the specific markets that they appear to dominate.*" (Chen and Rogoff 2003).

Still, some (few) studies challenge this view. First, Chen and Lee (2018) studied the impact of market shares on the strength of commodity currencies and found "*that as a country's market power increases, RER reacts less to a given COMP change.*" Second, Clements and Fry (2008) considered the case where a group of commodity-exporting countries have combined market power and found that "*spillovers from commodities to currencies contributed less than 1% to the volatility of the currency returns, while spillovers from currencies to commodities generally contributed between 2 and 5.2% to the commodities.*" They concluded that this spillover reflected the endogeneity of commodity prices induced by market power. Neutralizing countries with a potential market power should be included in robustness analyses of commodity currency papers.

Omitted Variable

Another source of endogeneity may arise from omitted variable. Some macroeconomic variables affect similarly the real exchange rate and the commodity prices. For example, interest rates influence (negatively) the commodity prices and (negatively) the (real) exchange rate of developing countries (see MacDonald and Nagayasu 2000). Similarly, business cycles are also common determinant. Some studies take these possibilities on board by accounting for unobserved global factors (Bodart et al. 2012) or discuss these potential biases (Chen and Rogoff 2003). According to Bai et al. (2009)'s approach, these global factors can be modelled by imposing a factor structure on $e_{i,t}$

$$\mathrm{RER}_{i,t} = \alpha_i + \beta\,\mathrm{CToT}_{i,t} + e_{i,t} \tag{10}$$

$$e_{i,t} = \sum_{r=1}^{R} \lambda_{i,r} f_{r,t} + \varepsilon_{i,t} \tag{11}$$

where $f_{r,t}$ is the, potentially non-stationary, r-th latent common factor, $\lambda_{i,r}$ is the loading on factor $f_{r,t}$, and $\varepsilon_{i,t}$ is the idiosyncratic error. Based on this approach, Bodart et al. (2012) showed that the estimations of the elasticity of the real exchange rate to commodity price fluctuations are smaller than those reported in the previous empirical literature.

3.6 Discussion

The commodity currency literature relies on the assumption that commodity prices are exogenous. Price variations are then viewed as quasi-natural experiments (Fuchs–Schuendeln and Hassan 2016). This assumption should not be taken as granted, as omitted variables and market power reverse causality are legitimate concerns that have been documented. These endogeneity concerns should be carefully discussed and addressed, when needed, by appropriate econometric responses.

Further relevance of using terms of trade measures based on fixed weights should be assessed country by checking the stability of the commodity-export weights.

We finally note that some recent studies showed that commodity currency exchange rates have surprisingly robust power in predicting global commodity prices, both in-sample and out-of-sample (Chen et al. 2010, but also Groen and Pessenti 2010; Ferraro et al. 2015). However, these findings on reverse causality differ from the commodity currency literature from two standpoints. First, they focus on the short-term dependence and not on the long-term one. Second, these studies rely on nominal exchange rates and not on real exchange rates.

4 Financialization of Commodity Markets

4.1 Introduction

Speculation in commodity markets is not a new phenomenon. The tulip mania in 1636–1637 gave rise to a dramatic rise and collapse of bulb prices that is comparable to the recent booms in oil (2008) or rare earths prices (2010). The recent commodity booms, however, are related to a specific development, referred to as the financialization of commodity markets, namely the large inflows of financial investors who have no commercial interest in the underlying commodities. The diversification power of commodity investments and their equity-like returns, as documented in Erb and Harvey (2006) and Gorton and Rouwenhorst (2006), have attracted a lot of investors, such as hedge funds and investment funds. According to BIS data, the notional value of outstanding OTC commodity derivatives has risen from USD 0.6 trillion in 2000 to USD 14.1 trillion, its peak, in 2008. According to the testimony of Masters (2008), a portfolio manager in a hedge fund active on the oil market, a new class of agents has come to the market and distorted the price discovery process by following a passive index investment.

Many papers addressed the question of assessing and measuring the genuine impact of the financialization. Recent surveys include those of Irwin and Sanders (2011), Cheng and Xiong (2014), Fattouh et al. (2013), Henderson et al. (2015), Bhardwaj et al. (2015), and Main et al. (2018). According to the typology of Haase et al. (2016), the financialization may have an impact on six categories of variables: (1) returns, (2) risk premia, (3) spreads (price differentials between long- and short-dated futures' contracts), (4) volatility, (5) spillover, and (6) spot or futures price levels.

As this is the thema of this paper, we focus here on this latter category, that is, spot or futures price levels. This focus variable is probably the key one in the debate on the financialization of commodity markets. Knowing if the financialization contributed to the rise of oil and rice prices in 2008 matters and potentially calls for policy responses.

We consider below different empirical approaches all designed to capture the effect of financialization on commodity prices. We first discuss the direct measures where we explicitly measure the positions held by financial investors. We then discuss indirect measures based on the detection of breaks around 2004. All these measures are unfortunately subject to endogeneity issues that we highlight in the next subsection, together with some identification strategies. We finally present the recent evolutions, which suggest that the presumed effects of financialization decrease.

4.2 How to Measure Financialization

4.2.1 Direct Measures

Few datasets are available to study the evolution of the share of financial commodity investors in the markets. The semi-annual BIS dataset on OTC derivative commodity contracts does not provide a breakdown by types of investors.

The US Commodity Futures Trading Commission (CFTC) on the contrary provides on a weekly basis reports that contribute to identifying different categories of investors. The Disaggregated Commitments of Traders (COT) covers 22 major physical commodity markets and reports the open interest positions by separating traders into the following four categories of traders: producer/merchant/processor/user; swap dealers; managed money; and other reportables. There are mainly two limits to these data. First, the breakdown does not inform on whether positions are taken on a speculation or hedging basis (Fattouh et al. 2013). Second, the rising role of institutional investors is not visible directly as such investors are split into the swap dealers, managed money, and other reportables categories.

As a response, the CFTC provides a report since January 2009 called the COT Supplement that covers 13 selected agriculturals with a breakdown now identifying explicitly the so-called Commodity Index Traders (CIT). This dataset has been widely used in the empirical literature and used to infer CIT positions on other commodities (Singleton 2013), despite the critics on the validity and representativeness of such inferred positions (Irwin and Sanders 2011).

Other papers go more granular by relying on specific proprietary data, such as Brunetti et al. (2016) who used individual daily positions of large market participants data from CFTC's large trader reporting system.

Finally, few papers rely on alternative data providers, as Henderson et al. (2015) that used commodity-linked notes issued by, and obligations of, financial institutions. Such notes are filed with the U.S. Securities and Exchange Commission (SEC) and made publicly available through the EDGAR database. These notes are typically purchased by non-informed traders and hedged via long positions on futures markets.

4.2.2 Indirect Measures

Although financialization relies on the inflow of financial investors, some studies do not use explicit dataset on investors' positions, but rather take 2003/2004 as an implicit break date where financialization takes effect. These papers provide evidence supporting a rise in comovement within commodity markets and with equity markets by relying on rolling window correlation (Bhardwaj et al. 2015), on variants of dynamic conditional correlation model (Silvennoinen and Thorp 2013; Zhang et al. 2017) or on the time-varying explanatory power of multifactor models (Christoffersen et al. 2019).

Whether direct measures dominate indirect measure remains an open question. The direct measures have a richer information set on the time-varying intensity of financial investor pressure. The indirect ones instead isolate from discussions on the relevant direct measures (CIT, swap dealers, and/or money managers) capturing financial investors' pressure and by taking the simple, clear, but arbitrary, view of a break in 2004.

4.3 *How to Measure Effects on Commodity Prices*

Financial investors are mainly active on the paper market. Most empirical studies therefore assess the impact of financialization on the futures prices. However, as discussed in Cheng and Xiong (2014), measuring the impact of financialization on futures is an intermediate step, as what ultimately matters is its impact on the spot market. There is then a second strand of literature that studies the mechanisms whereby the paper market (futures prices) impacts the real market (spot prices).

According to Cheng and Xiong (2014), futures prices are related to spot prices via three mechanisms. First, the theory of storage relates the futures and spot prices via an equilibrium relationship, as documented in Basak and Pavolva (2016). Second, the risk-sharing mechanism relates the futures and the expected future spot price via risk premia depending on the hedging pressure, as documented Acharya et al. (2013). Third, the informational role of futures markets takes futures prices as signals to guide commodity demand and thus spot prices, as documented in Dimpfl et al. 2017.

What the data providers call spot prices are not always spot prices, but often the nearest maturity futures contracts. Indeed, most spot trades occur over the counter and are not reported in harmonized datasets. Moreover, commodity spot prices are subject to substantial heterogeneity in data quality and commodity grades. Spot prices also reflect locations and specific transportation costs.

4.4 *Endogeneity Concerns*

4.4.1 Futures Prices and Hedging Demand

Most studies relating CIT trade positions to commodity prices presume that CIT (demand side) initiate the trades and Granger causes the futures price rises. However, CIT positions also reflect producers' hedging needs (supply side). We need here an identification strategy designed to identify a CIT demand shock in view to assess the genuine contribution of CIT investors to the price evolution. Cheng et al. (2015) used fluctuations in the VIX to isolate trades initiated by CITs and found a positive correlation between CIT position changes and futures prices. Henderson et al. (2015) used commodity-linked note (CLN) issuances to similarly identify trade initiated by financial traders. They found that financial traders "*have significantly positive and economically meaningful impacts on commodity futures prices around the pricing dates of the CLNs when the hedge trades are executed and significantly negative price impacts around the determination dates when the hedge trades are unwounded.*"

4.4.2 Spot Prices and Macro-Driven Boom

If large inflows of institutional investors on commodity markets can affect the commodity futures prices, the reverse is also true. Indeed, rising commodity prices also attract institutional investors. Most papers based on correlation measures are subject to this endogeneity concern.

Tang and Xiong (2012) studied the correlation of non-energy commodity returns with oil returns and propose a solution. They analyze separately the commodities included in the S&P GSCI and DJ-UBSCI (treatment group) and the commodities excluded from these indices (control group). They found that the commodities of the treatment group, which are presumed to be subject to commodity index traders' purchases, had a rise in their correlation with oil returns significantly larger than the one of the commodities in the control group.

As an alternative, Kilian and Murphy (2014) deal with reverse causality by relying on structural VAR modelling and sign restrictions as identification strategy. They use monthly "*the percent change in global oil production, a measure of cyclical variation in global real activity, the real price of crude oil, and the change in above-ground global crude oil inventories. The model is identified based on a combination of sign restrictions and bounds on the short-run price elasticities of oil demand and oil supply.*"

4.5 De-Financialization

Weekly futures open interest as reported in CFTC's CoT fell by 50% in 2008 but has then recovered and is currently far above its pre-crisis levels. However, all indicators do not support the belief of a constantly rising financialization. The BIS notional value of outstanding OTC commodity derivatives has collapsed from USD 14.1 trillion in 2008 to USD 2.1 trillion in 2019, now stable for more than five years. In addition, the composition of the open interest (in terms of producer, swap dealer, money managers, pother reportable) has remained remarkably stable since 2006 (see Fig. 9 in Bhardwaj et al. 2015).

Further, the presumed effects on financialization on inter-commodity correlation and equity-commodity correlations have vanished as documented via simple rolling correlations in Bhardwaj et al. (2015) and via the explanatory power of multifactor models in Christoffersen et al. (2019). Zhang et al. (2017) explicitly raise the question of "*de-financialization,*" measured as correlation between equity market and oil and gas markets. Based on a variance-threshold dynamic correlation model, they conclude that financialization persists since 2008.

4.6 Discussion

The literature on financialization of commodity markets is challenged by the difficulty to identify the exogenous contribution of financial investors to commodity prices. A clear rise of correlation among commodity prices and between commodity and equity prices has been documented by many from 2004 to around 2010, but only few papers were explicitly accounting for reverse causality (rising prices attract investors) or for hedging supply-demand determination (do financial investors go long because commodity hedgers are on the rise). Those that develop original identification strategy (Tang and Xiong (2012), Cheng et al. (2015), Henderson et al. (2015) among others) show that the debate on the persistent effects of financialization ten years after the financial crisis remains open.

5 Conclusion

Commodity prices are at the cross-road of many disciplines and raise a lot of relevant empirical challenges, about the handling of their long-term properties, about their potential exogeneity, viewed by some as quasi-natural experiments, and by their mixed status of financial asset on the one side and of consumption/intermediate good on the other side.

Their properties have given rise to a considerable amount of research on the PSH, and it seems that we still do not know for sure (cf. Table 1) if the long-term trend of their relative prices is negative or not. We might wonder if the applicability of new time series methodologies has not taken prevalence on the relevance of the research question. New research should better reflect the changing nature of imports, still made of manufactures but also of services and commodities.

The view of commodity prices being exogenous to real exchange rates, seen as quasi-natural experiments from the small economy perspective, is also to be mitigated. As discussed, beside the potential market power of some countries on some specific commodity markets, endogeneity also arises from the influence of common factors such as business cycles and interest rates. New research should reflect these interdependences, via for example structural VAR models or factor models.

Finally, the impact of financialization has been mainly investigated through diverse versions of dynamic conditional correlation models. Only few papers could find a convincing empirical design able to isolate the genuine contribution of financial investors inflows on the commodity prices and volatility, with results generally giving at best a small support to the hypothesis of speculator driven increases. The recent reduction of correlations supports these results and instead support the view of a price boom driven by the real economy.

This paper shows that the PSH, commodity currencies, and financialization literatures have surprisingly few connections. Shedding some light on these connections is certainly a promising path for research. For example, does the real exchange rate of commodity currency countries decline in the long run, as the PSH suggests? Does the higher commodity price volatility induced by financialization weaken the connection between commodity prices and real exchange rates? Should the PSH encourage financial investors to divest from commodity markets? There is no evidence that the empirical research on commodity prices will decline in the long run.

References

Acharya, V. V., Lochstoer, L. A., & Ramadorai, T. (2013). Limits to arbitrage and hedging: Evidence from commodity markets. *Journal of Financial Economics, 109*(2), 441–465.

Ahn, J. B., Mano, R., & Zhou, J. (2017). *Real exchange rate and external balance: How important are price deflators?* (Working Papers No. 17/81). International Monetary Fund.

Arezki, R., Hadri, K., Loungani, P., & Rao, Y. (2014). Testing the Prebisch-Singer hypothesis since 1650: Evidence from panel techniques that allow for multiple breaks. *Journal of International Money and Finance, 42,* 208–223.

Aslam, A., Beidas-Strom, S., Bems, R., Celasun, O., Çelik, S. K., & Koczan, Z. (2016). *Trading on their terms? Commodity exporters in the aftermath of the commodity boom.* (Working Papers No. 16/27). International Monetary Fund.

Bai, J., Kao, C., & Ng, S. (2009). Panel cointegration with global stochastic trends. *Journal of Econometrics, 149,* 82–99.

Bahmani-Oskooee, M., Chang, T., Elmi, Z., & Ranjbar, O. (2018). Re-testing Prebisch-Singer hypothesis: New evidence using Fourier quantile unit root test. *Applied Economics, 50*(4), 441–454.

Balassa, B. (1964). The purchasing power parity doctrine: A reappraisal. *Journal of Political Economy, 72,* 584.

Banerjee, A., & Carrion-I-Silvestre, J. L. (2015). Cointegration in panel data with structural breaks and cross-section dependence. *Journal of Applied Econometrics, 30*(1), 1–23.

Basak, S., & Pavlova, A. (2016). A model of financialization of commodities. *Journal of Finance, 71*(4), 1511–1556.

Bhagwati, J. N. (1984). Why are services cheaper in the poor countries? *Economic Journal, 94,* 279–286.

Bhardwaj, G., Gorton, G., & Rouwenhorst, K. G. (2015). Facts and fantasies about commodity futures ten years later, NBER 21243.

Bodart, V., Candelon, B., & Carpantier, J. F. (2012). Real exchange rates in commodity producing countries: A reappraisal. *Journal of International Money and Finance, 31,* 1485–1502.

Bodart, V., Candelon, B., & Carpantier, J. F. (2015). Real exchanges rates, commodity prices and structural factors in developing countries. *Journal of International Money and Finance, 51,* 264–284.

Bodart, V., & Carpantier, J. F. (2016). Real exchange rates and skills. *Journal of International Money and Finance, 67,* 305–319.

Breitung, J., & Pesaran, M. (2005). *Unit roots and cointegration in panels* (Working Paper No. 05–32). Institute of Economic Policy Research.

Brunetti, C., Buyuksahin, B., & Harris, J. H. (2016). Speculators, prices and market volatility. *Journal of Financial and Quantitative Analysis, 51*(5), 1545–1574.

Cashin, P., Cespedes, L. F., & Sahay, R. (2004). Commodity currencies and the real exchange rate. *Journal of Development Economics, 75*(1), 239–268.

Cashin, P., & McDermott, C. J. (2002). The long-run behavior of commodity prices: Small trends and big variability. *IMF Staff Papers, 49*(2), 175–199.

Chen, Y.-C., & Lee, D. (2018). Market power, inflation targeting, and commodity currencies. *Journal of International Money and Finance, 88*(C), 122–139.

Chen, Y.-C., & Rogoff, K. S. (2003). Commodity currencies. *Journal of International Economics, 60*(1), 133–160.

Chen, Y.-C., Rogoff, K. S., & Rossi, B. (2010). Can exchange rates forecast commodity prices? *Quarterly Journal of Economics, 125*(3), 1145–1194.

Cheng, I.-H., Kirilenko, A., & Xiong, W. (2015). Convective risk flows in commodity futures markets. *Review of Finance, 19*(5), 1733–1781.

Cheng, I.-H., & Xiong, W. (2014). The financialization of commodity markets. *Annual Review of Economic Financial Economics, 6,* 419–441.

Chinn, M. (2006). A primer on real effective exchange rates: Determinants, overvaluation, trade flows and competitive devaluation. *Open Economies Review, 17*(1), 115–143.

Clements, K. W., & Fry, R. (2008). Commodity currencies and currency commodities. *Resources Policy, 33*(2), 55–73.

Choudhri, E. U., & Khan, M. S. (2005). Real exchange rates in developing countries: Are Balassa-Samuelson effects present? *IMF Staff Papers, 52*(3), 1–2.

Christoffersen, P., Lunde, A., & Olesen, K. V. (2019). Factor structure in commodity futures return and volatility. *Journal of Financial and Quantitative analysis, 54*(3), 1083–1115.

Collier, P., & Goderis, B. (2012). Commodity prices and growth: An empirical investigation. *European Economic Review, 56*(6), 1241–1260.

Coudert, V., Couharde, C., & Mignon, V. (2011). Does euro or dollar pegging impact the real exchange rate. *World Economy, 34*(9), 1557–1592.

Cuddington, J. (1992). Long-run trends in 26 primary commodity prices: A disaggregated look at the Prebisch-Singer hypothesis. *Journal of Development Economics, 39,* 207–227.

Cuddington, J., & Urzua, C. (1989). Trends and cycles in the net barter terms of trade: A new approach. *Economic Journal, 99,* 426–442.

Darvas, Z. (2012. *Real effective exchange rates for 178 countries: A new database* (Working Paper No. 2012/06). Bruegel.

Dauvin, M. (2014). Energy prices and the real exchange rate of commodity-exporting countries. *International Economics, CEPII Research Center, 137,* 52–72.

Deaton, A., & Miller, R. (1996). International commodity prices, macroeconomic performance and politics in sub-saharan Africa. *Journal of African Economies, 5,* 99–191.

Di Iorio, F., & Fachin, S. (2018). The Prebish-Singer hypothesis in the post-colonial era: Evidence from panel cointegration. *Economics Letters, 166*(C), 86–89.

Dimpfl, T., Flad, M., & Jung, R. (2017). Price discovery in agricultural commodity markets in the presence of futures speculation. *Journal of Commodity Markets, 5,* 50–62.

De Gregorio, J., & Wolf, H. C. (1994). *Terms of trade, productivity, and the real exchange rate* (Working Papers No. 4807). National Bureau of Economic Research.

Edwards, S. (1986). Commodity export prices and the real exchange rate in developing countries: Coffee in Colombia. In *Economic Adjustment and Exchange Rates in Developing Countries, National Bureau of Economic Research* (pp. 233–266).

Erb, C. B., & Harvey, C. R. (2006). The strategic and tactical value of commodity futures. *Financial Analysts Journal, 62*(2), 66–97.

Fattouh, B., Kilian, L., & Mahadeva, L. (2013). The role of speculation in oil markets: What have we learned so far? *The Energy Journal, International Association for Energy Economics, 34*(3), 7–33.

Ferraro, D., Rogoff, K., & Rossi, B. (2015). Can oil prices forecast exchange rates? An empirical analysis of the relationship between commodity prices and exchange rates. *Journal of International Money and Finance, 54,* 116–141.

Fuchs-Schuendeln, N., & Hassan, T.A. (2016). Natural experiments in macroeconomics. In *Handbook of macroeconomics*. Elsevier.
Ghoshray, A. (2011). A re–examination of trends in primary commodity prices. *Journal of Development Economics, 95,* 242–251.
Gorton, G., & Rouwenhorst, K. G. (2006). Facts and fantasies about commodity futures. *Financial Analysts Journal, 62*(2), 47–68.
Grilli, E., & Yang, M. C. (1988). Primary commodity prices, manufactured goods prices, and the terms of trade of developing countries: What the long run shows. *The World Bank Economic Review, 2,* 1–47.
Groen, J., & Pesenti, P. (2010). *Commodity prices, commodity currencies, and global economic developments* (Working Papers No. 15743). National Bureau of Economic Research.
Gruss, B., & Suhaib, K. (2019). *Commodity terms of trade: A new database* (Working Paper No. 19/21). IMF.
Haase, M., Zimmermann, Y. S., & Zimmermann, H. (2016). The impact of speculation on commodity futures markets—A review of the findings of 100 empirical studies. *Journal of Commodity Markets, Elsevier, 3*(1), 1–15.
Habib, M. M., & Kalamova, M. M. (2007). *Are there oil currencies? The real exchange rate of oil exporting countries* (Working Paper No. 839). European Central Bank.
Hadri, K., Kurozumi, E., & Rao, Y. (2015). Novel panel cointegration tests emending for cross-section dependence with N fixed. *Econometrics Journal, 18,* 363–411.
Hallam, D. (2018). Revisiting Prebisch–Singer: What long–term trends in commodity prices tell us about the future of CDDCs. *Background paper to the UNCTAD-FAO Commodities and Development Report 2017.*
Harvey, D. I., Kellard, N. M., Madsen, J. B., & Wohar, M. E. (2010). The Prebisch—Singer hypothesis: Four centuries of evidence. *Review of Economics and Statistics, 92,* 367–377.
Henderson, B. J., Pearson, N. D., & Wang, L. (2015). New evidence on the financialization of commodity markets. *Review of Financial Studies, 28*(5), 1285–1311.
Hurlin, C., & Mignon, V. (2007). Second generation panel unit root tests https://ideas.repec.org/p/hal/wpaper/halshs-00159842.html.
Irwin, S., & Sanders, D. (2011). Index funds, financialization, and commodity futures markets. *Applied Economic Perspectives and Policy, 33,* 1–31.
Kellard, N., & Wohar, M. (2006). On the prevalence of trends in primary commodity prices. *Journal of Development Economics, 79,* 146–167.
Kilian, L., & Murphy, D. P. (2014). The role of inventories and speculative trading in the global market for crude oil. *Journal of Applied Econometrics, 29*(3), 454–478.
Kim, T.-H., Pfaffenzeller, S., Rayner, T., & Newbold, P. (2003). Testing for linear trend with application to relative primary commodity prices. *Journal of Time Series Analysis, 24,* 539–552.
Koranchelian, T. (2005). *The equilibrium real exchange rate in a commodity exporting country: Algerian experience* (Working Paper No. WP/05/135). IMF.
Lee, J., & Strazicich, M. (2003). Minimum LM unit root test with two structural breaks. *Review of Economics and Statistics, 85,* 1082–1089.
Leon, J., & Soto, R. (1997). Structural breaks and long–run trends in commodity prices. *Journal of International Development, 9,* 247–266.
Lumsdaine, R., & Papell, D. (1997). Multiple trend breaks and the unit root hypothesis. *Review of Economics and Statistics, 79,* 212–218.
MacDonald, R., & Nagayasu, J. (2000). The long-run relation between real exchange rates and real interest rate differentials: A panel study. *IMF Staff Papers, 47*(1), 116–128.
MacDonald, R., & Ricci, L. A. (2005). The real exchange rate and the Balassa-Samuelson effect: The role of the distribution sector. *Pacific Economic Review, 10,* 29–48.
Main, S., Irwin, S. H., Sanders, D. R., & Smith, A. (2018). Financialization and the returns to commodity investments. *Journal of Commodity Markets, 10*(C), 22–28.
Newbold, P., Pfaffenzeller, S., & Rayner, T. (2005). How well are long-run commodity price series characterised by trend components? *Journal of International Development, 17,* 479–494.

Ostry, J. D. (1994). Government purchases and relative prices in a two-country world. *The Economic Record, 70,* 149–161.

Pesaran, M. H. (2004). *General diagnostic tests for cross section dependence in panels* (Vol. 435). Cambridge Working Papers in Economics.

Pfaffenzeller, S., Newbold, P., & Rayner, A. (2007). A short note on updating the Grilli and Yang commodity price index. *World Bank Economic Review, 21,* 151–163.

Prebisch, R. (1950). *The economic development of Latin America and its principal problems.* New York: United Nations.

Ricci, L. A., Milesi-Ferretti, G. M., & Lee, J. (2013). Real exchange rates and fundamentals a cross-country perspective. *Journal of Money, Credit and Banking, 45*(5), 845–865.

Rose, A. K., Supaat, S., & Braude, J. (2009). Fertility and the real exchange rate. *Canadian Journal of Economics, 42,* 496–518.

Samuelson, P. (1964). Theoretical problems on trade problems. *Review of Economics and Statistics, 46,* 145–154.

Sapsford, D. (1985). The statistical debate on the net barter terms of trade between primary commodities and manufactures: A comment and some additional evidence. *Economic Journal, 95,* 781–788.

Schwert, G., & William, (1987). Effects of model specification on tests for unit roots in macroeconomic data. *Journal of Monetary Economics, 20,* 73.

Silvennoinen, A., & Thorp, S. (2013). Financialization, crisis and commodity correlation dynamics. *Journal of International Financial Markets, Institutions and Money, 24*(C), 42–65.

Singer, H. (1950). The distribution of gains between investing and borrowing countries. *American Economic Review, Papers and Proceedings, 11,* 473–485.

Singleton, K. (2013). Investor flows and the 2008 boom/bust in oil prices. *Management Science, 60,* 300–318.

Spatafora, N., & Tytell, I. (2009). *Commodity terms of trade: The history of booms and busts* (Working Paper No. 09205). IMF.

Tang, K., & Xiong, W. (2012). Index investment and financialization of commodities. *Financial Analysts Journal, 68*(6), 54–74.

Tashu, M. (2015). *Drivers of Peru's equilibrium real exchange rate; is the nuevo sol a commodity currency?* (Working Papers No. 15/26). International Monetary Fund.

Westerlund, J., & Edgerton, D. (2008). A simple test for cointegration in dependent panels with structural breaks. *Oxford Bulletin of Economics and Statistics, 70*(5), 665–704.

Yamada, H., & Gawon, Y. (2014). When Grilli and Yang meet Prebisch and Singer: Piecewise linear trends in primary commodity prices. *Journal of International Money and Finance, 42*(C), 193–207.

Zhang, Y.-J., Chevallier, J., & Guesmi, K. (2017). "De-financialization" of commodities? Evidence from stock, crude oil and natural gas markets. *Energy Economics, 68*(C), 228–239.

Zivot, E., & Andrews, D. (1992). Further evidence on the great crash, the oil price shock and the unit root hypothesis. *Journal of Business and Economic Statistics, 10,* 251–270.

Modeling Time-Varying Conditional Betas. A Comparison of Methods with Application for REITs

Marcel Aloy, Floris Laly, Sébastien Laurent, and Christelle Lecourt

JEL Classification C13 · C32 · C40 · C53 · C58 · G12 · R33

1 Introduction

More than fifty years after its birth, Sharpe and Lintner's Capital Asset Pricing Model (CAPM) is still widely used by academics and practitioners to measure the performance of managed portfolios or to estimate the cost of equity for companies. A common practice within the financial industry consists in estimating linear time series models via an ordinary least square (OLS) regression, so that the slope coefficients (the betas) are assumed to be constant over the estimation period.

However, both theoretically and empirically, many studies have shown that betas may vary over time, meaning that the assumption of constancy is erroneous and potentially misleading. As a result, models that assume the constancy of parameters

M. Aloy · S. Laurent (✉) · C. Lecourt
Aix-Marseille University (Aix-Marseille School of Economics), CNRS & EHESS, Marseille, France
e-mail: sebastien.laurent@univ-amu.fr

M. Aloy
e-mail: marcel.aloy@univ-amu.fr

C. Lecourt
e-mail: christelle.lecourt@univ-amu.fr

F. Laly
UCLouvain (Louvain School of Management), LFIN-LIDAM, Louvain-la-Neuve, Belgium
e-mail: floris.laly@uclouvain.be

S. Laurent
Aix-Marseille Graduate School of Management – IAE, Aix-en-Provence, France

G. Dufrénot and T. Matsuki (eds.), *Recent Econometric Techniques for Macroeconomic and Financial Data*, Dynamic Modeling and Econometrics in Economics and Finance 27, https://doi.org/10.1007/978-3-030-54252-8_9

tend to be misspecified, which can lead to poor estimations, irrelevant forecasts, and eventually bad financial decisions.

In order to model time-varying betas, three main alternatives to the simple OLS regression are typically used by academics and practitioners: using rolling-window OLS regressions, computing realized measures on sub-intervals of high-frequency data so as to obtain a realized beta, and using exogenous interaction variables.

On top of the three alternatives listed above, many more methodologies have flourished over the years such as state space models with time-varying slope coefficients or Markov-switching models. More recently, two statistical approaches intended to capture the dynamic aspects of time series data in the case of multiple betas have been developed. Engle (2016), extending the work of Bollerslev et al. (1988), introduces a new model called the Dynamic Conditional Beta (DCB) model, offering a way to indirectly retrieve the time-varying slope coefficients of the independent variables via an estimate of the full conditional covariance matrix (using a multivariate GARCH model for instance). This approach is very intuitive and has the advantage of being easily implementable because MGARCH models are now available in many econometrics softwares. However, the approach also presents some major drawbacks. First, testing and imposing the constancy of the conditional betas are not so practical. Second, it is impossible to introduce exogenous variables in the model and to identify precisely which ones influence the evolution of the different betas since conditional betas are retrieved after a nonlinear transformation of the elements that compose the estimated conditional covariance matrix instead of being modeled directly.

Darolles et al. (2018), extending the work of Pourahmadi (1999), take a different direction and offer a way to directly compute time-varying slope coefficients that depend on their lagged values and past shocks via a natural orthogonalization of the observed time series. Their model, called CHAR, which belongs to the class of MGARCH models, can also be used to obtain time-varying betas. The drawback of this approach is however that this method requires estimating the full multivariate system (like for the DCB model) even when one is interested in one equation only. To overcome this problem, building upon this method, Blasques et al. (2020) propose a new model, called autoregressive conditional beta (ACB) model, that allows a direct modeling of the conditional betas. This model differs from the CHAR model due to the fact the dynamics of the conditional betas does not require the estimation of a system of equations but only a univariate model, with GARCH errors for instance.

After reviewing different ways to estimate both static and time-varying betas, we compare the performance of the most advanced conditional beta modeling techniques, that is to say the state space, DCB, and ACB modeling techniques (with and without additional exogenous variables) to those of static betas (i.e., OLS and GARCH) in an empirical application focusing on the REIT market of the USA and developed Europe[1] using daily data over the period 2009–2019. To the best of our

[1] The term 'developed Europe' stands for the following countries: Austria, Belgium, Finland, France, Germany, Ireland, Italy, the Netherlands, Norway, Spain, Sweden, Switzerland, and the United Kingdom.

knowledge, we are the first to offer a comparison of these three conditional beta modeling techniques. In particular, we investigate the time variability of betas in a two-factor model, where $\beta_{B,t}$ and $\beta_{M,t}$ are respective measures of the sensitivity of the REIT index to changes in the bond market and the stock market. Results show that ACB models clearly outperform other competing models in-sample (when comparing the models on the basis of their log-likelihood) and that both the state space and ACB models outperform the other models out-of-sample (in a tracking exercise).

The remainder of the paper is organized as follows. Section 2 introduces the static beta model in a general framework and Sect. 3 presents different models used to estimate time-varying betas.[2] Section 4 describes the data and presents the results of the empirical application on REITs. Finally, Sect. 5 concludes.

2 Static Betas

2.1 One-Factor Model

The fundamental equation common to most asset pricing models states that the price of a given asset i at time t, denoted $P_{i,t}$, must be equal to the expected discounted value of its payoff (for more details see, for instance, Cochrane 2005; Ferson 2003; Smith and Wickens 2002):

$$P_{i,t} = E_t[m_{t+1}(P_{i,t+1} + D_{i,t+1})], \tag{1}$$

where $D_{i,t+1}$ is the payment (interest or dividend) received at time $t+1$, and m_{t+1} is a strictly positive random variable used to discount the future payoffs, called the *stochastic discount factor* (SDF). $E_t(.) = E(.|\Omega_t)$ denotes the time-t conditional expectation given the information set Ω_t.

Equation (1) assumes that the future payoff *and* discount factor[3] are stochastic: both are uncertain at date t and are contingent to future states of nature. It is worth noting that this equation holds for any investment horizon and type of asset (bond, share, option, real estate and so forth) and this without any specific assumption such as complete markets, financial market equilibrium, investor preference, or distribution of asset returns.

Equation (1) can be expressed in terms of returns:

$$1 = E_t(m_{t+1}R_{i,t+1}), \tag{2}$$

[2]Three main approaches have been used in the literature to evaluate asset pricing models (see for instance Cochrane, 2005): time series modeling, cross-sectional regressions, and calibration. This chapter only deals with time series models, excluding GMM estimates pioneered by Hansen (1982).

[3]The relationship between the discount factor m_{t+1} and the rate d_{t+1} at which future payoffs are discounted is: $m_{t+1} = \frac{1}{1+d_{t+1}}$. As a result, an increase in the discount factor m_{t+1} corresponds to a decrease in the stochastic discount rate d_{t+1}.

where $R_{i,t+1} = \frac{P_{i,t+1}+D_{i,t+1}}{P_{i,t}}$ is the gross return of the asset at time $t+1$.

In the case of a risk-free asset whose gross return R_f is known with certainty and assumed to be constant, Eq. (2) implies that the conditional expectation of the SDF is equal to the inverse of the risk-free rate:

$$E_t(m_{t+1}) = 1/R_f. \tag{3}$$

Equation (2) can be further developed as follows:

$$1 = E_t(m_{t+1})E_t(R_{i,t+1}) + \text{cov}_t(m_{t+1}, R_{i,t+1}), \tag{4}$$

where $\text{cov}_t(.)$ denotes the time-t expected conditional covariance given the information set Ω_t.

Substituting out Eq. (3) into Eq. (4) produces

$$1 = \frac{E_t(R_{i,t+1})}{R_f} + \text{cov}_t(m_{t+1}, R_{i,t+1}) \tag{5}$$

$$\Rightarrow E_t(R_{i,t+1}) - R_f = -R_f\text{cov}_t(m_{t+1}, R_{i,t+1}). \tag{6}$$

The risk premium $E_t(R_{i,t+1}) - R_f$ required to compensate risk-averse investors for holding risky assets only depends on the covariance of the payoffs with the discount factor, which is the only source of risk. This is due to the fact that investors have an incentive to pay more for assets with high payoffs in adverse conditions and the discount factor is precisely an index of adverse conditions (Cochrane and Culp 2003).

The various asset pricing models differ in the way they model the discount factor. Most of them are based on the implications of consumer/investor intertemporal optimization models.[4] For instance, in the basic theoretical *Consumption CAPM model* (CCAPM) with time-additive preferences and power utility, the SDF corresponds to the inverse of the marginal rate of substitution between consumption today and consumption in the next period[5]

$$m_{t+1} = \frac{\beta U'(C_{t+1})}{U'(C_t)}, \tag{7}$$

where $U'(C_s)$ is the marginal utility of consumption at time s, denoted (C_s).

The combination of Eqs. (6) and (7) sheds light on the mechanism underlying the determination of asset prices. Assets whose returns covariate negatively with the future marginal utility of consumption are hedging assets because their returns tend

[4]The first intertemporal CAPM model was developed by Merton (1973) in continuous time and led to a multi-factor model.

[5]Recall that the optimization of intertemporal utility in a deterministic model leads to the well-known Euler equation: $\frac{U'(C_t)}{\beta U'(C_{t+1})} = 1 + r$, where r is the interest rate and the LHS is the marginal rate of substitution.

to be high during times of market turbulence.[6] Consequently, the expected rate of return on these assets (and thus their risk premium) must be lower because investors have an incentive to buy them at a higher spot price.

A Taylor's series expansion of Eq. (7) leads to a linear relationship between m_{t+1} and the growth of consumption $(\Delta c_{t+1})^7$:

$$m_{t+1} \approx \beta - \beta\kappa_t \Delta c_{t+1}, \tag{8}$$

where $\kappa_t = -\dfrac{C_t U''(C_t)}{U'(C_t)}$ is the consumer/investor's subjective relative risk aversion coefficient.

Interesting results can be obtained by focusing on wealth rather than on consumption. Assume that wealth is held through the market portfolio,[8], i.e., the portfolio of all market assets weighted according to their relative value. Let $R_{M,t+1} = \dfrac{W_{t+1}}{W_t}$ be the gross return of the market portfolio and $r_{M,t+1} = R_{M,t+1} - 1$ the net return. The investor's optimization leads to[9]

$$m_{t+1} \approx 1 + \frac{W_t U''(W_{t+1})}{U'(W_t)} r_{M,t+1} \tag{9}$$

$$\Rightarrow m_{t+1} \approx (1 - \gamma_t) - \gamma_t R_{M,t+1}, \tag{10}$$

where $U'(W_s)$ is the marginal utility of W_s (the wealth at time s), and $\gamma_t = -\dfrac{W_t U''(W_{t+1})}{U'(W_t)}$ is the investor's subjective relative risk aversion coefficient.[10]

Dividing both terms of Eq. (6) by $\text{var}_t(R_{i,t+1}|\Omega_t)$, we get

$$\frac{E_t(R_{i,t+1}) - R_f}{\text{var}_t(R_{i,t+1})} = -R_f \frac{\text{cov}_t(m_{t+1}, R_{i,t+1})}{\text{var}_t(R_{i,t+1})}, \tag{11}$$

where $\text{var}_t(.) = \text{var}_t(.|\Omega_t)$ denotes the time-t expected conditional variance.

Using Eq. (10) and assuming for simplicity that $\text{Cov}_t(\gamma_t, R_{i,t+1}) = 0$, we obtain for $i = M$:

[6]Since the marginal utility of consumption decreases with the consumption level, an increase in the future marginal utility corresponds to an unfavorable state of nature.

[7]See Smith and Wickens (2002).

[8]The market portfolio may in theory include financial assets, consumer durables, real estate and human capital (Roll 1977; Fama and French 2004). Consequently, the gross market returns $R_{M,t+1}$ proxied by the return on an equity index is thus a rather narrow measure.

[9]See Harvey and Siddique (2000) as well as Phelan and Toda (2015). More to the point, Harvey and Siddique (2000) suggest a second-order approximation of Eq. (9) which leads to a nonlinear (quadratic) relation between m_{t+1} and $R_{M,t+1}$ in Eq. (10).

[10]As stated by Cochrane (2005), p. 464, γ_t in Eq. (10) represents aversion to bets on wealth while κ_t in Eq. (8) represents aversion to bets on consumption. γ_t is thus a more intuitive measure of risk aversion.

$$\frac{E_t(R_{M,t+1}) - R_f}{\text{var}_t(R_{M,t+1})} = -R_f \frac{\text{cov}_t(m_{t+1}, R_{M,t+1})}{\text{var}_t(R_{M,t+1})} \tag{12}$$

$$\Rightarrow \frac{E_t(R_{M,t+1}) - R_f}{\text{var}_t(R_{M,t+1})} = \gamma_t R_f. \tag{13}$$

The result given in Eq. (13) is fairly usual in conventional portfolio theories. Indeed, the risk premium required by an investor to hold the market portfolio is equal to his/her risk aversion multiplied by the conditional variance of the market portfolio (i.e., the risk): $E_t(R_{M,t+1}) - R_f = R_f \gamma_t \text{var}_t(R_{M,t+1})$.[11]

Substituting out Eqs. (10) and (13) into Eq. (6), the pricing of asset i gives

$$E_t(R_{i,t+1}) - R_f = -R_f \text{cov}_t(m_{t+1}, R_{i,t+1}) \tag{14}$$

$$= \gamma_t R_f \text{cov}_t(R_{M,t+1}, R_{i,t+1}) \tag{15}$$

$$= \frac{E_t(R_{M,t+1}) - R_f}{\text{var}_t(R_{M,t+1})} \text{cov}_t(R_{M,t+1}, R_{i,t+1}) \tag{16}$$

$$= \beta_{i,t} \left[E_t(R_{M,t+1}) - R_f \right], \tag{17}$$

where

$$\beta_{i,t} = \frac{\text{cov}_t(R_{M,t+1}, R_{i,t+1})}{\text{var}_t(R_{M,t+1})}. \tag{18}$$

Equation (17) can be expressed more conveniently in the following form:

$$E_t\left(\widetilde{r}_{i,t+1}\right) = \beta_{i,t} E_t\left(\widetilde{r}_{M,t+1}\right), \tag{19}$$

where $E_t\left(\widetilde{r}_{i,t+1}\right) = E_t(R_{i,t+1}) - R_f$, $E_t\left(\widetilde{r}_{M,t+1}\right) = E_t(R_{M,t+1}) - R_f$, i.e., respectively, the conditional expectation of the net excess return of asset i and of the market.[12]

Equation (19) is the *Conditional CAPM model*, i.e., a conditional version of the theoretical CAPM proposed by Sharpe (1964) and Lintner (1965). If $\beta_{i,t}$ is constant and conditional information plays no role in determining excess returns, then Eq. (19) becomes[13]

[11]Moreover, Eq. (13) states that if the risk is measured by the conditional variance, the market price of risk (LHS) is equal to the investor's subjective relative risk aversion (discounted by the risk-free rate).

[12]Recall that the net return of an asset i is linked to the gross return by the definition $r_{i,t+1} = R_{i,t+1} - 1$.

[13]The basic Sharpe and Lintner's unconditional CAPM rests on more restrictive assumptions than the CCAPM given in Eq. (17), since it results from the maximization of a single period mean-variance criterion. As stated by Merton (1973), the single-period utility function only coincides with intertemporal maximization when preferences and future investment opportunity sets are not state dependent.

As a consequence, the validity of the conditional CAPM does note imply the validity of the unconditional CAPM. As noted by Wang (1996), the unconditional CAPM can differ from the unconditional expectations of Eq. (17) if $\text{Cov}(\beta_{i,t}, R_{M,t+1}) \neq 0$, and thus the unconditional beta may differ from the expected conditional beta (Lewellen and Nagel 2006). However, if $\beta_{i,t}$ is a deter-

$$E(\widetilde{r}_i) = \beta E(\widetilde{r}_M), \tag{20}$$

where $E(\widetilde{r}_i)$ and $E(\widetilde{r}_M)$ are the unconditional expectations of net returns, and β (we omit the index i to simplify the notation) is the unconditional market beta of asset i.

The CAPM is the most widely studied asset valuation model and is used to estimate the required rate of return on an asset given a certain level of systematic risk (or market risk), expressed as the market beta β. While investors are facing two types of risks when investing—indiosyncratic and systematic risks—only systematic risk is priced since idiosyncratic risk can totally be offset through diversification. In the theoretical CAPM, market risk is the only source of systematic risk.

In this framework, the most basic way to estimate the beta of an asset is to run the simple bivariate OLS regression over the whole sample:

$$\widetilde{r}_{i,t} = \alpha + \beta \widetilde{r}_{M,t} + \varepsilon_t \tag{21}$$

$$\varepsilon_t \overset{\text{i.i.d.}}{\sim} D(0, \sigma^2), \tag{22}$$

where ε_t captures the idiosyncratic risk of asset i and follows a D distribution (e.g., Gaussian) with mean 0 and constant variance σ^2.

In this case, and from the market model regression, α is expected to be zero. The OLS estimate of β, is obtained as

$$\hat{\beta}^{\text{OLS}} = \frac{\text{cov}(\widetilde{r}_i, \widetilde{r}_M)}{\text{var}(\widetilde{r}_M)}, \tag{23}$$

where $\text{Cov}(\widetilde{r}_i, \widetilde{r}_M)$ is the unconditional covariance between the asset's excess return and the market excess return, and $\text{var}(\widetilde{r}_M)$ is the unconditional variance of the market excess return.

One way to relax the restrictive i.i.d. hypothesis on the residuals is to use a univariate GARCH model. The standard GARCH (1, 1) model offers the advantage of accounting for two features which are commonly observed in financial time series data, i.e., leptokurtosis (distribution's excess peakedness and fat tails) and volatility clustering (tendency for low/high volatility to persist and appear in bunches). In this case, the residuals and the conditional variance can be specified as

$$\varepsilon_t = \sigma_t z_t, \quad z_t \overset{\text{i.i.d.}}{\sim} D(0, 1), \tag{24}$$

$$\sigma_t^2 = \lambda_0 + \lambda_1 \sigma_{t-1}^2 + \lambda_2 \varepsilon_{t-1}^2, \tag{25}$$

where z_t is an i.i.d. random variable with mean 0 and unit variance while the conditional variance at time t, denoted σ_t^2, depends on both the lagged squared error term at time $t-1$ and the lagged variance term at time $t-1$.

---

ministic constant ($\beta_{i,t} = \beta, \forall t$), then the unconditional CAPM and the unconditional expectation of the conditional CAPM are equivalent.

2.2 Multiple Betas

The consumption CAPM and conditional CAPM are basically one-factor asset pricing models. Indeed, Eqs. (8) and (10) belong to the more general class of *linear pricing kernel*

$$m_{t+1} \approx a_{0,t} + a_{1,t}F_{1,t+1} + \cdots + a_{N,t}F_{N,t+1} \tag{26}$$

in which $F_{1,t+1}, \ldots, F_{N,t+1}$ are variables or factors that are good proxies for growth of marginal utility.

Many authors have considered the case where the stochastic discount factor can be represented as a linear function of *N* factors of the form given by Eq. (26). For instance, in Eq. (10), which leads to a conditional CAPM, the return on the market portfolio is proxied by the return on a stock market index $R_{M,t+1}$, but this assumption is criticized by Roll (1977), who argues that this approximation neglects the human capital component in total wealth. Wang (1996) and other authors suggest that the growth rate of labor income can be a good proxy for the return on human capital. Under this hypothesis, one can consider an SDF with two factors: the return of a stock market index $R_{M,t+1}$ and the growth rate of labor income Δy_{t+1}: $m_{t+1} \approx a_{0,t} + a_{1,t}R_{M,t+1} + a_{2,t}\Delta y_{t+1}$.

Considering Eq. (26), Ferson and Jagannathan (1996) show that in the case where the factors $F_{1,t+1}, \ldots, F_{N,t+1}$ are traded assets[14] and if the coefficients are defined as follows[15]

$$a_{j,t} = -\frac{E_t\left(\widetilde{f}_{j,t+1}\right)}{R_f \operatorname{var}_t(F_{j,t+1})}, \quad j = 1, \ldots, N \tag{27}$$

$$a_{0,t} = \frac{1}{R_f} - \sum_{j=1}^{N} a_{j,t}, \tag{28}$$

where $E_t\left(\widetilde{f}_{j,t+1}\right) = E_t(F_{j,t+1}) - R_f$ denotes the (conditional) expected risk premium of factor *j*, then we obtain a multi-factor representation of the conditional CAPM:

$$E_t(\widetilde{r}_{i,t+1}) = \sum_{n=1}^{N} \beta_{n,t} E_t(\widetilde{f}_{n,t}). \tag{29}$$

As we have done previously, if all $\beta_{n,t}$'s are constant and conditional information plays no role in determining excess returns, we obtain an unconditional factor model

---

[14] If one of the factors is not a traded asset return, then its expected risk premium is estimated by the conditional expectation of the excess return of the factor mimicking portfolio, i.e., a portfolio whose returns can be used instead of the factor itself. See, for example, Ferson (2003).

[15] If the *j*th factor is the market return, then $a_{j,t} = -\gamma_t$, i.e., the negative of the coefficient of time-varying relative risk aversion given in Eq. (13).

which looks like the equation used in the so-called arbitrage pricing theory (APT) initiated by Ross (1976), i.e.,

$$\widetilde{r}_{i,t} = \alpha + \sum_{n=1}^{N} \beta_n \widetilde{f}_{n,t} + \varepsilon_t. \tag{30}$$

Equation (30) can be written in a more compact and standard form as follows:

$$y_t = \mathbf{x}_t' \boldsymbol{\beta} + \varepsilon_t, \tag{31}$$

where $y_t = \widetilde{r}_{i,t}$, $\mathbf{x}_t = [1, \widetilde{f}_{1,t}, \ldots, \widetilde{f}_{N,t}]'$ and $\boldsymbol{\beta} = (\alpha, \beta_1, \ldots, \beta_N)'$.

The OLS estimator of the parameter vector $\boldsymbol{\beta} = (\alpha, \beta_1, \ldots, \beta_N)'$ is

$$(\hat{\alpha}, \hat{\beta}_1, \ldots, \hat{\beta}_N)' = (\mathbf{x}'\mathbf{x})^{-1}\mathbf{x}'\mathbf{y}. \tag{32}$$

In the specific case where the covariance matrix of the regressors is diagonal, (that is, if the different factors are mutually uncorrelated) this formula leads to a simple generalization of Eq. (23) since

$$\hat{\beta}_n^{OLS} = \frac{\text{cov}(\widetilde{f}_n, \widetilde{r}_i)}{\text{var}(\widetilde{f}_n)}, \quad \forall n = 1, \ldots, N. \tag{33}$$

According to the APT, the one-factor CAPM is not appropriate in a world with multiple risk factors represented by microeconomic or macroeconomic variables. Among possible risk factors, one can mention inflation, the spread between short-term and long-term bonds, industrial production growth or default risks (Brooks 2014). Many pricing models have been developed with additional risk factors such as the size and value risk factors (Fama and French 1993), the momentum risk factor (Carhart 1997), or the profitability and investment risk factors (Fama and French 2015).

However, multi-factor pricing models have been criticized for poor out-of-sample performance and for data snooping (see for instance Andersen et al. 2003; Harvey et al. 2015; Linnainmaa and Roberts 2018).

According to Harvey et al. (2015), more than 300 factors have been presented in the literature as important and significant in explaining the cross-sectional variation of stock returns. Many of these factors are however difficult to interpret from an economic perspective.

3 Time-Varying Betas

There is a large consensus in the literature about the fact that betas are actually time-varying. Such evidence has been pointed out by Fama and MacBeth (1973), Fabozzi and Francis (1977), Alexander and Chervany (1980), Sunder (1980), Ohlson and

Rosenberg (1982), DeJong and Collins (1985), Fisher and Kamin (1985), Brooks et al. (1992), Brooks et al. (1994) among others. The aim of this section is therefore to review various methods used to estimate time-varying betas (and potentially time-varying alphas) in a conditional model close to Eq. (29):

$$\widetilde{r}_{i,t} = \alpha_t + \sum_{n=1}^{N} \beta_{n,t} \widetilde{f}_{n,t} + \varepsilon_t, \tag{34}$$

or equivalently (31):

$$y_t = \mathbf{x}_t' \boldsymbol{\beta}_t + \varepsilon_t. \tag{35}$$

3.1 Rolling Betas

The first and simplest way to estimate time-varying betas is to estimate beta over moving sub-periods using a simple rolling-window OLS regression as proposed by Fama and MacBeth (1973) (see Van Nieuwerburgh 2019; Zhou 2013 for applications in the case of the REIT beta).

Let us assume we have historical data on a period spanning from t_0 to t_T. This method consists in selecting a window of h observations (for instance 120 daily returns) and then estimating Eq. (30) by OLS over the first h observations, that is from t_0 to t_{0+h}, so as to obtain $(\hat{\alpha}_h, \hat{\beta}_{1,h}, \ldots, \hat{\beta}_{N,h})'$. Then, the window is rolled one step forward by adding one new observation and dropping the most distant one. So, $(\hat{\alpha}_{h+1}, \hat{\beta}_{1,h+1}, \ldots, \hat{\beta}_{N,h+1})'$ are obtained in the same way over the period t_{0+1} to t_{0+h+1} and the sequence is reiterated until the end of the sample period.

This method can be considered as a quick and dirty time-varying regression model. When compared to the usual OLS regression model, the rolling-window OLS regression model offers the advantage of taking into account time variations in the alpha and the betas. However, the coefficients measured with this method only vary very slowly by construction due to the fact only one period is dropped and another is added between two successive estimates, which can lead to inaccurate estimates. The results also depend heavily on the size of the chosen window.[16]

[16]Some refinements can be made, for example, by introducing a weighting scheme giving less weight to observations from more distant periods (see for instance Nieto et al. 2014).

3.2 Realized Betas

An alternative estimation method of time-varying betas is to compute both the realized variance and realized covariance from intraday[17] data so as to estimate the beta using so called *realized measures*. The realized variance and covariance being computed from intraday data, they are much more accurate than standard measures (Hansen et al. 2014). This approach, based on the works of Andersen et al. (2003) and Barndorff-Nielsen and Shephard (2004),[18] offers a more accurate way of analyzing the dynamic behavior of beta than the rolling beta methodology (Patton and Verardo 2012). In the context of a single factor model, the realized market beta, denoted β^R, is defined as

$$\beta_t^R = \frac{\text{cov}^R(\widetilde{r}_i, \widetilde{r}_M)_t}{\text{var}^R(\widetilde{r}_M)_t} = \frac{\Sigma_{k=1}^{(s)} \widetilde{r}(t)_{i,k} \widetilde{r}(t)_{M,k}}{\Sigma_{k=1}^{(s)} \widetilde{r}(t)_{M,k}^2}, \tag{36}$$

where $\text{cov}^R(\widetilde{r}_i, \widetilde{r}_M)$ is the realized covariance between the asset's excess return and the market excess return, and $\text{var}^R(\widetilde{r}_M)$ is the realized variance of the market excess return. $\widetilde{r}(t)_{i,k}$ is the excess return on asset i during the kth intraday period on day t and s is the total number of intraday periods.[19] As a consequence, the realized beta is the ratio of an asset's sample covariance with the market to the sample variance of the market over several intraday periods.[20]

This method resembles the rolling beta approach but it relies on non-overlapping windows (of one day or one month for instance) and requires data sampled at a higher frequency within each window (e.g., 5-minute returns). The main limit of this methodology is, therefore, the need for a sufficient number of intraday data in order to be able to compute the realized variance and covariance. See Boudt et al. (2017) for an extension to multi-factor betas.

3.3 Time-Varying Betas with Interaction Variables

Rolling OLS and 'realized betas' do not rely on a parametric model to specify the dynamics of the time-varying betas. Shanken (1990) and Schadt (1996) introduce the hypothesis that the dynamics of the betas depends on a set of exogenous vari-

---

[17]Intraday data are used to estimate daily beta, variances or covariances, assuming they are fairly stable during the day. It is of course possible to use, for instance, daily data to compute beta, variances or covariances over a month if we assume that these parameters are fairly stable during the month. See for instance Andersen et al. (2006) and Lewellen and Nagel (2006).

[18]This methodology assumes an absence of jumps. In the case of jumps, see Todorov and Bollerslev (2010).

[19]For more details, see Patton and Verardo (2012).

[20]It is also possible, as in Lewellen and Nagel (2006), to directly estimate Eq. (30) for each intraday (or intra-monthly, intra-quarterly, etc.) period.

ables. This leads to an alternative way of modeling conditional betas, via the use of interaction variables (see also Gagliardini et al. 2016).

Indeed, interaction variables can be used to introduce dynamics in $\beta_{n,t}$ in the following way:

$$\beta_{n,t} = \beta_n + \sum_{k=1}^{K} \theta_{n,k} Z_{k,t-1}, \tag{37}$$

where each $Z_{k,t-1}$ variable ($k = 1, \ldots, K$) is an observable state variable, predetermined at the end of period $t-1$, and assumed to drive the dynamics of the beta of the n'th factor, and where $\theta_{n,k}$ is its associated coefficient. The above model can be rewritten as

$$\widetilde{r}_{i,t} = \alpha + \beta \widetilde{r}_{M,t} + \sum_{k=1}^{K} \theta_k Z_{k,t-1} \widetilde{r}_{M,t} + \varepsilon_t$$

and, therefore, corresponds to a multiple linear regression model with k interaction variables when the conditional variance is constant or a GARCH model with k interaction variables in the conditional mean when the conditional variance is assumed to follow a GARCH dynamics.

The advantage of this method is that the model can easily be estimated. However, this method requires the selection of suitable variables supposed to drive the dynamics of the betas. In addition, they must be able to generate a certain persistence that can be observed in the dynamic behavior of the betas.

Many studies have used interaction variables. It seems worth mentioning Schwert and Seguin (1990), in which betas are assumed to vary with the level of aggregate market volatility. The authors estimate the conditional market beta in the following way:

$$\beta_{i,t} = \beta_i + \theta_i \left(\frac{1}{\hat{\sigma}^2_{M,t}} \right), \tag{38}$$

where β_i and θ_i are constant parameters and $\hat{\sigma}^2_{M,t}$ is the time-varying volatility of the aggregate stock market. The model is mainly used to account for time-variation in stock betas or to study the relationship between firm size and time-varying betas (see for example Reyes 1999).

3.4 Indirect Dynamic Conditional Betas

As we have seen, the theoretical conditional CAPM expresses the conditional market beta as follows:

$$\beta_{i,t} = \frac{\text{cov}_t(R_{M,t+1}, R_{i,t+1})}{\text{var}_t(R_{M,t+1})}. \tag{39}$$

This expression opens up the possibility of obtaining time-varying betas from the estimation of conditional variances and covariances obtained, for instance, by a multivariate GARCH model (see for instance Bali 2010 for an application and Bauwens et al. 2006 for a survey on MGARCH models). In this case, the model imposes a minimal structure on the time-varying process, apart from the modeling of conditional variances and covariances in an autoregressive form. However, Eq. (39) does not apply in the multi-factor model when the factors are correlated.

Engle (2016) recently extended the multivariate GARCH approach to the case of a multi-factor model. Following Engle's (2016) methodology, the conditional betas are inferred from an estimate of the conditional covariance matrix Σ_t of $(\mathbf{x}_t, y_t)'$.

For ease of exposition, we assume in this section that $\mathbf{x}$ and y have been centered so that $\mathbf{x}$ does not contain a vector of ones (corresponding to α) and, therefore, one does not need to estimate the intercept in (31).

In order to obtain the coefficients of the multivariate regression of y_t (asset returns) on $\mathbf{x}_t$ (factors), Engle (2016) assumes that $(\mathbf{x}_t, y_t)'$ follows an $(N+1)$-dimensional normal distribution (conditional on the information set at time $t-1$, denoted $\mathcal{F}_{t-1}$), i.e.

$$\begin{pmatrix} \mathbf{x}_t \\ y_t \end{pmatrix} | \mathcal{F}_{t-1} \sim N\left(\begin{pmatrix} \mathbf{0}_{m-1} \\ 0 \end{pmatrix}, \Sigma_t \equiv \begin{pmatrix} \Sigma_{\mathbf{xx},t} & \Sigma_{\mathbf{x}y,t} \\ \Sigma_{y\mathbf{x},t} & \Sigma_{yy,t} \end{pmatrix} \right),$$

where subscripts embody natural partitions.

In order to derive an estimate of the conditional betas, Engle (2016) relies on the fact that the conditional distribution of y_t on $\mathbf{x}_t$ is

$$y_t|\mathbf{x}_t \sim N\left(\Sigma_{y\mathbf{x},t}\Sigma_{\mathbf{xx},t}^{-1}\mathbf{x}_t, \Sigma_{yy,t} - \Sigma_{y\mathbf{x},t}\Sigma_{\mathbf{xx},t}^{-1}\Sigma_{\mathbf{x}y,t}\right). \tag{40}$$

In more details, estimates of the time-varying coefficients inferred from the regression of y_t on $\mathbf{x}_t$ can be retrieved from Σ_t as follows:

$$\widehat{\boldsymbol{\beta}}_t^{\mathrm{DCB}} \equiv (\hat{\beta}_{1,t}^{\mathrm{DCB}}, \ldots, \hat{\beta}_{N,t}^{\mathrm{DCB}})' = \Sigma_{\mathbf{xx},t}^{-1}\Sigma_{\mathbf{x}y,t}. \tag{41}$$

When there is only one regressor, estimates of the time-varying coefficients inferred from the regression of y_t on x_t can simply be retrieved from $\Sigma_{xy,t}/\Sigma_{xx,t}$, i.e., the conditional covariance between y_t and x_t divided by the conditional variance of x_t. While any MGARCH model can be used to estimate Σ_t, Engle (2016) uses a dynamic conditional correlation (DCC) GARCH model on $(\mathbf{x}_t, y_t)'$, which relies on the following decomposition of Σ_t:

$$\Sigma_t = \mathbf{D}_t\mathbf{R}_t\mathbf{D}_t, \tag{42}$$

where $\mathbf{D}_t$ is a diagonal matrix containing the conditional volatilities (typically modeled with $N+1$ independent univariate GARCH models) while $\mathbf{R}_t$ is a conditional correlation matrix (usually obtained as a transformation of a scalar BEKK specification on the devolatilized series). Note that the constant conditional correlation (CCC) model proposed by Bollerslev et al. (1990) is obtained when $\mathbf{R}_t = \mathbf{R}$.

Based on the DCC–GARCH model, the DCB model presents several drawbacks, which we summarize here. First, the stationarity and ergodicity conditions of the DCC are not well known. Second, the model incorporates complicated constraints associated with correlation matrices. Third, the asymptotic properties of the Quasi Maximum Likelihood Estimator (QMLE) are unknown. And fourth, the effects of the DCC parameters on β_t are nearly impossible to interpret.

Note that out-of-sample forecasts of the betas can be obtained by adapting Eq. (41) to out-of-sample forecasts of the conditional covariance matrix.

3.5 Direct Dynamic Conditional Betas

In this section, we present two competing models that directly specify the dynamics of the conditional betas: the first model belongs to the general class of data-driven models while the second model belongs to the class of observation-driven models.

3.5.1 State Space Models

Adrian and Franzoni (2009) suggest a stylized model based on the conditional CAPM, in which beta changes over time and the investor's expectation of beta results from a learning process. This learning process is modeled via a Kalman process in which beta is treated as a latent variable. Adrian and Franzoni (2009) thus provide a theoretical foundation for the estimation of unobserved time-varying betas by state space modeling (see Choudhry and Wu 2008; Cisse et al. 2019; Faff et al. 2000; Huang and Hueng 2009; Mergner and Bulla 2008; Nieto et al. 2014 among others).

A useful state space representation of the multi-factor model is given by the following system of equations:

$$y_t = \mathbf{x}_t'\boldsymbol{\beta}_t + \varepsilon_t \tag{43}$$

$$\boldsymbol{\beta}_t - \overline{\boldsymbol{\beta}} = \Phi(\boldsymbol{\beta}_{t-1} - \overline{\boldsymbol{\beta}}) + \mathbf{u}_t, \tag{44}$$

where $y_t = \widetilde{r}_{i,t}$, $\mathbf{x}_t = [1, \widetilde{f}_{1,t}, \ldots, \widetilde{f}_{N,t}]'$, $\boldsymbol{\beta}_t = (\alpha_t, \beta_{1,t}, \ldots, \beta_{N,t})'$, Φ is a $\big((N+1) \times (N+1)\big)$ transition matrix which can be assumed to be diagonal: $\Phi = \text{diag}(\Phi_0, \Phi_1, \ldots, \Phi_N)$.

The $N+2$ residuals are assumed to be conditionally i.i.d and mutually independent, i.e.,

$$\begin{pmatrix} \mathbf{u}_t \\ \varepsilon_t \end{pmatrix} | \mathbf{Y}_{t-1}, \mathbf{X}_t \overset{\text{i.i.d}}{\sim} N\left(\begin{pmatrix} \mathbf{0} \\ 0 \end{pmatrix}, \begin{pmatrix} \mathbf{Q} & \mathbf{0} \\ \mathbf{0} & \sigma^2 \end{pmatrix} \right), \tag{45}$$

where $\mathbf{Q} = \text{diag}(\sigma_{u0}^2, \sigma_{u1}^2, \ldots, \sigma_{uN}^2)$, and we use the notations $\mathbf{Y}_t = \{y_t, y_{t-1}, \ldots, y_1\}$ and $\mathbf{X}_t = \{\mathbf{x}_t', \mathbf{x}_{t-1}', \ldots, \mathbf{x}_1'\}$ to denote observations available at time t.[21]

Equation (44) encompasses several specifications (see Chap. 2 of Moryson 1998, for more details).

If $\Phi = \mathbf{I}$, each conditional beta follows a *random walk* process, i.e.,

$$\beta_{n,t} = \beta_{n,t-1} + u_{n,t}, \qquad n = 1, \ldots, N. \tag{46}$$

In this case, the h-step-ahead out-of-sample forecast of the nth conditional beta at the end of the estimation period is given by $\beta_{n,T+h} = \beta_{n,T}, \forall h > 0$.

If $\Phi = \mathbf{0}$, we obtain the *random coefficient* model

$$\beta_{n,t} = \overline{\beta}_n + u_{n,t}, \qquad n = 1, \ldots, N, \tag{47}$$

where the deviation of $\beta_{n,t}$ from its unconditional mean $\overline{\beta}_n$ is caused solely by the noise $u_{n,t}$, and the h-step-ahead out-of-sample forecasts is $\beta_{n,T+h} = \overline{\beta}_n, \forall h > 0$.

Finally, Eq. (44) corresponds to the *mean reverting* model[22]

$$\beta_{n,t} = \overline{\beta}_n + \Phi_n(\beta_{n,t-1} - \overline{\beta}_n) + u_{n,t}. \tag{48}$$

The stochastic process $\beta_{n,t}$ reverts to its unconditional mean $\overline{\beta}_n$ after a shock and the parameter Φ_n controls the speed of reversion to the mean. The h-step-ahead out-of-sample forecast is given by $\beta_{n,T+h} - \overline{\beta}_n = \Phi_n^h(\beta_{n,T} - \overline{\beta}_n)$.

Let us write Eq. (44) in a more compact form as

$$\boldsymbol{\beta}_t = \boldsymbol{\mu} + \Phi\boldsymbol{\beta}_{t-1} + \mathbf{u}_t, \tag{49}$$

where $\boldsymbol{\mu} = (\alpha(1 - \Phi_0), \beta_1(1 - \Phi_1), \ldots, \beta_N(1 - \Phi_N))'$.

Estimation of the model's parameters can be achieved by the Kalman filter, an iterative algorithm producing at each time t an estimator of $\boldsymbol{\beta}_t$ denoted $\widehat{\boldsymbol{\beta}}_{t|t-1}$ based on the information up to time $t - 1$.

Given an estimate $\widehat{\boldsymbol{\beta}}_{t|t-1}$, the measurement Eq. (43) can be written as

$$y_t = \mathbf{x}_t'\widehat{\boldsymbol{\beta}}_{t|t-1} + \mathbf{x}_t'(\boldsymbol{\beta}_t - \widehat{\boldsymbol{\beta}}_{t|t-1}) + \varepsilon_t. \tag{50}$$

The one period-ahead conditional forecast is thus $E[y_t|\mathbf{Y}_{t-1}, \mathbf{X}_t] = y_{t|t-1} = \mathbf{x}_t'\widehat{\boldsymbol{\beta}}_{t|t-1}$ and the prediction error

[21] Several extensions can be accounted for, such as time-varying variances/covariances of the error terms.

[22] Since we assume $\Phi = \text{diag}(\Phi_0, \Phi_1, \ldots, \Phi_N)$, if all $(\Phi_0, \Phi_1, \ldots, \Phi_N)$ are inside of the unit circle, then the vector $\overline{\boldsymbol{\beta}}$ corresponds to the average value of $\boldsymbol{\beta}_{t+1}$.

$$\eta_{t|t-1} = y_t - y_{t|t-1} \\ = y_t - \mathbf{x}_t'\widehat{\boldsymbol{\beta}}_{t|t-1}. \tag{51}$$

From Eq. (50), the MSE of the prediction error can be easily computed as

$$f_{t|t-1} = E[(\eta_{t|t-1})^2|\mathbf{Y}_{t-1}, \mathbf{X}_t] = \mathbf{x}_t'\mathbf{P}_{t|t-1}\mathbf{x}_t + \sigma^2, \tag{52}$$

where $\mathbf{P}_{t|t-1} = E[(\boldsymbol{\beta}_t - \widehat{\boldsymbol{\beta}}_{t|t-1})(\boldsymbol{\beta}_t - \widehat{\boldsymbol{\beta}}_{t|t-1})'|\mathbf{Y}_{t-1}, \mathbf{X}_t]$ is the covariance matrix of $\boldsymbol{\beta}_t$ conditional on the information up to $t-1$. We can see from Eq. (52) that the MSE of the prediction error consists of two parts: the uncertainty associated with $\widehat{\boldsymbol{\beta}}_{t|t-1}$ and the variance of ε_t.

Under the gaussianity assumption, the sample log-likelihood is (Hamilton 1994; Durbin and Koopman 2001):

$$\sum_{t=1}^{T} \log f(y_t|\mathbf{Y}_{t-1}, \mathbf{X}_t) = -\left(\frac{T}{2}\right)\log(2\pi) - \left(\frac{1}{2}\right)\sum_{t=1}^{T}\log(f_{t|t-1}) \\ - \sum_{t=1}^{T}\frac{\left(\eta_{t|t-1}\right)^2}{f_{t|t-1}}, \tag{53}$$

where $\eta_{t|t-1}$ and $f_{t|t-1}$ are given respectively in Eqs. (51) and (52).

With initial condition $\widehat{\boldsymbol{\beta}}_{0|0}$ and $\mathbf{P}_{0|0}$, the conditional covariance matrix $\mathbf{P}_{t|t-1}$ and the conditional vector $\widehat{\boldsymbol{\beta}}_{t+1|t}$ are recursively computed according to the following prediction equations:

- One-step ahead forecast

$$\widehat{\boldsymbol{\beta}}_{t|t-1} = \boldsymbol{\mu} + \Phi\widehat{\boldsymbol{\beta}}_{t-1|t-1} \tag{54}$$

$$\mathbf{P}_{t|t-1} = \Phi\mathbf{P}_{t-1|t-1}\Phi' + \mathbf{Q} \tag{55}$$

  as well as Eqs. (51) and (52).
- Kalman gain

$$\mathbf{K}_t = \mathbf{P}_{t|t-1}\mathbf{x}_t'f_{t|t-1}^{-1}. \tag{56}$$

- Measurement update

$$\widehat{\boldsymbol{\beta}}_{t|t} = \widehat{\boldsymbol{\beta}}_{t|t-1} + \mathbf{K}_t\eta_{t|t-1} \tag{57}$$

$$\mathbf{P}_{t|t} = (\mathbf{I} - \mathbf{K}_t\mathbf{x}_t')\mathbf{P}_{t|t-1}. \tag{58}$$

The Kalman gain $\mathbf{K}_t$ is used to update $\widehat{\boldsymbol{\beta}}_{t|t}$ from $\widehat{\boldsymbol{\beta}}_{t|t-1}$. It determines the relative weighting of new information (given by the prediction error $\eta_{t|t-1}$) versus the current state estimate $\widehat{\boldsymbol{\beta}}_{t|t-1}$.

For given parameters of the model, the recursive Eqs. (54)–(58) provide the prediction error $\eta_{t|t-1}$ and its variance $f_{t|t-1}$. The unknown parameters are estimated through the maximization of the log-likelihood (53) with respect to these parameters, using $\eta_{t|t-1}$ and $f_{t|t-1}$ as inputs.

Finally, the smoothed estimates are obtained by iterating backward for $t = T - 1, T - 2, \ldots, 1$ the following equations:

$$\widehat{\boldsymbol{\beta}}_{t|T} = \widehat{\boldsymbol{\beta}}_{t|t} + \mathbf{P}_{t|t}\Phi'\mathbf{P}_{t+1|t}^{-1}\left(\widehat{\boldsymbol{\beta}}_{t+1|T} - \mu - \Phi\widehat{\boldsymbol{\beta}}_{t|t}\right) \tag{59}$$

$$\mathbf{P}_{t|T} = \mathbf{P}_{t|t} + \mathbf{P}_{t|t}\Phi'\mathbf{P}_{t+1|t}^{-1}\left(\mathbf{P}_{t+1|T} - \mathbf{P}_{t+1|t}\right)\left(\mathbf{P}_{t+1|t}^{-1}\right)'\Phi\left(\mathbf{P}_{t|t}\right)', \tag{60}$$

where $\widehat{\beta}_{T|T}$ and $\mathbf{P}_{T|T}$ are the initial value for the smoothing, obtained from the last iteration of the Kalman filter.

Adrian and Franzoni (2009) use a one-factor version of Eq. (44), i.e., they assume that the conditional beta follows a mean-reverting process, with and without conditioning variables such as the term spread. In addition, they also introduce a time-varying unobservable long-run beta $\overline{\boldsymbol{\beta}}_t$ and they consequently add an updating equation for this coefficient. According to Adrian and Franzoni (2009), the measurement update with the Kalman gain provides a realistic representation of the investor's learning process regarding the unknown beta.

From an empirical perspective, Choudhry and Wu (2008), Faff et al. (2000), Mergner and Bulla (2008), and Nieto et al. (2014) compare different methodologies for estimating time-varying betas. In particular, they compare different multivariate GARCH specifications and Kalman models (Random Walk, Random Coefficients, Mean-Reverting). Overall, the Kalman random walk model is considered as the best description of time-varying sectoral betas.

3.6 Autoregressive Conditional Betas

An alternative to the state space model presented above that also allows a direct specification of dynamic conditional betas has recently been proposed by Darolles et al. (2018). Their model, called CHAR, is a multivariate GARCH model based on the Cholesky decomposition of the $m \times m$ (with $m = N + 1$) conditional covariance matrix Σ_t of $(\mathbf{x}_t, y_t)'$.

As Pourahmadi (1999), let us consider the Cholesky decomposition of Σ_t, i.e.,

$$\boldsymbol{\Sigma}_t = \boldsymbol{L}_t \boldsymbol{G}_t \boldsymbol{L}_t',$$

where $\boldsymbol{G}_t = \text{diag}(g_{11,t}, \ldots, g_{mm,t})$ and $\boldsymbol{L}_t$ is a lower unitriangular matrix (i.e., triangular with 1's on the diagonal and 0's above the diagonal) with element $\ell_{ij,t}$ at the row i and column j for $i > j$.

Let us now illustrate this decomposition for $m = 3$.

$$\boldsymbol{L} = \begin{bmatrix} 1 & 0 & 0 \\ l_{21,t} & 1 & 0 \\ l_{31,t} & l_{32,t} & 1 \end{bmatrix} \quad \boldsymbol{G} = \begin{bmatrix} g_{11,t} & 0 & 0 \\ 0 & g_{22,t} & 0 \\ 0 & 0 & g_{33,t} \end{bmatrix}$$

and

$$\boldsymbol{\Sigma} = \begin{bmatrix} g_{11,t} & l_{21,t}g_{11,t} & l_{31,t}g_{11,t} \\ l_{21,t}g_{11,t} & l_{21,t}^2 g_{11,t} + g_{22,t} & l_{21,t}l_{31,t}g_{11,t} + l_{32,t}g_{22,t} \\ l_{31,t}g_{11,t} & l_{21,t}h_{31,t}g_{11,t} + l_{32,t}g_{22,t} & l_{31,t}^2 g_{11,t} + l_{32,t}^2 g_{22,t} + g_{33,t} \end{bmatrix}.$$

Darolles et al. (2018) show that if $\boldsymbol{w}_t \equiv (\mathbf{x}_t, y_t)'$ has mean $\mathbf{0}$,

$$\begin{aligned} w_{i,t} &= \sum_{j=1}^{i-1} \ell_{ij,t}\varepsilon_{j,t} + \varepsilon_{i,t} = \sum_{j=1}^{i-1} \ell_{ij,t}\left(w_{j,t} - \sum_{k=1}^{j-1} \ell_{jk,t}v_{k,t}\right) + \varepsilon_{i,t} \\ &= \sum_{j=1}^{i-1} \beta_{ij,t}w_{j,t} + \varepsilon_{i,t}. \end{aligned}$$

Interestingly, for $i = m$, the mth equation of the CHAR model is

$$y_t = \sum_{j=1}^{m-1} \beta_{ij,t}x_{j,t} + \varepsilon_{i,t}, \tag{61}$$

which corresponds to Eq. (35) when $\alpha_t = 0$, $N = m - 1$ and $\varepsilon_t = \varepsilon_{i,t}$.

Darolles et al. (2018) show that $g_{ii,t}$ is the conditional variance of $w_{i,t}$ and rely on a GARCH model to specify its dynamics. They also propose several specifications for the dynamics of the conditional betas and study the statistical properties of the MLE and Gaussian QML of this model. In their application, they retain the following specification of the conditional betas:

$$\beta_{ij,t} = \beta_{ij} + a_{ij}\varepsilon_{i,t-1}\varepsilon_{j,t-1} + b_{ij}\beta_{ij,t-1}. \tag{62}$$

The main drawback of this model is, therefore, that it requires estimating the system sequentially because $\beta_{ij,t}$ not only depends on $\varepsilon_{i,t-1}$ but also on $\varepsilon_{j,t-1}$, the error term of the j's (with $j < i$) equation in the Cholesky decomposition. Darolles et al. (2018) also derive stationarity conditions and prove the consistency and the asymptotic normality of the QML estimator of this model.

Building upon the CHAR model, Blasques et al. (2020) propose another model, called the autoregressive conditional beta (ACB) model, which does not require the estimation of the whole system and that outperforms the CHAR specification in the modeling of conditional betas.

Using the same notation as in Eq. (35), the ACB model is specified as

$$y_t = \beta_{0,t}x_{0,t} + \sum_{n=1}^{N} \beta_{n,t}x_{n,t} + \varepsilon_t \tag{63}$$

$$\varepsilon_t = \sigma_t z_t, \quad z_t \overset{\text{i.i.d.}}{\sim} N(0, 1) \tag{64}$$

$$\beta_{i,t} = \beta_i + \sum_{k=1}^{K} \theta_{i,k} Z_{k,t-1} + a_i x_{i,t-1}\varepsilon_{t-1} + b_i \beta_{i,t-1}, \forall i = 0, \ldots, N, \tag{65}$$

where $x_{0,t} = 1\ \forall t$ and σ_t^2 is a GARCH(1, 1) model as in (25). Note that $\beta_{0,t}x_{0,t} \equiv \alpha_t$ is a time-varying alpha when $a_0 \neq 0$ and $b_0 \neq 0$ but can be constrained to be a constant (like in the empirical application) by setting $a_0 = b_0 = 0$.

This model is very general because it nests the static model (31) when $\forall i = 0, \ldots, N$, $K = 0$ and $a_i = b_i = 0$ but also Model (37) with interaction variables when $\theta_{0,k} = 0$ and $\forall k = 0, \ldots, K$, $a_i = b_i = 0$.

Blasques et al. (2020) study the statistical properties of this model (stationarity and invertibility) but also those of the MLE and Gaussian Quasi-Maximum Likelihood estimators and prove convergence and asymptotic normality under mild conditions.

4 Empirical Application on REITs

Real Estate Investment Trusts[23] (REITs), which are publicly traded real estate companies that own and manage commercial or residential real estate, are attractive alternatives to the mainstream investment choice (e.g., stocks and bonds) since they allow investors to easily access real estate investments without directly owning or managing the underlying assets.[24] Moreover, the literature on real estate has shown that the inclusion of REITs within one's portfolio improves the risk-return profile of the portfolio. Compared to other asset classes such as bonds and stocks, they have the characteristics of offering more stable returns and a lower volatility historically. For the purpose of portfolio diversification, it is important to know how the level of exposure of REITs to both the bond market risk and to the stock market risk varies over time. The aim of this section is thus to perform a comparative analysis of the three most advanced modeling techniques (state space, DCB, and ACB) used in estimating the sensitivity of REIT indices to changes in both the bond market and the

[23]REITs were initially established in 1960 when the US Congress enacted the legislation authorizing their existence. Since then, REITs were authorized in many countries (even though REIT regimes vary country-by-country), including most European countries (e.g., the Netherlands in 1969, Belgium in 1995, France in 2003, Germany, Italy, the United Kingdom in 2007 or Portugal in 2019).

[24]Their revenues are mainly generated from the rents they receive and they do not have to pay any corporate tax in exchange for paying most of their taxable income to shareholders (US REITs must distribute 90% of their taxable income to shareholders through dividend payments for example).

stock market. Van Nieuwerburgh (2019) argues that a model with a bond market and stock market factor is both the most basic and most natural model of risk for REITs as the bond market beta measures how sensitive REITs are to changes in interest rates and the stock market beta measures how sensitive REITs are to changes in economic activity.[25] A similar model is used by Allen et al. (2000). Moreover, we note that the addition of three Fama-French risk factors (size, value, and momentum) to the original two-factor model in the study of Van Nieuwerburgh (2019) leaves the bond and stock market betas almost unchanged. As a consequence, we follow Van Nieuwerburgh (2019) and choose to perform our analysis on the following parsimonious two-factor model:

$$\widetilde{r}_{\text{REIT},t} = \alpha_t + \beta_{B,t}\widetilde{r}_{B,t} + \beta_{M,t}\widetilde{r}_{M,t} + \varepsilon_t, \tag{66}$$

where $\widetilde{r}_{\text{REIT}}$ is the excess return on the REIT market, measured by the daily excess return on the FTSE EPRA Nareit index, $\widetilde{r}_B$ is the excess return on the bond market, measured by the daily excess return on the sovereign bond index and $\widetilde{r}_M$ is the excess return on the stock market, measured by the daily excess return on the stock market index. Equation (66) corresponds to the conditional risk factor model that we use to estimate conditional betas on day t from a regression of the daily excess REIT index returns on the excess stock market and bond market returns. In the special case where $\boldsymbol{\beta}_t = (\beta_{B,t}, \beta_{M,t})' = \boldsymbol{\beta}$, the betas are restricted to be constant. Note also that for some models, the alpha is allowed to be time-varying as well. However, empirical results (not reported here to save space) suggest that $\alpha_t = \alpha$ $(\forall t)$ once allowing the conditional betas of this two-factor model to be time-varying.

Two strands of literature on REIT conditional betas are particularly relevant to our study. The first one considers the exposure of REITs to both interest-rate risk and stock market risk. Flannery and James (1984) put the emphasis on the fact that firms holding financial assets should be more sensitive to interest-rate risk. Allen et al. (2000) put forward four reasons why equity and mortgage REITs may be affected by changes in interest rates. First, REITs rely heavily on debt so that an increase in interest rates may dampen demand and have a negative impact on valuations (and vice versa). Second, an increase in interest rates may also translate into a higher cost of debt financing. Third, such an increase may result in a higher required rate of return by investors. And fourth, it may raise the cost of present development and refurbishment projects.

A second strand of literature considers potential regime shifts in market betas. Willard and Youguo (1991) find a decline in equity REIT betas over the period 1974–1983. Liang et al. (1995) in a similar study find that the market beta of equity REITs is rather stable over time while the market beta of mortgage REITS declined substantially over the period 1973–1989. However, Chiang et al. (2005) find that when using the Fama-French three-factor model, the declining trend in equity REIT

[25] We would probably formulate this a bit differently than Van Nieuwerburgh, arguing that market beta measures how sensitive REITs are to changes in economic activity *in the broad sense*, the variability of stock market returns being more related to the expectation of future profits than the real economic activity per se.

betas evaporates. Finally, Glascock (1991) tests for changes in the market beta of a REIT portfolio during bull and bear markets and finds that the beta behaves procyclically.

Despite the numerous empirical applications focusing on REITs, only a few papers examine how to best model the market beta of REITs. We can mention the papers by Zhou (2013) and Altınsoy et al. (2010). Zhou (2013) compares five modeling techniques in the estimation of the conditional beta of REITs: rolling regression, dynamic conditional correlation (DCC) GARCH model, Schwert and Seguin model, state space model, and Markov-switching model. In the same way, the study of Altınsoy et al. (2010) is based on the estimation of the conditional beta of Turkish REITs with a comparison of modeling techniques. Compared to the existing literature, the contribution of our study is four-fold : first we attempt to analyze how to model the conditional beta of REITs focusing on the two largest REIT markets (i.e., the US and European REIT markets), which allows us to compare both markets. Second, we extend the spectrum of modeling techniques by focusing on the most advanced conditional beta modeling techniques, that is to say the state space model, DCB and ACB modeling techniques. Third, we investigate the time variability of betas in a two-factor model and we introduce within this model exogenous variables that may affect the evolution of the bond market beta and the stock market beta. Finally, in addition to an in-sample beta estimation, we extend our analysis to an out-of-sample beta forecasting exercise.

4.1 Data and Sample

We focus on both the U.S. and developed Europe[26] REIT markets, REIT markets, proxied by the daily excess log-returns on the FTSE EPRA Nareit United States USD Total Return Index and the FTSE EPRA Nareit Developed Europe EUR Total Return Index. The FTSE EPRA Nareit data are from Thomson Reuters Eikon.

We explore the sensitivity of the two REIT indices to (1) the bond market factor as proxied by the daily excess log-returns on the S&P US Treasury Bond Index and the S&P Eurozone Developed Sovereign Bond Index respectively, and to (2) the stock market factor proxied by the daily excess log-returns on the S&P500 index and the EUROSTOXX600 index respectively. The bond market data are obtained from Standard and Poor's and the stock market data are from Thomson Reuters Eikon.

The risk-free rate is the one-month T-bill rate for the USA and the one-month Euribor rate for developed Europe computed on a daily basis. The data are from the Federal Reserve Economic Data (FRED) and the European Money Markets Institute (EMMI), respectively.

---

[26]Based on the FTSE EPRA Nareit Developed Europe EUR Total Return Index, developed Europe includes the following countries: Austria, Belgium, Finland, France, Germany, Ireland, Italy, the Netherlands, Norway, Spain, Sweden, Switzerland, and the United Kingdom.

Assuming that beta can be influenced by risk aversion, we use three risk aversion indicators commonly used in the literature (Bank 2007) as exogenous variables that may affect the evolution of the bond market beta and the stock market beta. In particular, we use implied volatility as an indicator of the volatility that is expected by the market, the Ted Spread as an indicator of credit risk in the interbank money market, and the High Yield Option-Adjusted Spread as an indicator of credit risk.

The volatility indices, i.e., the VIX for the USA and the VSTOXX for developed Europe, respectively measure the 30-day expected volatility of the US stock market based on the S&P500 option prices and the 30-day expected volatility of the European stock market based on the EUROSTOXX50 option prices. We hereafter both call them VIX for simplicity. The Ted Spread is the difference between the three-month Treasury bill and the three-month USD LIBOR rate.[27] The US High Yield Index Option-Adjusted Spread is the difference between a computed option-adjusted spread (OAS) index of all bonds in the Bank of BofAML US High Yield Master II Index and a spot Treasury curve. As for the EUR High Yield Index Option-Adjusted Spread of BofAML, it is the European equivalent of the US High Yield Index Option-Adjusted Spread.

The full sample runs from October 02, 2009, to October 01, 2019, which amounts to 2554 daily returns.

4.2 *Empirical Results*

The aim of our empirical study is to compare the performance of the competing models from two perspectives: in-sample beta estimates on the one hand and out-of-sample beta forecasts on the other hand. The in-sample analysis is meant to assess how well the different models fit the data while the out-of-sample analysis is useful in assessing what modeling technique provides the best beta forecasts in a tracking exercise. A better beta forecast can then be used as an input within many financial applications.

We use the first 2304 observations of our sample as the in-sample period and the remaining 250 observations as the out-of-sample period. Having performed a sensitivity analysis on the same sample by increasing the number of out-of-sample period observations (to 500 and 750 observations respectively) and having found that the results were qualitatively the same, we do not report them to save space.

4.2.1 In-Sample Estimates

Seven competing models have been estimated using the Ox programming language (Doornik 2012) and the G@RCH 8.0 software (Laurent 2018) to obtain the condi-

[27] While the Ted Spread originally measures interbank risk in the US, we believe that this indicator may also be relevant for Europe as well, the US being the leading engine of credit worldwide.

tional bond market betas ($\beta_{B,t}$) and the conditional stock market betas ($\beta_{M,t}$) for both the USA and developed Europe. We estimate the following two-factor model:

$$\widetilde{r}_{\mathrm{REIT},t} = \alpha_t + \beta_{B,t}\widetilde{r}_{B,t} + \beta_{M,t}\widetilde{r}_{M,t} + \varepsilon_t. \tag{67}$$

The seven competing models are the following:

1. OLS model:

$$\begin{aligned} \alpha_t &= \alpha,\ \beta_{B,t} = \beta_B,\ \beta_{M,t} = \beta_M\ \forall t, \\ \varepsilon_t &\overset{\text{i.i.d.}}{\sim} N(0, \sigma^2). \end{aligned}$$

2. Univariate GARCH model:

$$\begin{aligned} \alpha_t &= \alpha,\ \beta_{B,t} = \beta_B,\ \beta_{M,t} = \beta_M\ \forall t, \\ \varepsilon_t &= \sigma_t z_t, \quad z_t \overset{\text{i.i.d.}}{\sim} N(0, 1), \\ \sigma_t^2 &= \lambda_0 + \lambda_1 \sigma_{t-1}^2 + \lambda_2 \varepsilon_{t-1}^2. \end{aligned}$$

3. Univariate GARCH model with interaction variables (GARCH-Z):

$$\begin{aligned} \alpha_t &= \alpha\ \forall t, \\ \beta_{B,t} &= c_B + \theta_{B,\mathrm{TED}} \mathrm{TED}_{t-1} + \theta_{B,\mathrm{HY}} \mathrm{HY}_{t-1}, \\ \beta_{M,t} &= c_M + \theta_{M,\mathrm{VIX}} \mathrm{VIX}_{t-1}, \\ \varepsilon_t &= \sigma_t z_t, \quad z_t \overset{\text{i.i.d.}}{\sim} N(0, 1), \\ \sigma_t^2 &= \lambda_0 + \lambda_1 \sigma_{t-1}^2 + \lambda_2 \varepsilon_{t-1}^2, \end{aligned}$$

   where TED_t, HY_t, and VIX_t are, respectively, the TED spread, the High Yield Index Option-Adjusted Spread, and the VIX index, $\theta_{B,\mathrm{TED}}$ and $\theta_{B,\mathrm{HY}}$ are the coefficients of the two interaction variables entering into the conditional betas of the bond market, and $\theta_{M,\mathrm{VIX}}$ is the coefficient of the interaction variable entering into the conditional betas of the stock market.
4. State Space Model (SSM):

$$\begin{aligned} \alpha_t &= \alpha\ \forall t, \\ \beta_{B,t} &= \beta_{B,t-1} + u_{B,t}, \\ \beta_{M,t} &= \beta_{M,t-1} + u_{M,t}, \end{aligned}$$

$$\begin{pmatrix} u_{B,t} \\ u_{M,t} \\ \varepsilon_t \end{pmatrix} |\mathbf{Y}_{t-1}, \mathbf{X}_t \overset{\text{i.i.d}}{\sim} N\left(\begin{pmatrix} 0 \\ 0 \\ 0 \end{pmatrix}, \begin{pmatrix} \sigma_B^2 & 0 & 0 \\ 0 & \sigma_M^2 & 0 \\ 0 & 0 & \sigma_\varepsilon^2 \end{pmatrix}\right).$$

5. Dynamic Conditional Beta Model (DCB):

$$\begin{pmatrix}\mathbf{x}_t\\ y_t\end{pmatrix}|\mathcal{F}_{t-1} \sim N\left(\begin{pmatrix}\mathbf{0}\\ 0\end{pmatrix}, \Sigma_t \equiv \begin{pmatrix}\Sigma_{\mathbf{xx},t} & \Sigma_{\mathbf{x}y,t}\\ \Sigma_{y\mathbf{x},t} & \Sigma_{yy,t}\end{pmatrix}\right),$$
$$\alpha_t = \alpha \forall t,$$
$$\widehat{\boldsymbol{\beta}}_t^{DCB} \equiv (\hat{\beta}_{B,t}, \hat{\beta}_{M,t})' = \Sigma_{\mathbf{xx},t}^{-1}\Sigma_{\mathbf{x}y,t}.$$

where $\mathbf{x}_t = (\widetilde{r}_{B,t}, \widetilde{r}_{M,t})'$, $y_t = \widetilde{r}_{\mathrm{REIT},t}$ and Σ_t is specified as a DCC-GARCH(1, 1) model.

6. Autoregressive Conditional Beta (ACB):

$$\begin{aligned}\alpha_t &= \alpha\, \forall t,\\ \beta_{B,t} &= c_B + a_B x_{i,t-1}\varepsilon_{t-1} + b_B\beta_{B,t-1},\\ \beta_{M,t} &= c_M + a_M x_{i,t-1}\varepsilon_{t-1} + b_M\beta_{M,t-1},\\ \varepsilon_t &= \sigma_t z_t, \quad z_t \overset{\text{i.i.d.}}{\sim} N(0, 1).\end{aligned}$$

7. Autoregressive Conditional Beta with interaction variables (ACB-Z):

$$\begin{aligned}\alpha_t &= \alpha\, \forall t,\\ \beta_{B,t} &= c_B + a_B x_{i,t-1}\varepsilon_{t-1} + b_B\beta_{B,t-1},\\ \beta_{M,t} &= c_M + a_M x_{i,t-1}\varepsilon_{t-1} + b_M\beta_{M,t-1} + \theta_{M,\mathrm{VIX}}\mathrm{VIX}_{t-1},\\ \varepsilon_t &= \sigma_t z_t, \quad z_t \overset{\text{i.i.d.}}{\sim} N(0, 1),\end{aligned} \tag{68}$$

where $\theta_{B,\mathrm{VIX}}$ is the coefficient of the interaction variable entering into the conditional betas of the stock market.[28]

To ease the presentation of the results throughout the rest of the paper, we use GARCH, GARCH-Z, DCB, SSM, ACB, and ACB-Z to, respectively, denote the GARCH model, the GARCH model with interaction variables, the dynamic conditional beta model, the state space model, the autoregressive conditional beta model and the autoregressive conditional beta model with interaction variables.

The GARCH-Z model corresponds to a GARCH model with interaction variables in the conditional mean or equivalently an ACB model with no dynamics (i.e., $a_i = b_i = 0, \forall i$) but with explanatory variables. SSM is a state space model without dynamics in the intercept but in which the two slope coefficients follow a random walk. The ACB and ACB-Z models are two ACB models, but exogenous explanatory variables are only included in the latter model. Note also that a GARCH specification is included in the residuals of all the models except for the OLS and SSM.

[28]The other two coefficients entering into the conditional betas of the bond market (i.e., $\theta_{B,\mathrm{TED}}$ and $\theta_{B,\mathrm{HY}}$) are found to be insignificant so that the corresponding variables are removed from the model.

Estimation results of the different models are reported in Tables 1 and 2, respectively, for the USA and developed Europe.

We do report the estimated parameters for the sake of completeness but it is hard to draw any conclusion from them. To compare the models in-sample, we rely on both the log-likelihood and the Bayesian information criterion (BIC). Table 3 reports the number of parameters, the log-likelihood as well as the BIC of all the models at the exception of the DCB.[29] Results suggest that the GARCH-Z, ACB, and ACB-Z models always outperform the other models for both the USA and developed Europe. Interestingly, when comparing the GARCH-Z model to the GARCH model based on both the BIC and a likelihood ratio test, we find that results are clearly in favor of the GARCH-Z model, suggesting that the risk factors used to capture the dynamics in the conditional betas are relevant. Results reported in Tables 1 and 2 indeed suggest that these three variables help predict the dynamics of the two conditional betas in the GARCH-Z model. However, although the ACB model does not rely on these exogenous factors, it further improves the estimation of the dynamics of the two conditional betas.

According to the BIC, the best model is the ACB-Z model, i.e., an ACB model with the additional VIX explanatory variable in the conditional beta of the stock market. Note however that the SSM model imposes the variance of the error term to be homoscedastic, which certainly explains why it performs so badly according to the BIC. Extending the SSM model by accounting for GARCH effects in the residuals is, therefore, desirable but beyond the scope of this paper.

The estimated conditional betas $\beta_{B,t}$ and $\beta_{M,t}$ are plotted in Figs. 1 and 2, respectively, for the USA and developed Europe. Each graph contains the estimated betas for the seven competing models of our study. Both graphs reveal large fluctuations over time regarding the exposure of REITs to the two risk factors of our model.

The reading of these graphs leads to various observations. First, we observe that both the bond market beta and the stock market beta are not constant over time but time varying (contrary to the assumption made when using the OLS approach), which confirms the appropriateness of conditional CAPM modeling. Second, the stock market beta for the USA and to a lesser extent for developed Europe is on a declining track over the considered period, implying that the sensitivity of the REIT sector to the overall equity market is decreasing and can be interpreted as a sign of a maturing REIT market, which is consistent with the out-of-sample results of Zhou (2013), even though his study ends in 2011. The picture is however different regarding the bond market beta of both the USA and developed Europe since the bond market beta of both areas is rather on a rising track over the same period. This difference justifies both using a two-factor model and comparing the USA to Europe. Third, while we observe that all the dynamic methods (DCB, ACB, SSM) present similar dynamics, we also note that the conditional betas of the DCB model are far more erratic than those of the SSM and ACB models. Indeed, the conditional betas

[29]Recall that the DCB model requires estimating an MGARCH model (i.e., a DCC-GARCH model in our case) so that the obtained log-likelihood is for the joint distribution of the three series $(\widetilde{r}_{\mathrm{REIT},t}, \widetilde{r}_{B,t}, \widetilde{r}_{M,t})'$ and not for $\widetilde{r}_{\mathrm{REIT},t}$ given $\widetilde{r}_{B,t}$ and $\widetilde{r}_{M,t}$ as for the other six models.

Table 1 Parameter estimates for the USA

	α	β_B	β_M									
OLS	−0.018	1.243	1.081									
	(0.017)	(0.126)	(0.031)									
	λ_0	λ_1	λ_2	α	β_B	β_M						
GARCH	0.011	0.064	0.919	−0.014	1.463	0.970						
	(0.006)	(0.016)	(0.0232)	(0.015)	(0.121)	(0.030)						
	λ_0	λ_1	λ_2	c_B	c_M	a_B	a_M	b_B	b_M	$\theta_{B,\text{TED}}$	$\theta_{B,\text{HY}}$	$\theta_{M,\text{VIX}}$
GARCH-Z	0.007	0.041	0.947	2.887	0.524					−4.393	27.976	2.104
	(0.004)	(0.012)	(0.017)	(0.496)	(0.073)					(0.878)	(8.768)	(0.363)
ACB	0.007	0.034	0.950	0.004	0.002	−0.164	0.016	0.999	0.997			
	(0.005)	(0.012)	(0.020)	(0.000)	(0.000)	(0.007)	(0.001)	(0.000)	(0.000)			
ACB-Z	0.007	0.037	0.947	0.005	0.002	−0.156	0.015	0.997	0.990			0.032
	(0.005)	(0.012)	(0.021)	(0.000)	(0.000)	(0.004)	(0.001)	(0.000)	(0.000)			(0.000)
	σ_e	σ_B	σ_M	σ_η								
SSM	0.713	0.061	0.017	0.000								

The table reports the parameter estimates of the different models. Standard errors are in parentheses. *GARCH-Z* corresponds to the GARCH model with interaction variables. *ACB-Z* corresponds to the *ACB* model with both dynamics and an explanatory variable in the market beta. *SSM* corresponds to the state space model. Parameters of the DCC model used to get the conditional betas of the *DCB* are not reported

Table 2 Parameter estimates for developed Europe

	α	β_B	β_M									
OLS	0.008	0.478	0.808									
	(0.012)	(0.083)	(0.019)									
	λ_0	λ_1	λ_2	α	β_B	β_M						
GARCH	0.010	0.052	0.921	0.016	0.448	0.783						
	(0.007)	(0.022)	(0.041)	(0.012)	(0.075)	(0.021)						
	λ_0	λ_1	λ_2	c_B	c_M	a_B	a_M	b_B	b_M	$\theta_{B,\text{TED}}$	$\theta_{B,\text{HY}}$	$\theta_{M,\text{VIX}}$
GARCH-Z	0.007	0.043	0.936	0.515	0.487					−1.158	18.489	1.144
	(0.004)	(0.017)	(0.028)	(0.214)	(0.052)					(0.357)	(6.360)	(0.224)
ACB	0.009	0.058	0.913	0.007	0.021	0.201	0.023	0.983	0.971			
	(0.009)	(0.035)	(0.058)	(0.006)	(0.011)	(0.153)	(0.01)	(0.014)	(0.014)			
ACB-Z	0.008	0.046	0.930	0.009	0.058	0.263	0.015	0.979	0.863			0.192
	(0.005)	(0.020)	(0.032)	(0.005)	(0.031)	(0.151)	(0.007)	(0.011)	(0.070)			(0.107)
	σ_e	σ_B	σ_M	σ_η								
SSM	0.551	0.093	0.031	0.0000								

See the note of Table 1

Table 3 Comparison of the different models based on log-likelihood

	#para	United States		Developed Europe	
		Log-lik	BIC	Log-lik	BIC
OLS	4	−2726.81	2.3805	−2046.09	1.7896
GARCH	6	−2588.30	2.2670	−1972.63	1.7325
GARCH-Z	9	−2502.28	2.2033	−1930.44	1.7067
SSM	4	−2531.73	2.2010	−1996.49	1.7364
ACB	10	−2444.96	2.1560	−1939.16	1.7169
ACB-Z	11	−2442.61	2.1582	−1925.20	1.7089

The table reports the number of parameters (#para), log-likelihood (Log-lik) and Bayesian Information Criterion (BIC) of all the models (at the exception of the DCB) for both the United States and developed Europe

obtained with the DCB model are very choppy over the estimation period while those obtained with the SSM and ACB models are much smoother on average. Fourth, we observe periods where the betas obtained with the different models are very close to one another and periods where the betas are very far from one another so that one can expect differences between the different models in terms of performance. Finally, the conditional bond and stock market betas filtered with the SSM model are smoother than those obtained with the other models of our comparative study. However, this effect can be explained by the fact a random walk is imposed in this model. Finally, we find that the correlation between the betas of the SSM model and the betas of the ACB model is very high (0.89 for the U.S. bond market beta, 0.94 for the U.S. stock market beta, 0.77 for the developed Europe bond market beta, 0.71 for the developed Europe stock market beta) and we note that the correlation remains high when we add explanatory variables to the betas of the ACB model.

4.2.2 Out-of-Sample Estimates

The aim of this section is to illustrate the usefulness of the competing models in a portfolio and risk management exercise. Since realized betas are not observed, it is impossible to judge the quality of the models by looking at the forecasting errors of the conditional betas.

Instead, following Engle (2016) and Darolles et al. (2018), we perform a tracking exercise that consists in taking a position at time t in the two considered factors (bond and market) whose weights are the one-step ahead forecasts of the corresponding conditional betas.

For each model, the conditional betas forecasts are, therefore, used to construct a hedging portfolio. The returns on this portfolio are obtained using the conditional betas forecasts, i.e.,

$$z_{\text{REIT},t+1|t} = \beta_{B,t+1|t}\widetilde{r}_{B,t+1} + \beta_{M,t+1|t}\widetilde{r}_{M,t+1}, \tag{69}$$

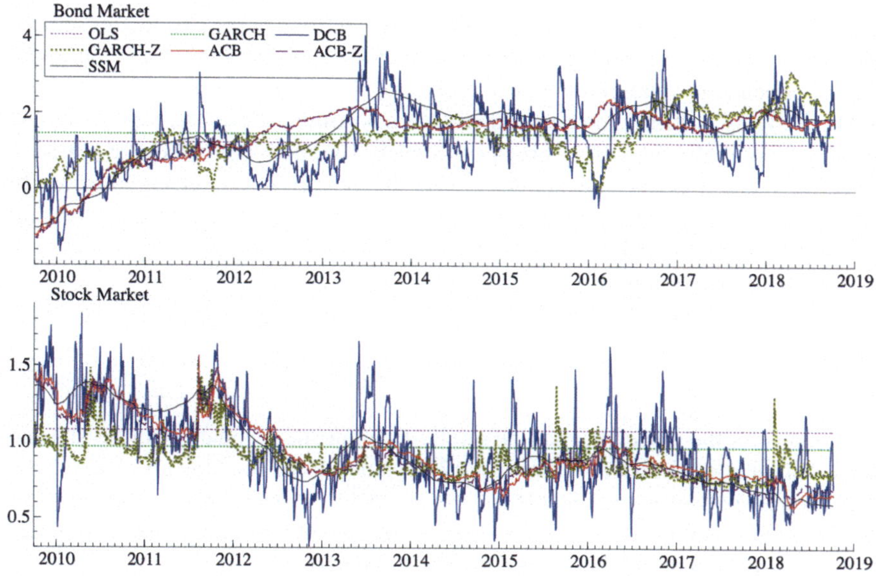

Fig. 1 In-sample conditional betas for the USA

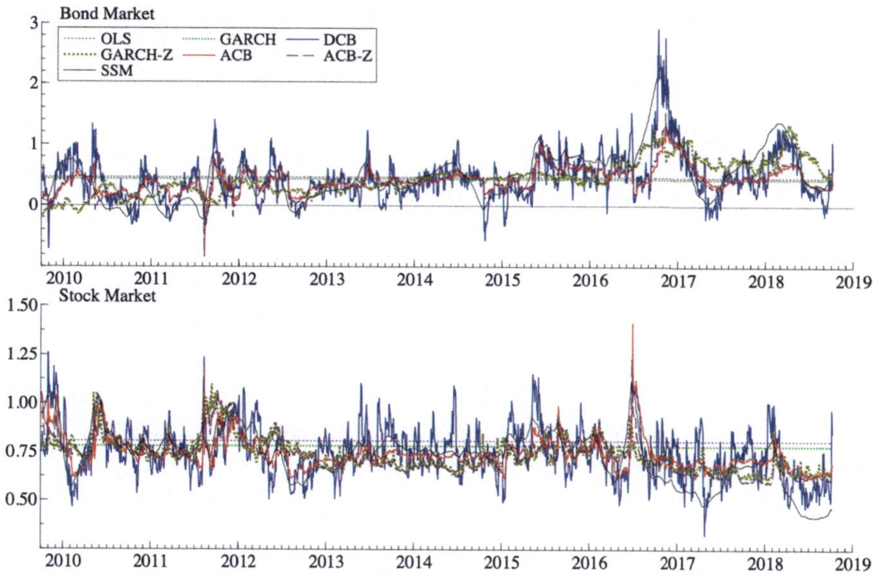

Fig. 2 In-sample conditional betas for developed Europe

where $\widetilde{r}_{B,t+1}$ and $\widetilde{r}_{M,t+1}$ are the realized excess log-returns of the two factors at time $t+1$ while $\beta_{B,t+1|t}$ and $\beta_{M,t+1|t}$ are the one-step-ahead forecasts of the conditional betas obtained at the end of day t.

This hedging portfolio can be interpreted as a portfolio invested in the risk factors and which optimally tracks the corresponding REIT returns. It is a hedging portfolio in the sense that it can be sold short to hedge the main risks of a given portfolio. In this asset pricing context, expected returns on any asset are linear in the betas and only depend upon the risk premiums embedded in the factors. In other words, there is no alpha or intercept in (69).

For both the USA and developed Europe, we compute the ex-post tracking errors as follows:

$$TE_{t+1|t} = \widetilde{r}_{\text{REIT},t+1} - Z_{\text{REIT},t+1|t} \tag{70}$$

and we look for the model that has the smallest sample mean square error (MSE) and mean absolute deviation (MAD) over the 250 values of the tracking errors using the model confidence set approach of Hansen et al. (2011). Models are reevaluated every 25 steps so that estimated parameters are kept constant to produce 25 one-step-ahead forecasts of the conditional betas before being updated.

The forecasted one-steap-ahead conditional betas $\beta_{B,t+1|t}$ and $\beta_{M,t+1|t}$ are plotted in Figs. 3 and 4, respectively, for the USA and developed Europe. Each graph contains the forecasted betas for the seven competing models. Both graphs again reveal large fluctuations over time regarding the exposure of REITs to the two risk factors of our model and we observe large discrepancies between the competing methods.

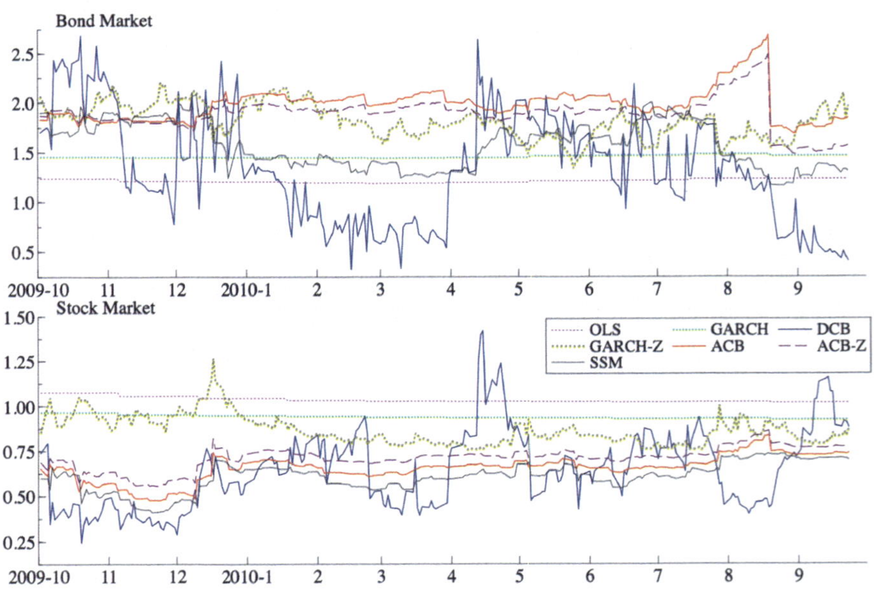

Fig. 3 Out-of-sample conditional betas for the USA

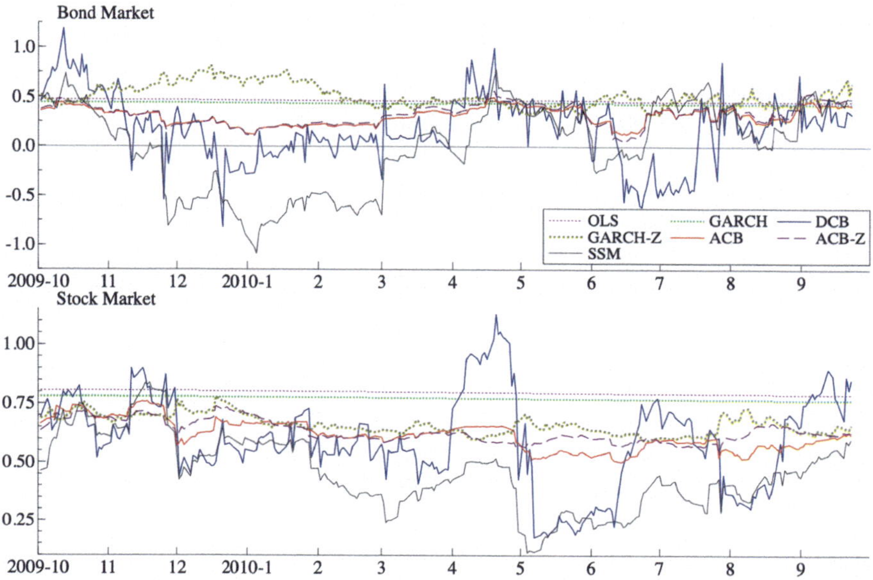

Fig. 4 Out-of-sample conditional betas for developed Europe

Table 4 reports the MSE and the MAD as well as the results of the MCS test with an MSE loss function, a significance level of 5%, and 10,000 bootstrap samples (with a block length of five observations). Models highlighted with the symbol ✓ are contained in the model confidence set (or set of superior models) when relying either on the MSE or MAD loss function (i.e., results are identical for both loss functions). The ACB, ACB-Z and SSM models clearly outperform the other models. The ACB model has the lowest MSE for both series although its MSE is not statistically different from those of the ACB-Z and SSM at the 5% nominal size according to the MCS test. Interestingly, the models with constant betas clearly underperform the models with time-varying betas. Furthermore, the conditional betas of the DCB model are found to be much more volatile than those of the ACB models and the SSM (which would imply higher transaction costs). Finally, it appears that the DCB model significantly underperforms in this tracking exercise.

5 Conclusion

In this chapter, we review the different time series models used to estimate static and time-varying betas and then compare the performance of the standard static beta models (i.e., OLS and GARCH models) to the most advanced conditional beta modeling techniques that are the state space model, the dynamic conditional beta model and the autoregressive conditional beta model (with or without additional

Table 4 Comparison of out-of-sample results

	MSE	MAD	MCS
Panel A: United States			
OLS	0.8463	0.6670	
GARCH	0.7381	0.6240	
GARCH-Z	0.7273	0.6175	
DCB	0.6808	0.6040	
SSM	0.6336	0.5694	✓
ACB	0.6071	0.5769	✓
ACB-Z	0.6018	0.5723	✓
Panel B: Developed Europe			
OLS	0.4758	0.5391	
GARCH	0.4663	0.5344	
GARCH-Z	0.4426	0.5219	
DCB	0.4504	0.5311	
SSM	0.4279	0.5201	✓
ACB	0.4063	0.5042	✓
ACB-Z	0.4074	0.5058	✓

The table reports three evaluation criteria: mean square error (MSE), mean absolute deviation (MAD), and model confidence set (MCS). Models highlighted with the symbol ✓ are contained in the model confidence set (or set of superior models)

exogenous variables). The analysis is performed on the two largest REIT markets in the world, that is to say the U.S. and developed Europe REIT markets, over the period 2009–2019. In particular, we investigate the time variability of betas in a two-factor model, where $\beta_{B,t}$ and $\beta_{M,t}$ are respective measures of the sensitivity of the REIT index to changes in the bond market and the stock market. Assuming that beta may depend on risk aversion, we use three risk aversion indicators as exogenous variables that may affect the evolution of the bond market beta and the stock market beta.

Based on the employed evaluation criteria, we evaluate the performance of the seven competing models both in terms of in-sample estimates and through an out-of-sample tracking exercise. Results reveal several meaningful findings. First, dynamic models clearly outperform static models both in- and out-of-sample, meaning that both the bond market beta and the stock market beta are not constant over time but time varying, which gives convincing arguments for modeling conditional, instead of static, betas. Second, the autoregressive conditional beta model with additional exogenous variables outperforms the other techniques for both the USA and developed Europe, followed by the autoregressive conditional beta model without additional variables and the state space model. The dynamic conditional beta model delivers an unsatisfactory out-of-sample predictive performance. Finally, the inclusion of risk aversion indicators as exogenous variables into the ACB model (but also into the GARCH model) helps improve the prediction of betas.

These results can be used in many financial situations, like for example, in estimating the cost of capital with the aim of capital budgeting involving REITs, in evaluating the performance of REIT portfolios or in deciding what asset allocation or portfolio diversification to target.

Acknowledgements The second author gratefully thanks the FNRS for financial support.

References

Adrian, T., & Franzoni, F. (2009). Learning about beta: Time-varying factor loadings, expected returns, and the conditional CAPM. *Journal of Empirical Finance*, *16*, 537–556.

Alexander, G. J., & Chervany, N. L. (1980). On the estimation and stability of beta. *Journal of Financial and Quantitative Analysis*, *15*(1), 123–137.

Allen, M. T., Madura, J., & Springer, T. M. (2000). Reit characteristics and the sensitivity of reit returns. *The Journal of Real Estate Finance and Economics*, *21*(2), 141–152.

Altınsoy, G., Erol, I., & Yıldırak, S. (2010). Time-varying beta risk of turkish real estate investment trusts. *Middle East Technical University Studies in Development*, *37*(2), 83–114.

Andersen, T. G., Bollerslev, T., Diebold, F. X., & Labys, P. (2003). Modelling and forecasting realized volatility. *Econometrica*, *71*(2), 579–625.

Andersen, T. G., Bollerslev, T., Diebold, F. X., & Wu, G. (2006). Realized beta: Persistence and predictability. In T. Fomby & D. Terrell (Eds.), *Econometric analysis of financial and economic time series*. Advances in econometrics (Vol. 20, pp. 1–39). Bingley: Emerald Group Publishing Limited.

Bali, T. G., & Engle, R. F. (2010). Resurrecting the conditional CAPM with dynamic conditional correlations. *SSRN Electronic Journal*.

Bank, E. C. (2007). Measuring investors' risk appetite. *Financial Stability Review*, 166–171.

Barndorff-Nielsen, O. E., & Shephard, N. (2004). Econometric analysis of realized covariation: High frequency based covariance, regression, and correlation in financial economics. *Econometrica*, *72*(3), 885–925.

Bauwens, L., Laurent, S., & Rombouts, J. (2006). Multivariate GARCH models: A survey. *Journal of Applied Econometrics*, *21*, 79–109.

Blasques, F., Francq, C., & Laurent, S. (2020). *Dynamic conditional betas*. Working paper.

Bollerslev, T., Engle, R. F., & Wooldridge, J. M. (1988). A capital asset pricing model with time-varying covariances. *Journal of Political Economy*, *96*(1), 116–131.

Bollerslev, T., et al. (1990). Modelling the coherence in short-run nominal exchange rates: A multivariate generalized arch model. *Review of Economics and Statistics*, *72*(3), 498–505.

Boudt, K., Lunde, A., Laurent, S., Quaedvlieg, R., & Sauri, O. (2017). Positive semidefinite integrated covariance estimation, factorizations and asynchronicity. *Journal of Econometrics*, *196*(2), 347–367.

Brooks, C. (2014). *Introductory econometrics for finance*. Cambridge: Cambridge University Press.

Brooks, R., Faff, R., & Lee, J. (1992). The form of time variation of systematic risk: Testing for beta instability. *Pacific-Basin Finance Journal*, *2*, 191–198.

Brooks, R. D., Faff, R. W., & Lee, J. H. (1994). Beta stability and portfolio formation. *Pacific-Basin Finance Journal*, *2*(4), 463–479.

Carhart, M. M. (1997). On persistence in mutual fund performance. *The Journal of Finance*, *52*(1), 57–82.

Chiang, K. C., Lee, M.-L., & Wisen, C. H. (2005). On the time-series properties of real estate investment trust betas. *Real Estate Economics*, *33*(2), 381–396.

Choudhry, T., & Wu, H. (2008). Forecasting ability of GARCH vs Kalman filter method: Evidence from daily UK time-varying beta. *Journal of Forecasting*, *27*, 670–689.

Cisse, M., Konte, M., Toure, M., & Assani, S. (2019). Contribution to the valuation of BRVM's assets: A conditional CAPM approach. *Journal of Risk and Financial Management, 12*, 27.

Cochrane, J. (2005). *Asset pricing*. Princeton: Princeton Univ. Press.

Cochrane, J. H., & Culp, C. L. (2003). Equilibrium asset pricing and discount factors: Overview and implications for derivatives valuation and risk management. In L. P. Hughston, (Ed.), Modern risk management: A history, introduced by Peter Field (Ch. 5, pp. 57–92). London, UK: Risk Books.

Darolles, S., Francq, C., & Laurent, S. (2018). Asymptotics of cholesky GARCH models and time-varying conditional betas. *Journal of Econometrics*, *204*(2), 223–247.

DeJong, D. V., & Collins, D. W. (1985). Explanations for the instability of equity beta: Risk-free rate changes and leverage effects. *Journal of Financial and Quantitative Analysis*, *20*(1), 73–94.

Doornik, J. (2012). Object-oriented matrix programming using Ox. London: Timberlake Consultants Press.

Durbin, J., & Koopman, S. J. (2001). *Time series analysis by state space methods*. Oxford statistical science series. Oxford: Oxford University Press.

Engle, R. F. (2016). Dynamic conditional beta. *Journal of Financial Econometrics*, *14*(4), 643–667.

Fabozzi, F. J., & Francis, J. C. (1977). Stability tests for alphas and betas over bull and bear market conditions. *The Journal of Finance*, *32*(4), 1093–1099.

Faff, R. W., Hillier, D., & Hillier, J. (2000). Time varying beta risk: An analysis of alternative modelling techniques. *Journal of Business Finance and Accounting*, *27*, 523–554.

Fama, E. F., & French, K. R. (1993). Common risk factors in the returns on stocks and bonds. *Journal of Financial Economics*, *33*(1), 3–56.

Fama, E. F., & French, K. R. (2004). The capital asset pricing model: Theory and evidence. *Journal of Economic Perspectives*, *18*(3), 25–46.

Fama, E. F., & French, K. R. (2015). A five-factor asset pricing model. *Journal of Financial Economics*, *116*(1), 1–22.

Fama, E. F., & MacBeth, J. D. (1973). Risk, return, and equilibrium: Empirical tests. *Journal of Political Economy*, *81*(3), 607–636.

Ferson, W. E. (2003). Tests of multifactor pricing models, volatility bounds and portfolio performance. In G. Constantinides, M. Harris, & R. Stulz (Eds.), *Handbook of the economics of finance*. Financial markets and asset pricing (Vol. 1, pp. 743–802). Amsterdam: Elsevier.

Ferson, W. E., & Jagannathan, R. (1996). Econometric evaluation of asset pricing models. In G. Maddala, & C. Rao (Eds.), *Handbook of statistics*. Statistical methods in finance (Vol. 14, pp. 1–33). Amsterdam: Elsevier.

Fisher, L., & Kamin, J. H. (1985). Forecasting systematic risk: Estimates of 'raw' beta that take account of the tendency of beta to change and the heteroskedasticity of residual returns. *Journal of Financial and Quantitative Analysis*, *20*(2), 127–149.

Flannery, M. J., & James, C. M. (1984). The effect of interest rate changes on the common stock returns of financial institutions. *The Journal of Finance*, *39*(4), 1141–1153.

Gagliardini, P., Ossola, E., & Scaillet, O. (2016). Time-varying risk premium in large cross-sectional equity data sets. *Econometrica*, *84*(3), 985–1046.

Glascock, J. L. (1991). Market conditions, risk, and real estate portfolio returns: Some empirical evidence. *The Journal of Real Estate Finance and Economics*, *4*(4), 367–373.

Hamilton, J. D. (1994). *Time series analysis*. Princeton: Princeton University Press.

Hansen, L. P. (1982). Large sample properties of generalized method of moments estimators. *Econometrica*, *50*(4), 1029–1054.

Hansen, P., Lunde, A., & Nason, J. (2011). The model confidence set. *Econometrica*, *79*, 453–497.

Hansen, P. R., Lunde, A., & Voev, V. (2014). Realized beta GARCH: A multivariate GARCH model with realized measures of volatility. *Journal of Applied Econometrics*, *29*(5), 774–799.

Harvey, C. R., Liu, Y., & Zhu, H. (2015). ...and the cross-section of expected returns. *The Review of Financial Studies*, *29*(1), 5–68.

Harvey, C. R., & Siddique, A. (2000). Conditional skewness in asset pricing tests. *The Journal of Finance*, *55*, 1263–1295.

Huang, P., & Hueng, C. J. (2009). Interest-rate risk factor and stock returns: A time-varying factor-loadings model. *Applied Financial Economics*, *19*, 1813–1824.

Laurent, S. (2018). *G@RCH 8. Estimating and forecasting GARCH models*. London: Timberlake Consultants Ltd.

Lewellen, J., & Nagel, S. (2006). The conditional CAPM does not explain asset-pricing anomalies. *Journal of Financial Economics*, *82*, 289–314.

Liang, Y., Prudential, W., & Webb, J. (1995). Intertemporal changes in the riskiness of reits. *Journal of Real Estate Research*, *10*(4), 427–443.

Linnainmaa, J. T., & Roberts, M. R. (2018). The history of the cross-section of stock returns. *The Review of Financial Studies*, *31*(7), 2606–2649.

Lintner, J. (1965). The valuation of risk assets and the selection of risky investments in stock portfolios and capital budgets. *Review of Economics and Statistics*, *47*, 13–37.

Mergner, S., & Bulla, J. (2008). Time-varying beta risk of Pan-European industry portfolios: A comparison of alternative modeling techniques. *European Journal of Finance*, *14*, 771–802.

Merton, R. C. (1973). An intertemporal capital asset pricing model. *Econometrica*, *41*(5), 867–887.

Moryson, M. (1998). *Testing for random walk coefficients in regression and state space models. Contributions to statistics*. Heidelberg: Physica-Verlag.

Nieto, B., Susan, O., & Ainoha, Z. (2014). Time-varying market beta: Does the estimation methodology matter? *SORT—Statistics and Operations Research Transactions*, *38*(1), 13–42.

Ohlson, J., & Rosenberg, B. (1982). Systematic risk of the CRSP equal-weighted common stock index: A history estimated by stochastic-parameter regression. *Journal of Business*, 121–145.

Patton, A. J., & Verardo, M. (2012). Does beta move with news? Firm-specific information flows and learning about profitability. *The Review of Financial Studies*, *25*(9), 2789–2839.

Phelan, G., & Toda, A. A. (2015). *On the robustness of theoretical asset pricing models*. Tech. rep. Department of Economics, Williams College.

Pourahmadi, M. (1999). Joint mean-covariance models with applications to longitudinal data: Unconstrained parameterisation. *Biometrika*, *86*(3), 677–690.

Reyes, M. G. (1999). Size, time-varying beta, and conditional heteroscedasticity in UK stock returns. *Review of Financial Economics*, *8*(1), 1–10.

Roll, R. (1977). A critique of the asset pricing theory's tests. Part I: On past and potential testability of the theory. *Journal of Financial Economics*, *4*,

Ross, S. A. (1976). The arbitrage theory of capital asset pricing. *Journal of Economic Theory*, *13*, 341–360.

Schadt, W. E. F. R. W. (1996). Measuring fund strategy and performance in changing economic conditions. *The Journal of Finance*, *51*, 425–461.

Schwert, G. W., & Seguin, P. J. (1990). Heteroskedasticity in stock returns. *The Journal of Finance*, *45*(4), 1129–1155.

Shanken, J. (1990). Intertemporal asset pricing: An empirical investigation. *Journal of Econometrics*, *45*, 99–120.

Sharpe, W. F. (1964). Capital asset prices: A theory of market equilibrium under conditions of risk. *The Journal of Finance*, *19*(3), 425–442.

Smith, P., & Wickens, M. (2002). Asset pricing with observable stochastic discount factors. *Journal of Economic Surveys*, *16*, 397–446.

Sunder, S. (1980). Stationarity of market risk: Random coefficients tests for individual stocks. *The Journal of Finance*, *35*(4), 883–896.

Todorov, V., & Bollerslev, T. (2010). Jumps and betas: A new framework for disentangling and estimating systematic risks. *Journal of Econometrics*, *157*(2), 220–235.

Van Nieuwerburgh, S. (2019). Why are reits currently so expensive? *Real Estate Economics*, *47*(1), 18–65.

Wang, R. J. Z. (1996). The conditional CAPM and the cross-section of expected returns. *The Journal of Finance*, *51*, 3–53.

Willard, M., & Youguo, L. (1991). An examination of the small-firm effect within the reit industry. *Journal of Real Estate Research*, *6*(1), 9–17.
Zhou, J. (2013). Conditional market beta for reits: A comparison of modelling techniques. *Economic Modelling*, *30*, 196–204.

Revisiting the Glick–Rogoff Current Account Model: An Application to the Current Accounts of BRICS Countries

Yushi Yoshida and Weiyang Zhai

1 Introduction

Understanding what drives the changes in current accounts is one of the most important macroeconomic issues for developing countries. Excessive surplus in a current account can trigger trade wars, and excessive deficits in the current account can lead to currency crises. For example, Brazil's current account has frequently fallen into a deficit since the 1990s and India experienced a current account deficit of 91 billion US dollars in 2012, whereas China's current account surplus has become the world's largest (see Fig. 1).

The current trade war between the USA and China began under the administration of President Trump following a decade in which the US's bilateral trade deficit with China remained large. Against the backdrop of the worldwide effort being made to reduce tariff and nontariff barriers, the USA raised the tariffs for steel products and thousands of products in other industries from China in consecutive sequences starting in 2018, and China responded with retaliatory tariff increases.

Currency crises are, in many cases, preceded by a current account deficit. Obstfeld (2012) documents that many crises have been preceded by a large current account deficit, pointing out the crises of Chile in 1981, Finland in 1991, Mexico in 1994, and Thailand in 1997, which subsequently led to the outbreak of the Asian currency crisis. It should be noted that current account deficits are not a prerequisite for currency crises. However, as shown in previous studies, a current account deficit is considered as an important warning signal of consequent crises. Roy and Kemme (2011) and Catao and Milesi-Ferretti (2014) find that current accounts are a powerful predictor of crisis; a higher current account deficit position is associated with a higher risk of crisis. Zorzi et al. (2012) conculde that current accounts are not aligned with economic

Y. Yoshida (✉) · W. Zhai
Faculty of Economics, Shiga University, Shiga, Japan
e-mail: yushi.yoshida@biwako.shiga-u.ac.jp

G. Dufrénot and T. Matsuki (eds.), *Recent Econometric Techniques for Macroeconomic and Financial Data*, Dynamic Modeling and Econometrics in Economics and Finance 27, https://doi.org/10.1007/978-3-030-54252-8_10

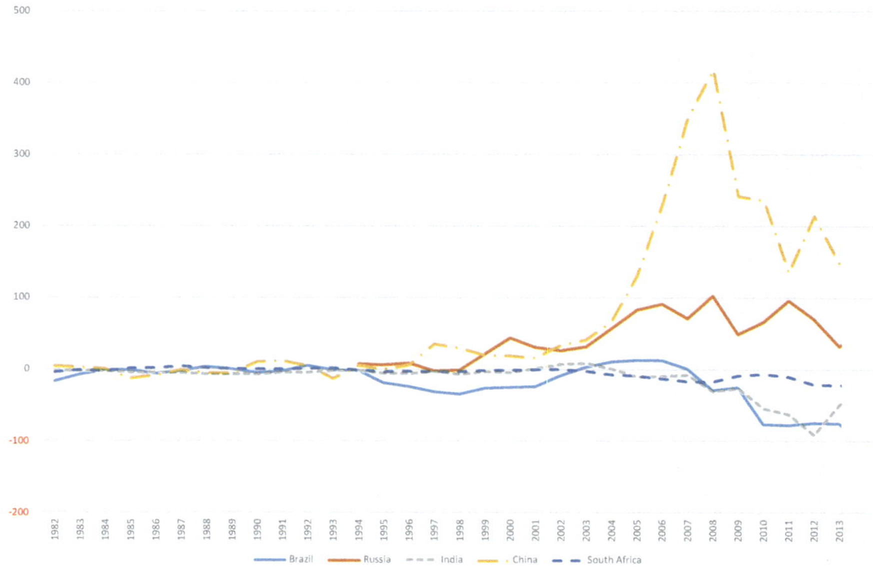

Fig. 1 Current accounts of BRICS countries. *Note* Current accounts are in terms of current billion US dollars. *Source* World development indicators, the World Bank

fundamentals prior to the financial crisis. During the Asian currency crisis, Corsetti et al. (1999) pointed to Taiwan's current account surplus as what prevented contagion from neighboring countries. Davis et al. (2016) showed that a higher private sector debt increases the probability of a crisis, especially when the current account is in a sizable deficit. Observing trade balance as a key determinant of current accounts, Kaminsky and Reinhart (1999) also concluded that exports often decrease just before a crisis.

Among other current account models, Glick and Rogoff (1995) developed an empirical model of current accounts to highlight the relationship of productivity with investment and current accounts. Current account changes are explained by country-specific productivity shocks, global productivity shocks, and lagged investments. The model performs surprisingly well with G7 data during 1975–1990. Their results show that current accounts had a negative response only to country-specific shocks, whereas investment showed a positive response to both global and country-specific productivity shocks.

In this study, we revisit Glick–Rogoff's model, in which productivity shocks act as a key driver of current account changes, and apply the model to the BRICS countries. This study aims to contribute to the literature by having the following goals. First, a model that emphasizes productivity shocks should be tested against fast-growing countries such as the BRICS countries. The BRICS countries experience much more volatile productivity shocks than developed countries in the G7 do. Second, understanding the current account changes of the largest economies is important for the surveillance of the global economy. In addition to have knowledge of the G7 countries, understanding the BRICS countries' current account movements, including the world's second largest economy, is essential for policymakers to adopt appropriate macroeconomic policies.

The results of the empirical application of the Glick–Rogoff model to BRICS countries show that the empirical model with productivity shocks works relatively well for developing countries except for Russia. However, the empirical Glick–Rogoff model collapses when the sample is extended to cover the post-crisis period. The fitness of regression in terms of adjusted R-squared becomes close to zero. Following the recent development of the empirical current account literature, we extended the Glick–Rogoff model with five macroeconomic variables, namely financial deepening, old dependency ratio, young dependency ratio, net foreign assets, and trade openness. The results of the extended model improve the fitness of regression for the pre-crisis period in India, China, and South Africa by more than 10% and that in Brazil by twofold.[1] Interestingly, the modified model even works well during the sample including the post-crisis period for Brazil, China, and Russia.

From the empirical investigations in this paper, we obtained the following conclusions for developing countries. (i) Productivity is only important in non-turbulent environments. The Glick–Rogoff model performs well in the period prior to the global financial crisis but loses all explanatory power in the sample period, which includes the global financial crisis. (ii) Other macroeconomic variables are important determinants regardless of the inclusion of the crisis in the sample period. Additional five macroeconomic variables in the modified Glick–Rogoff model improved the fitness of regression in both samples.

In comparison with developed countries, we also find the following implications. (iii) Productivity shocks explain the movements of the current account better for developing countries than for developed countries. (iv) However, productivity shocks retain some explanatory power for the current account of developed countries even in the post-crisis period, whereas productivity shocks have no effect on the current account of developing countries in the post-crisis period. (v) Inclusion of macroeconomic variables sometimes deteriorates the performance of the current account regression of developed countries, especially for Euro countries.

The construction of the rest of this paper is as follows. Section 2 reviews the theoretical model of Glick and Rogoff (1995) and the subsequent developments of both theoretical and empirical studies. Section 3 examines the empirical application of the Glick–Rogoff model and its modified model with macroeconomic variables as controls to the BRICS countries. Section 4 compares the results of the BIRCS countries with those of the G7 countries and discusses the characteristics of the current account for fast-growing emerging economies. Section 5 discusses how the five macroeconomic variables used in this empirical study are related to other macroeconomic variables used in the literature. Section 6 concludes the paper.

[1]The extended model cannot be applied to Russia for the pre-crisis period due to the lack of data availability.

2 Productivity Shocks and Current Accounts

Glick and Rogoff (1995) introduced a theoretical small-country model in which productivity shocks play crucial roles in determining current account movements. The next subsection briefly reviews the assumptions and underlying structure of their model. We discuss the applicability of the model assumptions that may lead to the misspecification of the empirical model for developing countries in Sect. 2.2.

2.1 Glick–Rogoff Small-Country Model with Adjustment Costs to Investment[2]

A small-country faces both country-specific productivity shocks and global productivity shocks. Global productivity shock can be mitigated by trading global bonds in the world capital market at the riskless gross world interest rate r. However, the representative agent in each country cannot diversify country-specific shocks. The representative firm chooses the path of investments to maximize the present discounted value of future profits under the given aggregate output (1).[3] Taking a linear approximation to the first-order condition yields Eqs. (2) and (3).

$$Y_t = A_t^c K_t^\alpha \left[1 - \frac{g}{2}\left(\frac{I_t^2}{K_t}\right)\right] \tag{1}$$

$$Y_t \cong \alpha_I I_t + \alpha_K K_t + \alpha_A A_t^c \tag{2}$$

$$I_t \cong \beta_1 I_{t-1} + \eta \sum_{s=1}^{\infty} \lambda^s \left(E_t A_{t+s}^c - E_{t-1} A_{t+s-1}^c\right) \tag{3}$$

In Eq. (3), the first term captures the past investment (or lagged productivity shock) on the current investment, and the second term captures the impact of revisions in expectations about the future path of productivity.

The representative agent chooses the path of consumption to maximize the present discounted utility (4).

$$\mathrm{E}_t \sum_{s=0}^{\infty} \beta^s U(C_{t+s}), \text{ where } U_t = C_t - \frac{h}{2} C_t^2, \text{ s.t.} F_{t+1} = rF_t + Y_t - I_t - C_t, \tag{4}$$

---

[2]This subsection closely follows the work of Glick and Rogoff (1995), with special focus on productivity shocks on current accounts. For the complete derivations of Eq. (8), please refer to the appendix of their original work. See also Marquesz (2004) and Bussiere et al. (2010) for another extension of the Glick–Rogoff model.

[3]Global productivity, A_t^W, is introduced multiplicatively to the aggregate output in a similar manner as country-specific productivity.

where r is assumed to be equal to β. The solution to the maximization for consumer yields (5), and the ex-post rate of change of consumption depends only on the unanticipated movement in permanent net income (6).

$$C_t = \frac{r-1}{r}\left(F_t + E_t \sum_{s=0}^{\infty} \frac{Y_{t+s} - I_{t+s}}{r^s}\right) = \frac{r-1}{r} F_t + \overline{Y}_t - \overline{I}_t \tag{5}$$

$$\Delta C_t = (E_t - E_{t-1})\frac{r-1}{r}\left(E_t \sum_{s=0}^{\infty} \frac{Y_{t+s} - I_{t+s}}{r^s}\right) = \left(\overline{Y}_t - \overline{I}_t\right) - E_{t-1}\left(\overline{Y}_t - \overline{I}_t\right) \tag{6}$$

Differencing the accounting identity for the current account, we obtain the following equation.

$$\Delta CA_t = (r-1)\Delta F_t + \Delta Y_t - \Delta I_t - \Delta C_t \tag{7}$$

Combining the equations obtained from maximization for ΔI_t, ΔY_t, ΔC_t with Eq. (7) yields the estimating equation for the current account[4].

$$\Delta CA_t = \gamma_1 I_{t-1} + \gamma_2 \Delta A_t^c + (r-1) CA_{t-1}, \tag{8}$$

where

$$\gamma_1 \equiv (\alpha_I - 1)(\beta_1 - 1) + \alpha_K > 0,$$
$$\gamma_2 \equiv -\beta_2[(\alpha_I - 1)(\beta_1 - 1) + \alpha_K]/(r - \beta_1) < 0$$

If all countries are symmetric in terms of preferences, technology, initial capital stocks, and zero initial net foreign asset positions, then the global shock should not affect an individual country's current account because the global shock affects all countries in the same manner; therefore, we obtain the final version of the basic Glick–Rogoff model.

$$\Delta CA_t = \gamma_1 I_{t-1} + \gamma_2 \Delta A_t^c + \gamma_3 \Delta A_t^W, \tag{9}$$

where γ_3 is assumed to be zero.

In this paper, leaving the investment equation aside, we focus on the effect of both country-specific and global productivity shocks on current accounts. The regression model derived from the theoretical result of Glick and Rogoff (1995) is as follows.

---

[4] α_I and α_K are marginal production of investment and capital as in Eq. (2). β_1 is the autoregressive coefficient of investment in Eq. (3). The first appearance of β_2 is omitted in this study, but it is equal to $\eta[\lambda/(1-\lambda)] > 0$.

$$\Delta CA_t = \gamma_1 I_{t-1} + \gamma_2 \Delta A_t^c + \gamma_3 \Delta A_t^W + \varepsilon_t, \tag{10}$$

where CA_t is the current account of the home country, A_t^c is the total factor productivity of the home country, A_t^W is the total factor productivity of the rest of the world, and I_t is the lagged investment in the home country.

2.2 Where Can the Glick–Rogoff Model Go Wrong in an Application to the BRICS Countries?

The effect of global productivity shocks is assumed to have no effect on a change in current accounts, as shown in Eq. (10). However, there are at least two strong arguments against this assumption. For the application to the BRICS countries, there are problems with assuming a zero initial net foreign asset position and assuming that global productivity shocks affect developed and developing countries equally.

First, Glick and Rogoff (1995) state that "zero initial net foreign asset positions ... is a reasonable empirical approximation for the G7 countries over the sample period." However, if positive (or negative) net foreign asset positions are at a level as high as was observed prior to global financial crisis in China, then the global productivity shocks affect these countries with nonzero net foreign asset positions asymmetrically.

Second, we followed Glick and Rogoff (1995) in measuring global productivity based on the largest economies of the world. In an application to the G7 countries as in Glick and Rogoff (1995), each sample country's productivity also contributes to the global productivity. Therefore, observed country productivity (from the original data) is decomposed into a country-specific component and a global component. The effect of the global component of productivity shock is nil because it affects both home country and the rest of the world similarly. However, for an application to the developing countries as in this study, observed country productivity (from the original data) does not constituent global productivity. The global factor should affect the BRICS countries and the rest of the world asymmetrically.

3 Domestic and Global Productivity

In this section, we apply the Glick–Rogoff model to the fast-growing emerging economies, namely the BRICS countries, Brazil, China, India, Russia, and South Africa. The key determinants of current account change in the Glick–Rogoff model are global and country-specific productivity shocks. To account for the severe negative shocks experienced during the global financial crisis, we estimate the model in two sample periods; one ending in 2008 and the other ending in 2017. In the second subsection, we also apply the extended model with additional macroeconomic variables after we obtain the base results from the original Glick–Rogoff model. In next section, we apply the same model to developed countries, namely Canada, France, Germany, Italy, Japan, the UK, and the USA. We discuss similarities and differences in current account determinants between the BRICS and G7 countries.

3.1 Estimation Results of the Basic Glick–Rogoff Model

Global productivity is constructed from the weighted average of the productivities of the G7 countries, namely Canada, France, Germany, Italy, Japan, the UK, and the USA. Alternatively, the first principal component of the productivities of the G7 countries is also used as a measure of global productivity.[5] The regression model of Eq. (10) is restated here.

$$\Delta CA_t = \gamma_1 I_{t-1} + \gamma_2 \Delta A_t^c + \gamma_3 \Delta A_t^W + \varepsilon_t,$$

From the Glick–Rogoff model, the expected sign of the past investment is positive, that of the first difference of each country's productivity is negative, and that of the first difference of worldwide productivity is zero; that is, $\gamma_1 > 0$, $\gamma_2 < 0$, and $\gamma_3 = 0$. The dynamic optimization model of Glick and Rogoff (1995) integrates the endogenous decisions of producers and consumers; therefore, the derived parameters of the model are affected by several sources. However, if we simply decompose the dependent variable, which is the first difference of the current account in terms of private saving and investment, and leave aside the government role, we can observe (in the first equality) the first-degree importance of the current investment and the past investment on the dependent variable. Adjusted for marginal production with respect to investment and capital stock, i.e., α_I and α_K, and the impact of past investment shock on the current investment, i.e., β_1, the coefficient of unity in the equation remains positive, γ_1, as shown in Eq. (8).

$$\Delta CA_t \equiv CA_t - CA_{t-1} = (S_t - I_t) - (S_{t-1} - I_{t-1}) = \Delta S_t - \Delta I_t \qquad (11)$$

It is also clear that a change in a country's productivity negatively affects a change in its current account, γ_2, via a change in investment through the second equality.

The empirical results of estimating the Glick–Rogoff model for the BRICS countries during 1983–2008 are provided in panel 1 of Table 1a, b.[6] The differences in the two tables arise from the way the global productivity index is constructed. The global

---

[5] Gregory and Head (1999) used dynamic factor analysis to construct a measure of common economic activity for the G7 countries. They find that the common economic activity has substantial impact on productivity but not on current account. İşcan (2000) further disintegrates overall productivity into traded good productivity and nontraded good productivity. He finds that the most influential of all on current account is country-specific traded good productivity.

[6] The sample period for Russia only begins in 1995 due to the availability of data.

Table 1 **a** Basic Glick–Rogoff regression with TFP as country-specific shock and weighted average as global shock

		Russia	Brazil	India	China	South Africa
Panel 1	Period	1995–2008	1983–2008	1983–2008	1983–2008	1983–2008
Country-specific		−1.12(1.72)	−0.01(0.04)	0.04(0.07)	0.13(0.08)	−0.06(0.05)
Global		5.14(10.09)	0.91(0.40)**	−0.03(0.20)	0.40(0.26)	−0.38(0.22)*
Investment		0.10(0.09)	0.01(0.03)	−0.02(0.01)	0.06(0.01)***	−0.10(0.02)***
Number of obs		14	26	26	26	26
Adj R-squared		− 0.14	0.29	0.20	0.60	0.58
Panel 2	Period	1995–2017	1983–2017	1983–2017	1983–2017	1983–2017
Country-specific		0.11(1.33)	−0.18(0.12)	0.27(0.15)*	0.17(0.18)	−0.14(0.09)
Global		15.17(9.56)	0.15(0.45)	−0.29(0.33)	0.62(0.78)	−0.20(0.70)
Investment		−0.02(0.05)	0.00(0.03)	−0.01(0.01)	−0.02(0.01)	0.00(0.02)
Number of obs		23	35	35	35	35
Adj R-squared		−0.05	−0.01	−0.02	0.02	−0.06

(continued)

Table 1 (continued)

b Basic Glick–Rogoff regression with TFP as country-specific shock and principal component as global shock

		Russia	Brazil	India	China	South Africa
Panel 1	Period	1995–2008	1983–2008	1983–2008	1983–2008	1983–2008
Country-specific		−1.23(1.58)	−0.00(0.04)	0.02(0.07)	0.14(0.09)	−0.08(0.05)
Global		0.16(0.28)	0.02(0.01)**	0.00(0.01)	0.01(0.01)	0.00(0.01)
Investment		0.11(0.08)	−0.01(0.02)	−0.02(0.01)	0.06(0.01)***	−0.07(0.02)***
Number of obs		14	26	26	26	26
Adj R-squared		−0.13	0.20	0.23	0.58	0.53
Panel 2	Period	1995–2017	1983–2017	1983–2017	1983–2017	1983–2017
Country-specific		−0.21(1.39)	−0.18(0.12)	0.23(0.13)	0.16(0.18)	−0.15(0.10)
Global		0.30(0.25)	0.00(0.10)	0.00(0.01)	0.01(0.02)	0.01(0.02)
Investment		−0.02(0.05)	0.00(0.02)	−0.01(0.01)	−0.01(0.07)	0.00(0.02)
Number of obs		23	35	35	35	35
Adj R-squared		−0.08	−0.01	−0.04	−0.01	−0.05

Note Global shock is calculated as the weighted average (the first principal component) of total factor productivity of the G7 in Table 1a (Table 1b) and country-specific shocks as the country's total factor productivity. Heteroscedasticity-robust standard errors are in parentheses. *, **, and *** represent the 10, 5, and 1% statistical significance levels, respectively

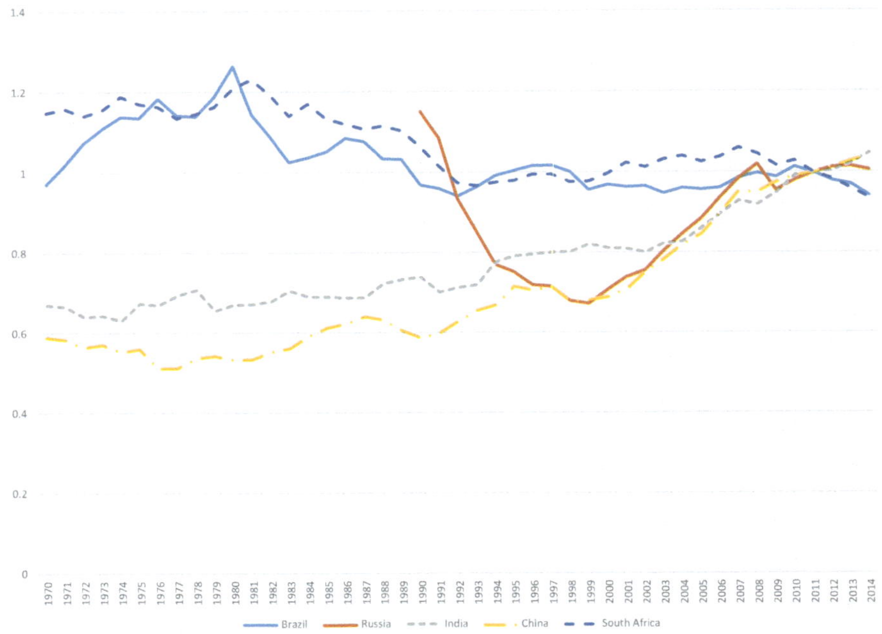

Fig. 2 Total factor productivity of BRICS countries. *Note* Total factor productivity is normalized for unity in 2011 for each country. *Source* World development indicators, the World Bank

productivity index is simply a GDP-weighted average of the G7 countries in Table 1a and the first principal component of the G7 countries in Table 1b. Country-specific productivity is based on total factor productivity, as shown in Fig. 2.

First, by comparing Table 1a and 1b, we find that the results are quite similar in terms of both the size of the coefficients and the statistical significance except for the coefficient size of global productivity shock. The result of the basic model is robust regardless of how global productivity is measured.

Second, country-specific productivity shock is not statistically significant for all BRICS countries, although theory predicts a negative effect on change in the current account. This result is quite different from the results obtained for the G7 countries during 1961–1990 in Glick and Rogoff (1995, Table 3); the estimated coefficients of the country-specific productivity shock are negative and statistically significant for five countries, namely the USA, Japan, Italy, the UK, and Canada.

Third, except for Brazil and South Africa (only in Table 1a), the results for global productivity shock are consistent with the Glick and Rogoff model. Under the assumption that global productivity shock symmetrically affects all countries in the world, the effect of global productivity shock on current account change must be equal to zero. The estimated coefficients for Russia, India, China, and South Africa (only in Table 1b) are not significantly different from zero at the conventional significance level. For the case of Brazil, the estimated coefficient is positive and statistically significant. The positive sign is consistent if the global productivity shock represents

Table 2 **a** Basic Glick–Rogoff regression with TFP residual as country-specific shock and weighted average as global shock

		Russia	Brazil	India	China	South Africa
Panel 1	Period	1995–2008	1983–2008	1983–2008	1983–2008	1983–2008
Country-specific		−1.12(1.72)	−0.01(0.04)	0.04(0.07)	0.13(0.08)	−0.06(0.05)
Global		4.10(10.52)	0.90(0.39)**	0.01(0.22)	0.51(0.27)*	−0.44(0.20)**
Investment		0.10(0.09)	0.01(0.03)	−0.02(0.01)	0.06(0.01)***	−0.10(0.02)***
Number of obs		14	26	26	26	26
Adj R-squared		−0.14	0.29	0.20	0.60	0.58
Panel 2	Period	1995–2017	1983–2017	1983–2017	1983–2017	1983–2017
Country-specific		0.11(1.33)	−0.18(0.12)	0.27(0.15)*	0.17(0.18)	−0.14(0.09)
Global		15.31(9.32)	−0.04(0.51)	−0.03(0.34)	0.76(0.76)	−0.35(0.73)
Investment		−0.02(0.05)	0.00(0.03)	−0.01(0.01)	−0.00(0.01)	0.00(0.02)
Number of obs		23	35	35	35	35
Adj R-squared		−0.05	−0.01	−0.02	0.02	−0.06

(continued)

Table 2 (continued)

b Basic Glick–Rogoff regression with TFP residual as country-specific shock and principal component as global shock

		Russia	Brazil	India	China	South Africa
Panel 1	Period	1995–2008	1983–2008	1983–2008	1983–2008	1983–2008
Country-specific		−1.24(1.58)	−0.00(0.04)	0.02(0.07)	0.14(0.08)	−0.08(0.06)
Global		−0.32(0.71)	0.02(0.01)**	0.01(0.01)	0.02(0.01)*	−0.00(0.01)
Investment		0.11(0.08)	−0.01(0.02)	−0.02(0.01)	0.06(0.01)***	−0.07(0.02)***
Number of obs		14	26	26	26	26
Adj R-squared		−0.13	0.20	0.23	0.58	0.53
Panel 2	Period	1995–2017	1983–2017	1983–2017	1983–2017	1983–2017
Country-specific		−0.22(1.39)	−0.18(0.12)	0.23(0.13)*	0.16(0.18)	−0.15(0.10)
Global		0.21(0.54)	−0.00(0.01)	0.02(0.02)	0.03(0.02)	−0.01(0.02)
Investment		−0.02(0.05)	0.00(0.02)	−0.01(0.01)	−0.00(0.01)	0.00(0.02)
Number of obs		23	35	35	35	35
Adj R-squared		−0.08	−0.01	−0.04	−0.01	−0.05

Note Global shock is calculated as the weighted average (the first principal component) of total factor productivity of the G7 in Table 2a (Table 2b) and country-specific shocks as the residual of each country's total factor productivity on global shock. Heteroscedasticity-robust standard errors are in parentheses. *, **, and *** represent the 10, 5, and 1% statistical significance levels, respectively

Table 3 **a** Modified Glick–Rogoff regression with TFP residual as country-specific shock and weighted average as global shock

		Russia	Brazil	India	China	South Africa
Panel 1	Period		1983–2008	1983–2008	1983–2008	1983–2008
Country-specific			0.01(0.03)	0.02(0.06)	0.02(0.08)	−0.13(0.04)**
Global			0.86(0.26)***	−0.01(0.22)	−0.26(0.43)	−0.43(0.27)
Investment			−0.21(0.07)***	0.00(0.08)	−0.19(0.15)	−0.14(0.10)
fdeep			−0.01(0.01)	−0.01(0.05)	−0.05(0.05)	−0.03(0.05)
reldepo			0.87(0.98)	2.76(3.20)	1.38(2.69)	−0.79(0.82)
reldepy			−0.15(0.08)	0.02(0.21)	0.02(0.12)	−0.05(0.03)
nfa/GDP			2.87(1.66)	1.50(13.68)	18.58(7.88)**	
open			8.08(7.73)	−11.96(11.37)	8.12(5.10)	7.49(1.84)***
Number of obs			26	26	26	26
Adj R-squared			0.60	0.29	0.69	0.68
Panel 2	Period	2001–2017	1983–2016	1983–2015	1983–2017	1983–2017
Country-specific		7.07(4.46)	−0.23(0.12)*	0.10(0.10)	−0.21(0.14)	0.05(0.13)
Global		40.35(16.75)**	−0.44(0.65)	−0.25(0.34)	0.54(037)	−0.56(0.78)
Investment		0.31(0.26)	−0.26(0.08)**	0.03(0.06)	0.02(0.03)	0.10(0.07)
fdeep		−0.03(0.03)	0.01(0.01)	−0.04(0.09)	−0.01(0.03)	−0.03(0.08)
reldepo		−0.21(0.22)	5.09(1.76)***	0.95(7.51)	0.45(1.18)	−0.70(2.26)
reldepy		−0.37(0.13)**	0.40(0.17)**	−0.09(0.47)	0.43(0.28)	0.08(0.06)
nfa/GDP		4.30(1.87)*	1.70(2.11)	5.69(21.77)	2.56(5.46)	
open		13.77(12.05)	8.81(20.97)	−14.10(13.64)	24.99(7.79)***	−6.68(7.64)
Number of obs		17	34	33	35	35
Adj R-squared		0.31	0.22	−0.02	0.24	−0.04

(continued)

Table 3 (continued)

b Modified Glick–Rogoff regression with TFP residual as country-specific shock and principal component as global shock

		Russia	Brazil	India	China	South Africa
Panel 1	Period		1983–2008	1983–2008	1983–2008	1983–2008
Country-specific			0.01(0.04)	0.02(0.07)	0.02(0.08)	−0.14(0.04)***
Global			0.01(0.01)**	0.01(0.01)	0.00(0.01)	−0.00(0.01)
Investment			−0.26(0.08)***	0.01(0.08)	−0.12(0.14)	−0.05(0.10)
fdeep			−0.01(0.01)	−0.01(0.05)	−0.02(0.05)	−0.06(0.04)
reldepo			1.65(1.00)	2.26(3.24)	0.12(2.64)	−1.70(1.26)
reldepy			−0.12(0.08)	−0.02(0.21)	0.02(0.12)	−0.05(0.04)
nfa/GDP			3.69(1.95)*	0.38(13.35)	16.13(7.25)**	
open			3.64(7.63)	−12.22(11.27)	6.38(5.18)	8.09(2.16)***
Number of obs			26	26	26	26
Adj R-squared			0.45	0.31	0.68	0.66

(continued)

Table 3 (continued)

b Modified Glick–Rogoff regression with TFP residual as country-specific shock and principal component as global shock

		Russia	Brazil	India	China	South Africa
Panel 2	Period	2001–2017	1983–2016	1983–2015	1983–2017	1983–2017
Country-specific		4.58(4.07)	−0.23(0.12)*	0.06(0.07)	−0.19(0.15)	0.00(0.13)
Global		2.51(1.78)	−0.01(0.01)	0.01(0.05)	−0.01(0.01)	−0.01(0.02)
Investment		0.31(0.26)	−0.25(0.10)***	0.03(0.08)	0.02(0.03)	0.10(0.09)
fdeep		−0.02(0.03)	0.01(0.01)	−0.04(0.49)	−0.02(0.04)	−0.04(0.08)
reldepo		−0.15(0.21)	5.03(1.77)***	−0.34(3.24)	0.36(1.21)	−0.75(2.63)
reldepy		−0.38(0.14)**	0.40(0.17)**	−0.18(0.21)	0.30(0.28)	0.06(0.05)
nfa/GDP		3.96(1.88)*	1.85(2.10)	8.69(13.35)	−0.38(5.68)	
open		16.74(13.05)	9.93(23.14)	−14.77(11.27)	23.92(7.52)***	−5.76(7.32)
Number of obs		17	34	33	35	35
Adj R-squared		0.29	0.22	−0.04	0.20	−0.05

Note Global shock is calculated as the weighted average (the principal component) of total factor productivity of the G7 in Table 3a (Table 3b)and country-specific shocks as residual of each country's total factor productivity on global shock. Heteroscedasticity-robust standard errors are in parentheses. *, **, and *** represent the 10, 5, and 1% statistical significance levels, respectively the rest of the world instead of the world with Brazil included (see the discussion in Sect. 2.2). Foreign productivity shock should positively affect home current account change because home productivity shock negatively affects home current account change, as in γ_2.

The basic Glick–Rogoff model is also estimated for the extended sample of 1983–2017, including the post-global financial crisis period, and the results are provided in panel 2 of Table 1a, b. The surprising result is that nothing in the Glick–Rogoff model works for the BRICS countries if the worldwide turbulent period is included in the sample. For all BRICS countries, the fitness of the regression in terms of adjusted R-squared is literally zero.

The productivity shocks implemented in the preceding empirical approach need to meet the requirement assumed in the theoretical model; country-specific productivity and global productivity are independent. In the original Glick and Rogoff (1995) study, global productivity is constructed from the same countries in the sample; therefore, the independent assumption is more likely to be violated if no adjustment is made. Following the methodology implemented in Glick and Rogoff (1995), we regressed the original country productivity on the global productivity and used the residual as country-specific productivity. By design, the independence of country-specific and global productivity shock is guaranteed. The results using the residual as country-specific productivity are shown in Table 2a, b.

The estimated results in Table 2a, b are very similar to those of Table 1a, b in terms of both the size of coefficients and the statistical significance. The only noteworthy point is that the global productivity shock becomes statistically significant for an additional country in the shorter sample. For China, in the period between 1983 and 2008, as seen in Table 2b, the global productivity shock is positive and statistically significant at the 10% level. This is similar to the case of Brazil, in which the positive sign indicates foreign productivity shock rather than global productivity shock. As for the results of South Africa in Table 2a, it is puzzling that the global productivity shock is negative and statistically significant.[7]

3.2 *Extended Models with Other Macroeconomic Variables*

Not all empirical models of current account movements emphasize productivity shocks. The advantage of the Glick–Rogoff regression model is its concrete derivation based on the theoretical dynamic model. However, many researchers have continued to explore the possibility of many other macroeconomic variables to explain current account movements, frequently without theoretical models.

Chinn and Prasad (2003) investigated the medium-term determinants of current accounts for a large sample of developed and developing countries. They find that current account balance is positively correlated with government budget balance

[7]Smit et al. (2014) provided an explanation of the irregular movement of South Africa's current account deficit as being driven by substantial net capital inflows and their reversals afterward.

and the initial level of net foreign assets. Among developing countries, financial deepening is positively associated with current account balance, while trade openness is negatively correlated with current account balance.

Cudre and Hoffmann (2017) and Romelli et al. (2018) also showed that trade openness is a significant driver of current accounts. Romelli et al. (2018) investigated the impact of trade openness on the relationship between the current account and the real exchange rate. They find that during the balance of payment distress episodes, currency depreciations are associated with larger improvements in the current accounts of countries that are more open to trade, and the magnitude of exchange rate depreciations over the adjustment process of current accounts is related to the degree of openness to trade. Cudre and Hoffmann (2017) also find that trade openness is an important factor even across regions within a nation.

Following the recent development of the empirical current account literature, we extended the Glick–Rogoff model with five macroeconomic variables: financial deepening, old dependency ratio, young dependency ratio, net foreign assets, and trade openness.[8] First, the fitness of regression substantially improved for Brazil, India, and China. In the shorter sample between 1983 and 2008, the adjusted R-squared increased from 0.29 to 0.60 for Brazil, from 0.60 to 0.69 for India, and from 0.58 to 0.68 for China. In the longer sample that included the post-crisis period, the adjusted R-squared values were 0.31 for Russia, 0.21 for Brazil, and 0.24 for China; all of these values increased from zero or even negative values of the adjusted R-squared in the basic model estimations.

Second, we obtained estimation results that are consistent with the theoretical prediction for country-specific productivity shock although none of the estimates were statistically significant in the base model; negative responses are obtained for South Africa in the shorter sample and Brazil for the longer sample. In addition, by assuming global productivity as foreign productivity, we find a positive association with statistical significance between the current account and global productivity for Russia in the longer sample in addition to those for Brazil.

Third, the importance of additional macroeconomic variables for explaining current accounts varies among the BRICS countries. Financial deepening has no explanatory power for all countries. The old dependency ratio has a positive effect only for Brazil in the longer sample, while the young dependency ratio exerts opposite effects on Russia and Brazil in the longer sample. Net foreign assets have a positive effect for China in the shorter sample and for Russia in the longer sample. Trade openness has a positive effect for South Africa in the shorter sample and China in the longer sample.

[8]The definitions and sources of macroeconomic variables are provided in the Appendix A.

4 Can the Same Current Account Model be Applied to Both Developed and Developing Countries?

For the BRICS countries, except for Russia, we find that the Glick–Rogoff model can explain approximately 20–60% of the changes in the current account for the period between 1983 and 2008; however, productivity shocks completely lose explanatory power when the sample is extended to cover the post-crisis period. The modified model extended with macroeconomic variables improves the fitness of regression for both samples. To conclude, whether these results are general or specific to fast-growing developing countries, we need to compare the results with those of developed countries. Therefore, we also estimated the same regression models for the G7 countries. The results are shown in appendix Tables 5 and 6.[9]

First, the explanatory power of the original Glick–Rogoff model does not work, or at least does not work better for G7 countries than for BRICS countries in the pre-crisis period. The degree of fitness in terms of adjusted R-squared is 0.16, 0.15, 0.08, and 0.28 for France, Germany, Italy, and Japan, respectively. It is less than zero for Canada and the UK. The only exception is the USA; 43% of current account movements in the USA are explained by both global and local productivity shocks and past investment.

Second, similar to the BRICS countries, the basic model fits less well for G7 countries in the post-crisis period. However, unlike the BRICS countries, even in the post-crisis period, the Glick–Rogoff model retains some explanatory power, at least for Canada, Germany, Japan, and the USA. This is a surprising finding because developed countries, especially the USA, are the most affected countries in the world by the global financial crisis. We may need to adjust our understanding so that the financial crisis per se does not break the relationship between productivity shocks and current account changes.

Third, unlike in the case of the BRICS countries, the modified model does not necessarily improve the fitness of regressions in terms of the adjusted R-squared. More interestingly, the decline in the explanatory power of the overall regression model is found only for European countries, more precisely Euro countries. For the pre-crisis sample, additional macroeconomic variables in France, Germany, and Italy reduced the adjusted R-squared from 0.16, 0.15, and 0.08 to 0.12, 0.09, and 0.00, respectively.

---

[9]These conclusions are drawn from comparing the results in Table 2a, b and Table 3a, b. However, the sample periods for the modified model do not exactly match those in the basic model due to the exclusion of a few years for missing macroeconomic variables. The estimation results in appendix Table 4a, b with the sample period adjusted to match those of Table 3a, b confirm that the qualitative results do not change.

By comparing the estimated results of the BRICS and G7 countries, we can draw the following implications for the movement of current accounts. (i) Productivity shocks as determinants of current account movements are more important in developing countries. (ii) However, productivity shocks lose their explanatory power for developing countries in the midst of the financial crisis. (iii) Demography, net foreign assets, and trade openness contribute to the movements of the current account for the BRICS countries, but this is less so for the G7 countries and even the opposite for euro countries.

5 Discussions

In this study, five macroeconomic variables are selected for the modified Glick–Rogoff model, which examines the effects of productivity shocks on current account changes. In the current account literature, researchers tested many more variables as determinants of current account movements. In this section, we discuss the possibility of other macroeconomic variables that may contribute to an increase in the fitness of the current empirical model. Some of the discussions in this section are meant to help improve future works.

First, in addition to trade openness, financial openness is also an important determinant of current account movements. By interpreting financial openness as unrestricted international capital flow, Yan and Yang (2012) find a shift in causality between the current account and capital inflows after the global financial crisis. Chin and Ito (2007) find that financial market development causes developed countries to have smaller savings and thus a current account deficit, while the opposite is true for Asian countries. Furthermore, Tan et al. (2015) examined the effect of the structure of the financial system on current accounts. They find that a country with a fully developed capital market is more likely to run a current account deficit. In this study, the financial deepening variable, i.e., the ratio of broad money to GDP, is most closely related to financial openness; however, this variable is not statistically significant for most of the cases in either the BRICS or the G7 countries. Our study confirmed the finding in the literature that development in capital markets is more relevant for current account adjustment than the growth in bank lending or money supply is.

Second, the income distribution within a country matters for current accounts. Income inequality raises national savings and thus increases current accounts if the savings rate of the richer individuals is higher than that of the poorer individuals. However, there are a variety of theoretical models that can generate the reduction in the current account associated with income inequality; see Behringer and van Treeck (2018). Belabed et al. (2018) suggested that the US current account deficit can be explained in part by rising income inequality in the USA. From investigating a sample of 20 countries, Behringer and van Treeck (2018) also find that income inequality leads to a decline in current accounts. In our study, the dependency ratio of young and elderly individuals may capture the part of the mechanism by which income distribution affects current accounts. The national savings rate should decline if the dependency ratio of elder (or dissaving) cohorts increases. Puzzlingly, in this study, the dependency ratio of elder cohorts on current accounts is positive for both BRICS and G7 countries whenever the ratio is statistically significant. One possible explanation for this contradiction between the theoretical predictions and empirical results in this paper may come from regressions based on the time series of a single country rather than the panel framework used in Chin and Ito (2007). As often experienced in many other applied works, demographical characteristics can only be captured in the difference in cross sections of countries.

Third, real exchange rates, or terms of trade, are not considered in this study. The effects of the terms of trade, i.e., relative price of exports and imports or relative price of tradable and non-tradable goods, on the current account is a classic issue in international macroeconomics. The so-called HLM effect works through the real income effect by which the terms of trade deterioration decreases the current account balance (Harberger, 1950 and Laursen and Metzler, 1950). Svensson and Razin (1983) examined the effect of terms of trade changes on a small country's current account under perfect international capital mobility. A temporary terms of trade deterioration implies a deterioration of the current account, whereas a permanent terms of trade deterioration has an uncertain effect on the current account. Gervais et al. (2016) analyzed a large set of emerging countries over the period from 1975 to 2008. They indicate that real exchange rate adjustment contributed to reducing current account imbalances. Focusing on the non-tradable sector, as in the Balassa–Samuelson effect on the real exchange rate, Hoffmann (2013) claims that the present-value model with non-tradable goods explains more than 70% of China's current account variability over the period 1982–2007.

6 Conclusion

Current account adjustment became a classic macroeconomic issue in the 1950s and is still one of the important macroeconomic policy objectives today. Especially, for fast-growing developing counties such as Brazil, China, India, Russia, and South Africa, a large current account imbalance can lead to an economic crisis for the worst case. In this study, we investigated the determinants of current account changes for these BRICS countries between 1983 and 2017. As an empirical model, we selected the Glick–Rogoff model, which emphasizes productivity shocks at home and in the world and fits well with developed economies in the 1970s and 1980s (Glick and Rogoff 1995). However, the Glick–Rogoff model fits poorly when it is applied to fast-growing BRICS countries for the period including the global financial crisis.

Productivity shocks are important determinants of current account movements; however, a set of global and country-specific productivity shocks alone cannot explain a country's current account[10]. A set of macroeconomic variables help to improve

[10] Attanasio and Weber (2010) question the validity of strong assumption of economic agents being able to solve the intertemporal optimization problem as in the standard macroeconomic models

the fitness of regression for developing countries but can worsen the adjusted R-squared for Euro countries. It is not surprising that different mechanisms of current account adjustment work for different groups of countries, i.e., developed and developing countries, because there are many differences in terms of monetary policy, exchange rate systems, tariffs, and trade regulations between the two groups. This result suggests that policymakers should search for a framework in which the current account adjusts through its own country-specific mechanism.

Acknowledgements Yoshida is grateful for financial support from JSPS KAKENHI 19K01673.

Appendix A. List of variables

Variable	Source	Description
tfp	PWT9.1	TFP at constant national prices (2011=1)
GDP	WDI	gross domestic product (current LCU)
ca	WDI	current account balance (BoP, current US$) times official exchange rate (LCU per US$, period average)
investment	WDI	gross fixed capital formation (current LCU)
		plus changes in inventories (current LCU)
aw1		weighted average of G7 countries' TFP
aw2		principal component of G7 countries' TFP
fdeep	WDI	financial deepening: broad money (% of GDP)
reldepo	WDI	age dependency ratio, old (% of working age population)
reldepy	WDI	age dependency ratio, young (% of working-age population)
nfa	WDI	net foreign assets (current LCU): Net foreign assets are the sum of foreign assets held by monetary authorities and deposit money banks minus their foreign liabilities. Data are in current local currency.
open	WDI	[exports of goods and services (current LCU) + imports of goods and services (current LCU)]/GDP (current LCU)

Appendix B

See Tables 4, 5, and 6.

Table 4 **a** Basic Glick–Rogoff regression with the TFP residual as country-specific shock and weighted average as global shock; sample adjusted to coincide with Table 3a

		Russia	Brazil	India	China	South Africa
Panel 2	Period	2001–2017	1983–2016	1983–2015	1983–2017	1983–2017
Country-specific		0.01(3.12)	−0.19(0.11)	0.26(0.15)*	0.17(0.18)	−0.14(0.09)
Global		14.62(13.29)	−0.15(0.48)	0.12(0.37)	0.76(0.76)	−0.35(0.73)
Investment		−0.02(0.06)	−0.00(0.03)	−0.00(0.02)	−0.00(0.01)	0.00(0.02)
Number of obs		17	34	33	35	35
Adj R-squared		−0.13	0.00	−0.02	0.02	−0.06

b Basic Glick–Rogoff regression with the TFP residual as country-specific shock and the principal component as global shock; sample adjusted to coincide with Table 3b

		Russia	Brazil	India	China	South Africa
Panel 2	Period	2001–2017	1983–2016	1983–2015	1983–2017	1983–2017
Country-specific		−1.45(2.95)	−0.19(0.11)*	0.23(0.13)*	0.16(0.18)	−0.15(0.10)
Global		−0.29(1.27)	−0.01(0.11)	0.26(0.17)	0.03(0.02)	−0.01(0.02)
Investment		−0.03(0.06)	−0.01(0.03)	0.00(0.02)	−0.00(0.01)	0.00(0.02)
Number of obs		17	34	33	35	35
Adj R-squared		−0.15	0.00	−0.02	−0.01	−0.05

Note Global shock is calculated as the weighted average (the first principal) of total factor productivity of the G7 in Table 4a (Table 4b) and country-specific shocks as the residual of each country's total factor productivity on global shock. Heteroscedasticity-robust standard errors are in parentheses. *, **, and *** represent the 10, 5, and 1% statistical significance levels, respectively

Table 5 Basic Glick–Rogoff regression for G7 countries with the TFP residual as country-specific shock and the principal component as global shock

		Canada	France	Germany	Italy	Japan	United Kingdom	United States
Panel 1	Period	1983–2008	1983–2008	1983–2008	1983–2008	1983–2008	1983–2008	1983–2008
Country-specific		0.18(0.17)	−0.01(0.18)	−0.39(0.18)*	0.07(0.10)	−0.43(0.17)**	0.05(0.12)	−0.57(0.15)***
Global		0.49(0.39)	−0.44(0.30)	0.43(0.59)	−0.28(0.56)	0.67(0.39)*	−0.08(0.27)	−0.43(0.16)**
Investment		0.02(0.03)	−0.07(0.02)**	0.07(0.07)	−0.05(0.03)	0.01(0.03)	−0.05(0.03)	−0.02(0.02)
Number of obs		26	26	26	26	26	26	26
Adj R-squared		−0.02	0.16	0.15	0.08	0.28	−0.02	0.43
Panel 2	Period	1983–2017	1983–2017	1983–2017	1983–2017	1983–2017	1983–2017	1983–2017
Country-specific		0.34(0.15)**	−0.06(0.17)	−0.33(0.18)*	0.03(0.14)	−0.30(0.14)**	0.01(0.14)	−0.25(0.23)
Global		0.76(0.45)	−0.12(0.17)	0.46(0.29)	−0.59(0.38)	0.63(0.25)**	−0.05(0.35)	−0.76(0.27)***
Investment		−0.01(0.02)	−0.02(0.02)	0.07(0.04)*	−0.02(0.03)	0.01(0.03)	−0.01(0.04)	−0.01(0.01)
Number of obs		35	35	35	35	35	35	35
Adj R-squared		0.21	−0.04	0.14	0.01	0.13	−0.10	0.34

Note Global shock is calculated as the first principal component of the total factor productivity of the G7 and country-specific shocks as the residual of each country's total factor productivity on global shock. Heteroscedasticity-robust standard errors are in parentheses. *, **, and *** represent the 10, 5, and 1% statistical significance levels, respectively

Table 6 Modified Glick–Rogoff regression for G7 countries with the TFP residual as country-specific shock and the principal component as global shock

		Canada	France	Germany	Italy	Japan	United Kingdom	United States
Panel 1	Period	1983–2008	1983–2008	1983–2008	1983–2008	1983–2008	1983–2008	1983–2008
Country-specific		0.13(0.11)	0.05(0.17)	−0.38(0.37)	0.11(0.15)	−0.24(0.31)	0.18(0.14)	−0.36(0.11)***
Global		1.08(0.37)***	−0.25(0.25)	0.58(0.70)	−0.38(0.57)	1.42(0.52)**	0.07(0.32)	−0.29(0.22)
Investment		0.17(0.15)	0.02(0.14)	0.04(0.09)	−0.12(0.19)	0.25(0.19)	0.05(0.16)	0.01(0.04)
fdeep						−0.01(0.01)	−0.01(0.01)	0.05(0.04)
reldepo		0.50(0.28)*	0.34(0.17)*	0.10(0.47)	0.34(0.35)	0.27(0.36)	−0.06(0.82)	0.73(0.41)*
reldepy		0.77(0.43)*	0.56(0.53)	−0.14(0.62)	0.05(0.17)	0.49(0.57)	0.33(0.37)	0.11(0.44)
nfa/GDP		0.72(2.71)	3.33(4.82)	1.91(14.65)	5.04(5.66)	4.96(4.31)	−16.09(9.41)	10.41(17.13)
open		1.88(3.20)	−7.68(7.80)	−0.03(11.06)	−7.72(8.30)	−4.29(8.91)	1.40(9.40)	−1.64(9.15)
Number of obs		26	26	26	26	26	26	26
Adj R-squared		0.24	0.12	0.09	0.00	0.33	0.13	0.53
Panel 2	Period	1983–2017	1983–2017	1983–2017	1983–2017	1983–2017	1983–2017	1983–2017
Country-specific		0.23(0.15)	−0.01(0.13)	−0.29(0.20)	−0.06(0.23)	−0.28(0.17)	−0.05(0.23)	−0.22(0.16)
Global		0.95(0.39)**	−0.15(0.15)	0.39(0.39)	−0.80(0.37)**	0.88(0.37)**	−0.03(0.42)	−0.71(0.25)***
Investment		0.08(0.14)	−0.00(0.09)	0.03(0.07)	−0.16(0.13)	0.14(0.08)*	0.01(0.15)	0.00(0.04)
fdeep						−0.02(0.01)	0.00(0.02)	0.09(0.06)
reldepo		0.07(0.21)	0.17(0.10)*	−0.10(0.30)	0.25(0.11)**	0.12(0.07)	0.16(0.68)	0.08(0.09)
reldepy		0.34(0.39)	0.25(0.26)	−0.03(0.32)	0.06(0.15)	0.19(0.11)	0.36(0.38)	0.53(0.43)
nfa/GDP		−1.74(1.58)	1.66(4.35)	−4.53(5.81)	−1.63(5.88)	3.92(4.03)	−2.13(11.73)	6.33(13.42)
open		6.20(2.63)**	−5.28(5.41)	3.32(7.47)	−2.77(7.17)	−1.77(6.54)	2.27(10.22)	−0.74(3.99)
Number of obs		35	35	35	35	35	35	35

(continued)

Table 6 (continued)

		Canada	France	Germany	Italy	Japan	United Kingdom	United States
Adj R-squared		0.33	0.00	0.11	0.07	0.21	−0.23	0.57

Note Global shock is calculated as the first principal component of the total factor productivity of the G7 and the country-specific shocks as the residual of the country's total factor productivity on global shock. Heteroscedasticity-robust standard errors are in parentheses. *, **, and *** represent the 10, 5, and 1% statistical significance levels, respectively

References

Attanasio, O. P., & Weber, G. (2010). Consumption and Saving: Models of intertemporal allocation and their implications for public policy. *Journal of Economic Literature, 48,* 693–751.

Behringer, J., & van Treeck, T. (2018). Income distribution and the current account. *Journal of International Economics, 114,* 238–254.

Belabed, C. A., Theobald, T., & van Treeck, T. (2018). Income distribution and current account imbalances. *Cambridge Journal of Economics, 42,* 47–94.

Bussière, M., Fratzscher, M., & Müller, G. J. (2010). Productivity shocks, budget deficits and the current account. *Journal of International Money and Finance, 29,* 1562–1579.

Catao, L. A. V., & Milesi-Ferretti, G. M. (2014). External liabilities and crises. *Journal of International Economics, 94*(1), 18–32.

Chinn, M. D., & Ito, H. (2007). Current account balances, financial development and institutions: Assaying the world "saving glut". *Journal of International Money and Finance, 26,* 546–569.

Chinn, M. D., & Prasad, E. S. (2003). Medium-term determinants of current accounts in industrial and developing countries: an empirical exploration. *Journal of International Economics, 59,* 47–76.

Corsetti, G., Pesenti, P., & Roubini, N. (1999). What caused the Asian currency and financial crisis? *Japan and the World Economy, 11*(3), 305–373.

Cudre, S., & Hoffmann, M. (2017). A provincial view of global imbalances: Regional capital flows in China. *Review of World Economics, 153,* 573–599.

Davis, J. S., Mack, A., Phoa, W., & Vandenabeele, A. (2016). Credit booms, banking crises, and the current account. *Journal of International Money and Finance, 60,* 360–377.

Gervais, O., Schembri, L., & Suchanek, L. (2016). Current account dynamics, real exchange rate adjustment, and the exchange rate regime in emerging-market economies. *Journal of Development Economics, 119,* 86–99.

Glick, R., & Rogoff, K. (1995). Global versus Country-specific productivity shocks and the current account. *Journal of Monetary Economics, 35,* 159–192.

Gregory, A. W., & Head, A. C. (1999). Common and country-specific fluctuations in productivity, investment, and the current account. *Journal of Monetary Economics, 44,* 423–451.

Harberger, A. C. (1950). Currency depreciation, income, and the balance of trade. *Journal of Political Economy, 58*(1), 47–60.

Hoffmann, M. (2013). What drives China's current account? *Journal of International Money and Finance, 32,* 856–883.

İşcan, T. B. (2000). The terms of trade, productivity growth and the current account. *Journal of Monetary Economics, 45,* 587–611.

Kaminsky, G. L., & Reinhart, C. M. (1999). The twin crises: The causes of banking and balance-of-payments problems. *American Economic Review, 89*(3), 473–500.

Laursen, S., & Metzler, L. A. (1950). Flexible exchange rates and the theory of employment. *Review of Economics and Statistics, 32,* 281–299.

Marquez, J. (2004). Productivity, investment, and current accounts: Reassessing the evidence. *Review of World Economics, 104,* 282–301.

Obstfeld, M. (2012). Does the current account still matter? *American Economic Review, 102,* 1–23.

Romelli, D., Terra, C., & Vasconcelos, E. (2018). Current account and real exchange rate changes: The impact of trade openness. *European Economic Review, 105,* 135–158.

Roy, S., & Kemme, D. M. (2011). What is really common in the run-up to banking crises? *Economics Letters, 113*(3), 211–214.

Sevensson, L. E. O., & Razin, A. (1983). The terms of trade and the current account: The Harberger-Laursen-Metzler effect. *Journal of Political Economy, 91,* 97–125.

Smit, B., Grobler, C., & Nel, C. (2014). Sudden stops and current account reversals: Potential macroeconomic consequences for South Africa. *South African Journal of Economics, 82*(4), 616–627.

Tan, Z. B., Yao, Y., & Wei, S. J. (2015). Financial structure, corporate savings and current account imbalances. *Journal of International Money and Finance, 54,* 142–167.
Yan, H. D., & Yang, C. L. (2012). Are there different linkages of foreign capital inflows and the current account between industrial countries and emerging markets? *Empirical Economics, 43,* 25–54.
Zorzi, M. C., Chudik, A., & Dieppe, A. (2012). Thousands of models, one story: current account imbalances in the global economy. *Journal of International Money and Finance*, 31, 1319–1338.

Cycles and Long-Range Behaviour in the European Stock Markets

Guglielmo Maria Caporale, Luis A. Gil-Alana, and Carlos Poza

1 Introduction

Understanding the behaviour of asset prices is crucial for both investors and monetary authorities to design effective portfolio management strategies and stabilisation policies, respectively. However, there is still no agreement on the most appropriate modelling framework to use. Whilst the early literature used specifications based on the classical dichotomy between $I(0)$ and $I(1)$, more recently the possibility of fractional integration and long-memory behaviour has also been taken into account. For instance, a long-memory specification was adopted by Caporale and Gil-Alana (2002) for US stock prices. In a subsequent paper, the same authors advocated an approach incorporating both long-run and cyclical components (see Caporale and Gil-Alana 2014). The present study uses a similar modelling framework which includes two singularities (or poles) in the spectral density function, one corresponding to the long-run (zero) frequency and the other to the cyclical (nonzero) frequency. The adopted specification is very general, since it allows for fractional integration with stochastic patterns at the zero and cyclical frequencies and includes both long-memory and short-memory components. The cyclical patterns are modelled using Gegenbauer processes. A motivation for this type of specification is that it is sufficiently general to include as particular cases the efficient-market hypothesis, mean reversion with different speeds of adjustment towards the long-run

G. M. Caporale (✉)
Brunel University London, London, UK
e-mail: Guglielmo-Maria.Caporale@brunel.ac.uk

L. A. Gil-Alana
University of Navarra, Pamplona, Spain

L. A. Gil-Alana · C. Poza
Universidad Francisco de Vitoria, Madrid, Spain

G. Dufrénot and T. Matsuki (eds.), *Recent Econometric Techniques for Macroeconomic and Financial Data*, Dynamic Modeling and Econometrics in Economics and Finance 27, https://doi.org/10.1007/978-3-030-54252-8_11

equilibrium level, financial cycles, explosive patterns, etc. This model is then estimated in the case of five European stock market índices using monthly data for five European stock market indices (DAX30, FTSE100, CAC40, FTSE MIB40, IBEX35) from January 2009 to January 2019.

Therefore, our contribution is twofold: we incorporate trends in a general framework with both long-memory and short-memory processes at both zero and nonzero frequencies and then provide new evidence on the behaviour of various European stock markets obtained by following this approach. The remainder of the paper is structured as follows. Section 2 reviews the relevant literature. Section 3 outlines the modelling approach. Section 4 describes the data and presents the empirical results. Section 5 offers some concluding remarks.

2 Literature Review

According to Borio (2014), the term "financial cycle" refers to the self-reinforcing interactions among perceptions of value and risk, risk-taking and financing constraints. Typically, rapid increases in credit boost property and asset prices, which in turn increases collateral values and the amount of credit the private sector can obtain, but the process subsequently tends to go into reverse. As highlighted by Borio et al. (2019), this mutually reinforcing interaction between financing constraints and perceptions of value and risks has historically been likely to generate severe macroeconomic imbalances.

The financial cycle can be approximated in different ways. The empirical literature suggests that a reasonable strategy is to capture it through fluctuations in credit and property prices, but also by means of the debt service ratio, defined as interest payments plus amortisation divided by GDP. Drehmann et al. (2018) find a robust relationship between debt accumulation and subsequent debt service (i.e., interest payments plus amortisation), which has a large negative effect on economic growth. All these series may be used individually or combined, as a composite financial cycle proxy similar to that constructed by Drehmann et al. (2012).

Borio et al. (2019) point out that previous literature has identified two important features of the financial cycle. First, its peaks generally coincide with banking crises or considerable financial stress. During expansions, the interaction between asset prices and risk-taking can overstretch balance sheets, making them more fragile and generating the consequent financial tightening. Second, financial cycles can be much longer than business cycles: most of the former have lasted around 15 to 20 years since the early 1980s, whilst the latter have typically lasted up to eight years. Therefore, a financial cycle can span more than one business cycle, which is the reason why peaks in a financial cycle are generally followed by downturns, whilst not all recessions are preceded by one of those peaks.

A number of recent studies have provided more evidence on financial cycles. Specifically, Oman (2019), using a frequency-based filter, details the existence of a Eurozone financial cycle and high-amplitude and low-amplitude national financial cycles. Applying concordance and similarity analysis to business and financial cycles, he provides evidence of several empirical regularities: the aggregate Eurozone credit-to-GDP ratio behaved pro-cyclically in the years preceding euro-area recessions; financial cycles are less synchronised than business cycles; business cycle synchronisation has risen whilst financial cycle synchronisation has decreased; financial cycle desynchronisation was more pronounced between high-amplitude and low-amplitude countries; high-amplitude countries experienced divergent leverage dynamics after 2002. Filardo et al. (2018) explore financial conditions (120 years of data) over time in order to improve our understanding of financial cycles. They find that financial cycles are characterised by recurrent, endogenous swings in financial conditions, which result in booms and busts. Yet the recurrent nature of such swings may not appear so obvious when looking at conventionally plotted time-series data. Using the pioneering framework developed by Stock (1987), they offer a new statistical characterisation of the financial cycle based on a continuous-time autoregressive (AR) model subject to time deformation.

Iacoviello (2015), using Bayesian methods, estimates a DSGE model where a recession is initiated by losses suffered by banks and exacerbated by their inability to extend credit to the real sector. Claessens et al. (2011) provide a wide-ranging analysis of financial cycles using a large database covering 21 advanced countries over the period 1960:1-2007:4. They study cycles in credit, house prices and equity prices. The main results are the following: (1) financial cycles tend to be long and severe, especially those in housing and equity markets; (2) financial cycles are highly synchronised within countries, especially with credit and house price cycles and (3) financial cycles magnify each other, especially when the downturns in credit and housing markets coincide. DePenya and Gil-Alana (2006) propose a method for testing nonstationary cycles in financial time-series data. They develop a procedure that enables the researcher to test unit root cycles in raw time series. Their test has several distinguishing features compared with alternative ones. In particular, it has a standard null limit distribution and is the most efficient test against the fractional alternatives. In addition, it allows the researcher to test unit root cycles at each of the frequencies, and, thus, to approximate the number of periods per cycle. Finally, as already mentioned, Caporale and Gil-Alana (2014) propose a general framework including linear and segmented time trends and stationary and non-stationary processes based on integer and/or fractional degrees of differentiation; moreover, the spectrum is allowed to contain more than a single pole or singularity, occurring at both zero but nonzero (cyclical) frequencies. They find that US dividends, earnings, interest rates and long-term government bond yields exhibit fractional integration with one or two poles in the spectrum; further, a model with a segmented trend and fractional integration outperforms rival specifications over long horizons in terms of its forecasting properties. A similar approach is taken in the present study (see the next section for details).

3 The Model

The adopted model is the following:

$$(1 - L)^{d_1}(1 - 2\cos w_r L + L^2)^{d_2} x_t = u_t, \quad t = 1, 2, \ldots, \tag{1}$$

where x_t is the observed time series; d_1 and d_2 are the orders of integration corresponding to the long-run (zero) and the (cyclical) (nonzero) frequency, respectively, and u_t is an $I(0)$ process, defined as a covariance-stationary process with a spectral density function that is positive and finite at all frequencies in the spectrum. The first polynomial in Eq. (1) refers to the standard case of fractional integration or $I(d)$ that basically imposes a singularity or pole in the spectrum at the long-run or zero frequency. The literature includes plenty of papers with such a specification and testing for unit or fractional degrees of differentiation (for the unit root case, see, e.g., Fama and French 1988a, b; Poterba and Summers 1988; for the fractional case see instead Baillie 1996; Gil-Alana and Robinson 1997; Abbritti et al. 2016; and others).

The second polynomial refers to the case of integration at a frequency away from zero and uses Gegenbauer processes, where $w_r = 2\pi r/T$ and $r = T/s$. Thus, s indicates the number of time periods per cycle, whilst r refers to the frequency with a pole or singularity in the spectrum of x_t. In this context, if $r = 0$ ($s = 1$), the second polynomial in (2) becomes $(1-L)^{2d_2}$, and therefore, the whole process corresponds to the classical fractional integration model widely studied in the literature. Andel (1986) introduced this process for values of r different from 0 and fractional values of d_2 and Gray et al. (1989, 1994) showed that, by denoting $\mu = \cos w_r$, one can express the polynomial in terms of the orthogonal Gegenbauer polynomial $C_{j,d_2}(\mu)$, so that, for all $d_2 \neq 0$,

$$(1 - 2\mu L + L^2)^{-d_2} = \sum_{j=0}^{\infty} C_{j,d_2}(\mu) L^j,$$

where we can define $C_{j,d_2}(\mu)$ recursively as follows:

$$C_{0,d_2}(\mu) = 1, \quad C_{1,d_2}(\mu) = 2\,\mu\, d_2,$$

and

$$C_{j,d_2} = 2\mu\left(\frac{d_2 - 1}{j} + 1\right)C_{j-1,d_2}(\mu) - \left(2\frac{d_2 - 1}{j} + 1\right)C_{j-2d_2}(\mu),\ j = 2, 3, \cdots$$

Authors such as Giraitis and Leipus (1995), Chung (1996a, b), Gil-Alana (2001) and Dalla and Hidalgo (2005), among others, subsequently examined these processes;

a recent empirical application using UK inflation can be found in Gil-Alana and Trani (2019).

In this paper, we combine these two approaches in a single framework testing simultaneously for the orders of integration at both the zero and a nonzero frequency. This type of model has been already employed to analyse US inflation by Canarella et al. (2019), but to date there have been no applications to stock prices.

4 Data Description and Empirical Results

We use closing prices of the following five European stock market indices: DAX30 (Germany), FTSE100 (UK), CAC40 (France), FTSE MIB40 (Italy) and IBEX35 (Spain). The frequency is monthly, and the sample period goes from January 2009 to January 2019. The data source is Thomson Reuters Eikon. Plots of the series are shown in Fig. 1. Visual inspection suggests that DAX30, FTSE100 and CAC40 exhibit an upward trend, whilst FTSE MIB40 and IBEX35 fluctuate around their mean.

As a first step, we compute the periodogram for the five series under examination. This is an asymptotically unbiased estimator of the spectral density function and can be used to obtain some preliminary evidence about the peaks in the spectrum of the series.

Table 1 displays the first five values of the periodogram for each series. It can be seen that for the stock markets of France, Germany and UK, the highest value corresponds to the smallest frequency, following by frequency 3; however, for France and Spain, it occurs at frequency 2, followed by frequency 1 and frequency 3, respectively.

In order to avoid deterministic terms, we use the demeaned series and estimate the model given by Eq. (1), testing the null hypothesis:

$$H_0 : d = d_0, \tag{2}$$

where $d = (d_1, d_2)^T$, with both values ranging from -2.00 to 2.00 with 0.01 increments. Thus, the estimated model under the null is:

$$(1 - L)^{d_{1o}}(1 - 2\cos w_r L + L^2)^{d_{2o}} x_t = u_t, \quad t = 1, 2, \ldots, \tag{3}$$

where u_t is assumed to be in turn an uncorrelated (white noise) process and an autocorrelated one, for the latter the exponential spectral approach of Bloomfield (1973) being used for the disturbances u_t; this is a non-parametric method that only requires specifying the spectral density function, which is given by:

$$f(\lambda; \tau) = \frac{\sigma^2}{2\pi} \exp\left(2 \sum_{r=1}^{m} \tau_r \cos(\lambda r)\right), \tag{4}$$

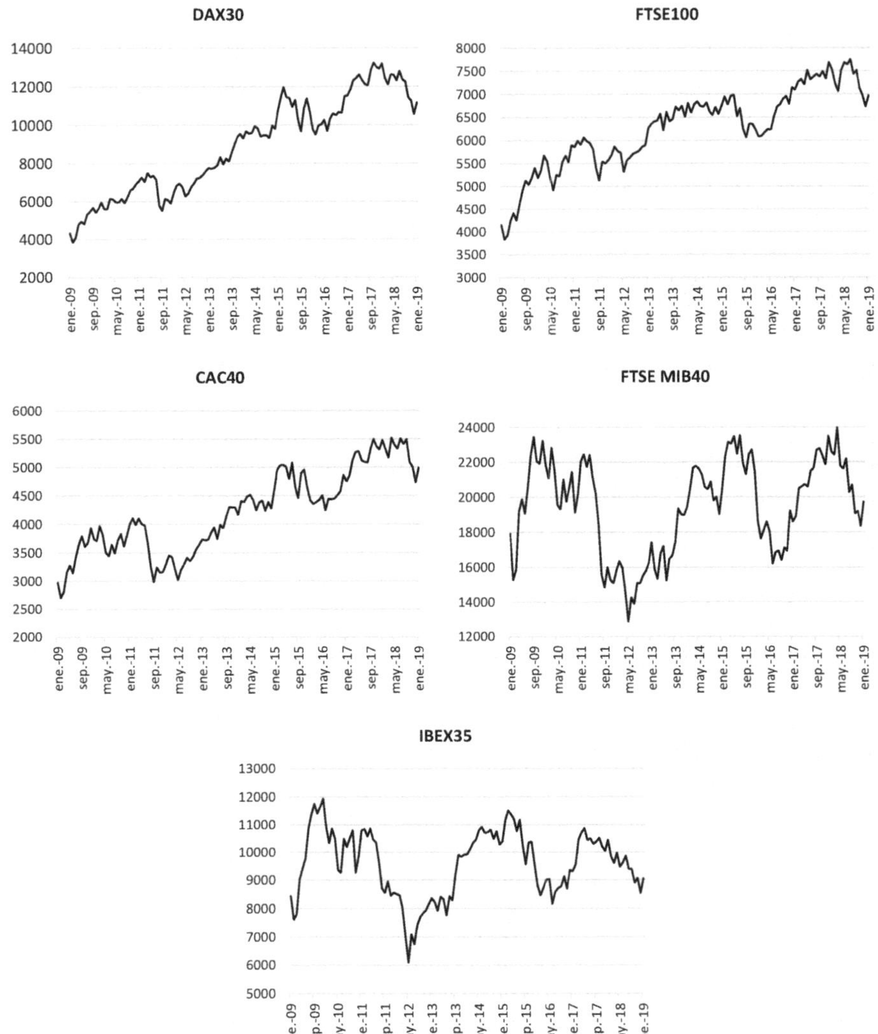

Fig. 1 Closing prices of five European stock market indices

Table 1 First five values in the periodogram of the series

Country	1	2	3	4	5
France	**0.17205**[a]	0.00407	**0.04472**	0.02301	0.00021
Germany	**0.55260**[a]	0.06697	**0.10928**	0.04837	0.00627
Italy	**0.04423**	**0.04806**[a]	0.04228	0.01590	0.00905
Spain	0.01546	**0.05215**[a]	**0.04449**	0.00507	0.01091
U.K.	**0.08426**[a]	0.02807	**0.04186**	0.01064	0.00331

[a] refers to the largest value and in bold the largest two values

where σ^2 is the variance of the error term and m indicates the short-run dynamic components. Bloomfield (1973) showed that this function approximates fairly well the behaviour of highly parameterised ARMA models and performs well in the context of fractional integration (Gil-Alana 2004).

For the sake of generality, we do not restrict the first polynomial to be constrained at the zero frequency and therefore consider initially a model with two factors of the Gegenbauer polynomial of the form:

$$\prod_{j=1}^{2} (1 - 2\cos w_r^{(j)} L + L^2)^{d_o^{(j)}} x_t = u_t, \quad t = 1, 2, \ldots, \tag{5}$$

where $d_o^{(1)}$ becomes $d_{1o}/2$ if $w_r^{(1)} = 0$ (or $j_1 = 1$). The estimated value of j is equal to 1 in all cases, which supports the existence of a pole or singularity in the spectrum at the zero frequency. Thus, in what follows we focus exclusively on the model given (3), estimating simultaneously d_{1o} (the order of integration at the long-run or zero frequency, d_{2o} (the order of integration at the cyclical frequency) and j_2 (the frequency in the spectrum that goes to infinity and that is related to the number of periods per cycle in the cyclical structure, i.e., $r_2 = j_2/T$).

Table 2 focuses on the case of white noise errors. It can be seen that that the frequency j_2 is equal to 2 for France, Italy and Spain, and to 3 for the UK and Germany. This implies that the number of periods per cycle is approximately 60 (five years) for the stock markets in the former three countries and 49 ($T = 121$)/3 $\approx$ 40 months (3.3 years) for the latter two. Concerning the estimates of the differencing parameters, d_1 is smaller than 1 in the case of France, though the unit-root null hypothesis cannot be rejected, whilst for the other countries the $I(1)$ hypothesis is rejected in favour of values of d_1 above 1. As for the estimates of d_2, the highest is for France (0.33), and only for this country and Germany (0.08), the values are significantly positive. In the other three cases, they are positive but very close to zero, and the $I(0)$ null cannot be rejected.

Table 3 displays the results for the case of weak autocorrelation using the model of Bloomfield (1973). The values of j_2 are now 2 for Italy and Spain and 3 for the other three countries; d_1 is substantially smaller than in the previous table, its estimates ranging between 0.58 (UK) and 0.71 (Spain), and evidence of mean reversion with respect to this frequency is only obtained in the UK case. In all other cases, the

Table 2 Estimates of d based on a model with white noise disturbances

Country	j_1	j_2	d_1	d_2
France	1	2	0.89 (0.73, 0.96)	0.33 (0.17, 0.65)
Germany	1	3	1.36 (1.11, 1.44)	0.08 (0.01, 0.25)
Italy	1	2	1.24 (1.01, 1.39)	0.02 (−0.08, 0.14)
Spain	1	2	1.38 (1.14, 1.52)	0.08 (−0.03, 022)
U.K.	1	3	1.34 (1.10, 1.53)	−0.05 (−0.17, 0.15)

Table 3 Estimates of d based on a model with autocorrelated disturbances

Country	j_1	j_2	d_1	d_2
France	1	3	0.65 (0.27, 1.09)	0.05 (−0.27, 0.11)
Germany	1	3	0.66 (0.49, 1.18)	0.04 (−0.09, 0.18
Italy	1	2	0.64 (0.45, 1.03)	0.01 (−0.18, 0.20)
Spain	1	2	0.71 (0.56, 1.24)	0.05 (−0.14, 0.26)
U.K.	1	3	0.58 (0.31, 0.99)	0.02 (−0.11, 0.21)

intervals indicate that the unit root null cannot be rejected. Finally, the estimates of d_2 are all positive, but the null $d_2 = 0$ cannot be rejected in any country.

On the whole, our results indicate high persistence at the long-run frequency, but they are not very supportive of the existence of cyclical stochastic structures in the European financial markets. The only clear evidence of a stochastic cycle is obtained in the case of France under the assumption of white noise disturbances; in all other cases, although d_2 is found to be positive, the confidence intervals are such that the null $d_2 = 0$ cannot be rejected, and therefore, there is no evidence of cycles.

5 Conclusions

In this paper, we have examined the possible presence of stochastic cycles in financial series. For this purpose, we have proposed a model that allows simultaneously for both long-run and cyclical patterns in the data using a method based on long-memory processes. For the zero frequency, the standard $I(d)$ approach is followed, whilst for the cyclical structure a Gegenbauer polynomial is used which also allows for fractional degrees of differentiation. Therefore, the chosen specification contains two singularities in the spectrum corresponding to the long-run (zero) and the cyclical (nonzero) frequencies, respectively.

Using monthly data for five European stock market indices (namely DAX30 (Germany), FTSE100 (UK), CAC40 (France), FTSE MIB40 (Italy) and IBEX35 (Spain)) over the period from January 2009 to January 2019, we find that the order of integration at the long-run or zero frequency is significantly higher than the one at the cyclical frequency, the latter being insignificantly different from zero in the majority of cases. The cycles seem to have a periodicity between three and five years.

However, these results should be taken with a degree of caution given the relatively short sample period. Specifically, with 121 monthly observations as in our cas,e the smallest possible frequency apart from $j_1 = 1$ (that corresponds to the long-run frequency) is 2, which implies cycles of $T/2$ at most, i.e., 60 months or five years. Analysing much longer series, possibly spanning decades, would be much more informative about the possible existence of stochastic cycles. This is left for further research.

Acknowledgement Luis A. Gil-Alana gratefully acknowledges financial support from the Ministerio de Ciencia y Tecnología ((ECO2017-85503-R).

References

Abbritti, M., Gil-Alana, L. A., Lovcha, Y., & Moreno, A. (2016). Term structure persistence. *Journal of Financial Econometrics, 14*(2), 331–352.

Andel, J. (1986). Long memory time series models. *Kybernetika, 22,* 105–123.

Baillie, R. T. (1996). Long memory processes and fractional integration in econometrics. *Journal of Econometrics, 73,* 5–59.

Bloomfield, P. (1973). An exponential model in the spectrum of a scalar time series. *Biometrika, 60,* 217–226.

Borio, C. (2014). The financial cycle and macroeconomics. What have we learnt? *Journal of Banking & Finance, 45,* 182–198.

Borio, C., Drehmann, M., & Xia, D. (2019). The financial cycle and recession risk. *BIS Quarterly Review*, December 2018.

Canarella, G., Gil-Alana, L. A., Gupta, R., & Miller, S. (2019). Modeling U.S. historical time-series prices and inflation using alternative long-memory approaches. *Empirical Economics*, forthcoming.

Caporale, G. M., & Gil-Alana, L. A. (2002). Fractional integration and mean reversion in stock prices. *Quarterly Review of Economics and Finance, 42,* 599–609.

Caporale, G. M., & Gil-Alana, L. A. (2014). Long run and cyclical dynamics in the US stock market. *Journal of Forecasting, 33,* 147–161.

Chung, C.-F. (1996a). A generalized fractionally integrated autoregressive moving-average process. *Journal of Time Series Analysis, 17,* 111–140.

Chung, C.-F. (1996b). Estimating a generalized long memory process. *Journal of Econometrics, 73,* 237–259.

Claessens, S., Ayhan, K., & Terrones, M. (2011). Financial cycles. What? How', When? *NBER International Seminar on Macroeconomics, 7*(1), 303–334.

Dalla, V., & Hidalgo, J. (2005). A parametric bootstrap test for cycles. *Journal of Econometrics, 129,* 219–261.

DePenya, F., & Gil-Alana, L. (2006). Testing of nonstationary cycles in financial time series data. *Review of Quantitative Finance and Accounting, 27*(1), 47–65.

Drehmann, M., Borio, C., Tsatsaronis, K. (2012). Characterising the financial cycle, don't lose sight of the medium term. BIS Working Paper No. 380.

Drehmann, M., Juselius, K., & Korinke, A. (2018). Going with the flows, new borrowing debt service and the transmission of credit booms. NBER Working Paper 24549.

Fama, E. F., & French, K. R. (1988a). Dividend yields and expected stock returns. *Journal of Financial Economics, 22,* 3–25.

Fama, E. F., & French, K. R. (1988b). Permanent and temporary components of stock prices. *Journal of Political Economy, 96,* 246–273.

Filardo, A., Lombardi, M., & Raczko, M. (2018). Measuring financial cycle time. BIS Working Papers 755, Monetary and Economic Department.

Gil-Alana, L. A. (2001). Testing stochastic cycles in macroeconomic time series. *Journal of Time Series Analysis, 22,* 411–430.

Gil-Alana, L. A. (2004). The use of the Bloomfield (1973) model as an approximation to ARMA processes in the context of fractional integration. *Mathematical and Computer Modelling, 39,* 429–436.

Gil-Alana, L. A., & Robinson, P. M. (1997). Testing of unit roots and other nonstationary hypotheses in macroeconomic time series. *Journal of Econometrics, 80,* 241–268.

Gil-Alana, L. A., & Trani, T. (2019). The cyclical structure of the UK inflation rate. 1210–2016. *Economics Letters, 181*, 182–185.
Giraitis, L., & Leipus, R. (1995). A generalized fractionally differencing approach in long memory modeling. *Lithuanian Mathematical Journal, 35*, 65–81.
Gray, H. L., Yhang, N., & Woodward, W. A. (1989). On generalized fractional processes. *Journal of Time Series Analysis, 10*, 233–257.
Gray, H. L., Yhang, N., & Woodward, W. A. (1994). On generalized fractional processes. A correction. *Journal of Time Series Analysis, 15*, 561–562.
Iacoviello, A. (2015). Financial business cycles. *Review of Economic Dynamics, 18*(1), 140–163.
Oman, W. (2019). The synchronization of business cycles and financial cycles in the Euro area. *International Journal of Central Banking, 15*, 1.
Poterba, J. M., & Summers, L. H. (1988). Mean reversion in stockprices: Evidence and implications. *Journal of Financial Economics, 22*, 27–59.
Stock, J. (1987). Measuring business cycle time. *Journal of Political Economy, 95*(6), 1240–1261.

A Non-linear Approach to Measure the Dependencies Between Bitcoin and Other Commodity Markets

Stéphane Goutte and Benjamin Keddad

JEL Classification C32 · G15 · Q02

1 Introduction

The objective of the paper is to study the interactions and dependencies across volatility of the main cryptocurrency Bitcoin and other commodity and energy markets. The Bitcoin is an electronic currency based on a vast peer-to-peer network, totally decentralized. New Bitcoins are introduced to the market via a process called mining. The miners receive rewards as soon as they validate new transactions after solving an optimisation problem by the way of a Proof of Work which needs intensive computation. The first Bitcoin was created in 2009. Since its origin, we observe a high volatility of its price and specific features that need to be understood for investments' objectives. The potential of cryptocurrencies as a means of payment and store of value has also been addressed by many stakeholders such as central banks and governments, but their high price volatility continues to reduce their scope on a larger scale.

We have observed recently a huge amount of papers on cryptocurrencies with the aim to predict their price, volatility, or to compare their trends regarding other markets. A lot of literature focuses on the prediction of the volatility and nearly all the models existing in the linear and nonlinear time series have been applied to predict prices from one hand or volatility in the other. Indeed, the Bitcoin had an

S. Goutte (✉)
CEMOTEV, UVSQ, Paris-Saclay, France
e-mail: stephane.goutte@univ-paris8.Fr

B. Keddad
PSB—Paris School of Business, 59 rue Nationale, 75013 Paris, France
e-mail: b.keddad@gmail.com

G. Dufrénot and T. Matsuki (eds.), *Recent Econometric Techniques for Macroeconomic and Financial Data*, Dynamic Modeling and Econometrics in Economics and Finance 27, https://doi.org/10.1007/978-3-030-54252-8_12

explosiveness period in 2017, a drop dawn and a quite stable period in 2018 and a new increasing trend during the first six months of 2019. All of these behaviors argue the interest of studying cryptocurrencies and their interactions with other assets.

Motivated by the growth of the Bitcoin market, and the recent interest of market participants and academics, this study focuses on the cross-comparisons of the Bitcoin against a panel of commodities to identify dependencies and common volatility regimes. This study is innovative since there is no research dealing with this issue. Moreover, as soon as Bitcoin is used also for diversification in portfolios, we need to analyze its behavior taking into account the behavior of other financial assets and specially the commodity and energy markets. This issue is even more important as academics and professionals suggest linking the value of crypotcurrencies to a basket of reference assets that would include commodities, fiat currencies or government securities, in order to stabilizing their value (see, Bis 2019).

A lot of paper has investigated the existence of cross-correlation between cryptocurrencies and different classes of stocks and bonds. For instance, in a recent paper, Aslanidis et al. (2019) detected that correlation of traditional assets against Monero is even closer to zero than against other cryptocurrencies. This result could mean that Monero behavior is different from other cryptocurrencies. They conclude that a small portion of cryptocurrencies will dominate the stochastic dynamic of a whole portfolio if they are integrate it. Spillover effects are also studied in several papers (Gillaizeau et al. 2019). Using realized volatility of the cryptocurrencies, some authors (Kurka 2019) detected asymmetries in shock transmissions between the cryptocurrencies and traditional assets. Informational efficiency of Bitcoin has also been studied using high-frequency data in Zargar and Kumar (2019).

Dynamic interdependence of Bitcoin prices in a cross-market context has been studied in Cheah et al. (2018) or Alvarez-Ramirez et al. (2018). Associated to the possible long memory behavior, models with jumps have been used to investigated both the behavior of the returns and the volatility. Some references are Chevallier et al. (2019), Phillip et al. (2019) or Mensi et al. (2019).

Philippas et al. (2019) uses dual process diffusion model to identify jumps attributed to informative signals on Bitcoin prices. The informative signals are derived from Twitter and Google trends. The authors show that the signals justify a sentimental appetite for information demand.

At the same time, researchers begins to question the issues of the Bitcoin for investment (CME, Futures). Looking at the evolution of the price on economic and financial specific periods, they try to show that Bitcoins can be associated to a commodity. Due to the specific evolution of the Bitcoin market, it is crucial to continuously investigate the future of this money to understand if its valuation has a unique behavior and the links it has with commodity markets.

Thus, in this paper, we try to analyse the behavior of the Bitcoin and its futures with a Markov-switching model approach. More precisely, the aim of our research design is to detect common episodes of volatility between the Bitcoin and commodities and quantify their regime-dependent connectedness. We employ a vector autoregressive version of the well-known Markov-switching model (MS-VAR) and compute the regime-dependent impulse response function (IRF) in regimes of high and low volatility. This methodology will be details in the next section.

The paper is organized as follows: in Sect. two, we present the data; in Sect. 3, the methodology used will be stated and in the last section, economic and financial interpretations and policies will be given.

2 Empirical Strategy and Data

We work on a two year dataset: 01/01/2017 to 18/12/2018 which corresponds to an economic cycle for the cryptocurrencies like the Bitcoin, even if we do not know if the trend in 2019 will become positive. We selected data sourced from Coinbase and Datastream. Along with the Bitcoin, the variables we used are the futures of Ethanol, Natural Gas, Lead, Wheat, Coffee, Cotton, Silver, Copper, Zinc, and Lead. All data are expressed in USD. This large panel of energy and commodity markets allows us to compare the impact, relationship, and dependency of Bitcoin against these markets.

We propose a bivariate MS(m)-VAR(p) model to link the returns of the Bitcoin and commodities and to 1\ detect common regimes among these two classes of asset and 2\ compute regime-dependent IRFs to detect asymmetric spillovers. Here, m stands for the number of regimes an p for the number of autoregressive lags within the model. We define $m = 2$ implying that the dependent variable can visit two different regimes, while the number of lags is chosen according to HQ information criterion.

We choose the less restrictive MS-VAR specification where all parameters are allowed to switch between regimes. We use the following regime-dependent specification:

$$Y_t = \begin{cases} \mu_1 + \sum_{i=1}^{p} B_{i,1} Y_{t-i} + A_1 u_t & \text{if } s_t = 1, \\ \vdots & \\ \mu_m + \sum_{i=1}^{p} B_{i,m} Y_{t-i} + A_m u_t & \text{if } s_t = m. \end{cases} \tag{1}$$

where Y_t is a $Kx1$ vector of the asset returns and u_t is a $K \times 1$ fundamental disturbance vector with $u_t \sim N(0, I_K)$. Each regime is characterized by an intercept μ_i, autoregressive parameter matrices $B_{1,i}, \ldots, B_{p,i}$. We define $\sum_i$ as the variance–covariance matrix of the residuals $A_i u_t$.

All these parameters are conditioned on an unobserved variable s_t that follows an ergodic two-states Markov process. Accordingly, the realization of the state $s_t \in 1, 2$ takes the form of a probability and is governed by a hidden Markov chain process with the following conditional transition probabilities:

$$p_{ij} = \Pr(s_{t+1} = i | s_t = j), \quad \sum_{j=1}^{m} p_{ij} = 1, \quad \text{for all } i, j \in \{1, \ldots, m\}. \tag{2}$$

The transition probability between the two states could be expressed as the following 2×2 probability matrix

$$P_{ij}(s_{t+1} = i|s_t = j) = \begin{pmatrix} P_{11} & 1 - P_{22} \\ 1 - P_{11} & P_{22} \end{pmatrix}, \tag{3}$$

The model in Eqs. (1–3) notably assumes varying heteroskedasticity according to the volatility state in the Bitcoin and commodity markets. The retained specification is entirely fitting with the purpose of the study. Indeed, in our two states model, regimes are defined as low and high volatility regimes. The first regime is labeled as the high volatility and the second as the low volatility regime. Accordingly, we assume that volatility of both assets should be higher in the second regime.

Here, the main advantages to use this nonlinear framework are two-fold. First, the MS-VAR has the advantage to infer regime probabilities and then episodes that overlap common dynamics among two or more time series. This implies, we can detect specific events where these series display some commonalities in their behavior such as returns or volatility. Second, we can investigate in each regime the connectedness across the variables and estimate potential asymmetries in their relationship.

In this study, we compute the regime-dependent IRFs to measure bidirectional spillovers between the Bitcoin and commodities during episodes of financial stress. Accordingly, we define spillovers as the response of one specific asset during common episodes of high volatility and we contrast the responses with those observed during normal times.

The reduced form error covariance matrix and switches in autoregressive parameters are then exploited in a Markov-switching context to identify structural shocks and compute IRFs. The next section describes how regime-dependent IRFs are computed.

2.1 *Impulse Response Functions*

As explained above, we compute the IRFs obtained from the MS-VAR. The impulse responses are then regime-dependent. The Bitcoin and commodities display regime-dependent responses after we simulate an unit deviation shock to a fundamental disturbance in each regime. We define the impulse response in regime i as in Ehrmann et al. (2003).

$$\left. \frac{\partial E_t Y_{t+h}}{\partial u_{k,t}} \right|_{s_t = \cdots = s_{t+h} = i} = \theta_{ki,h} \quad \text{for } h \geq 0. \tag{4}$$

Equation (4) shows the expected changes in endogenous variables at time $t + h$ when a unit shock occurs in the kth fundamental disturbance at time t, conditional on regime i. Estimates of the impulse responses can be calculated by combining the parameter estimates of the unrestricted MS-VAR with the regime-dependent matrix

$\hat{A}_i$ obtained through Cholesky restrictions. The relationship between the estimated response vectors and estimated parameters is obtained as follows:

$$\hat{\theta}_{ki,0} = \hat{A}_i u_0 \tag{5}$$

$$\hat{\theta}_{ki,h} = \sum_{j=1}^{\min(h,p)} \hat{B}_{ji}^{h-j+1} \hat{A}_i u_0 \text{ for } h \geq 0. \tag{6}$$

where $u_0 = (0, \ldots, 0, 1, 0, \ldots, 0)'$ is the initial disturbance vector, in which only the kth element is 1.

3 Results

Figures 6 and 7 display the regime-dependent IRFs between the Bitcoin and the commodities. For each commodity, the two first graphs show the Bitcoin response to a unit commodity shock in both regimes while the other graphs show the opposite. We also plot the common probabilities of being in the high volatility regime in Figs. 4a, 5 and 6b.

The common regime probabilities allow to detect episodes where both assets co-move through different regimes. We are particularly interested in detecting common high volatility episodes to assess how the underlying risk in both markets are connected in each regime. For purpose of illustration we plot Coffee and Bitcoin absolute returns as well as the common regime of high volatility (Fig. 1). The result clearly shows the ability of the MS-VAR to capture the co-regime between both assets as regime probabilities seem to overlap episodes of financial stress in each market. This is particularly obvious in the mid-2017 and mid-2018 when both markets have been particularly under stress.

For each commodity, we then compute these probabilities with the aim to compute regime-dependent IRFs and bidirectional spillovers between both markets. The smoothed probabilities are displayed for each estimated bivariate MS-VAR in Figs. 4, 5 and 6. When the probability is higher than 0.5, it more likely that the Bitcoin and the commodity evolve simultaneously in the high volatility regime. For all commodities, we observe that common episodes of high volatility are rather short-lived suggesting that both markets are dominated by idiosyncratic risk factors although they seems to be connected during some specific events. The only two exceptions are Silver and Wheat for which the high volatility regime is found to be more persistent. This is mainly explained by the lack of commonalities between both assets leading regime probabilities to be mainly driven by the asset with the highest volatility, i.e., the Bitcoin (see Fig. 2). Accordingly, for Silver and Wheat, we should be careful when interpreting the results.

Regarding the regime-dependent IRFs, the responses clearly show two striking results. The first important observation is that responses to commodity shocks are

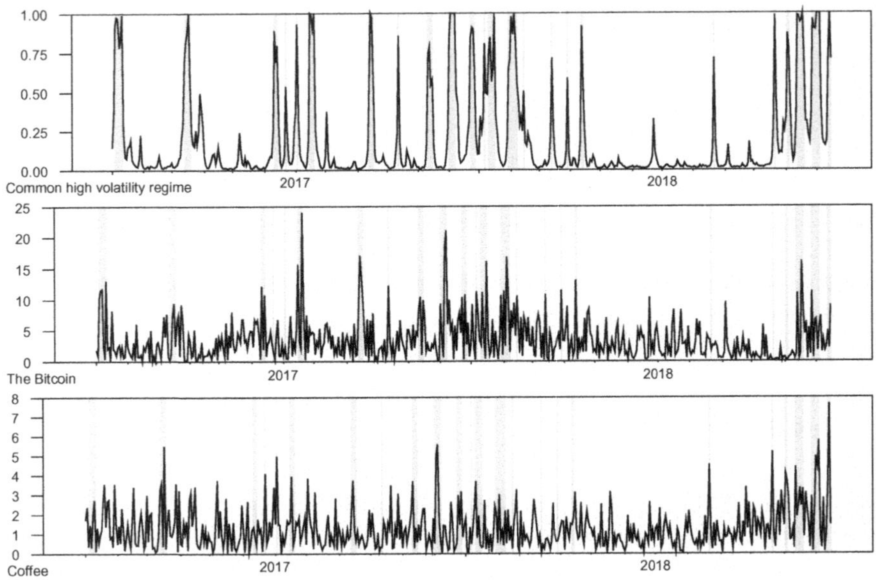

Fig. 1 Coffee, Bitcoin, and common regimes

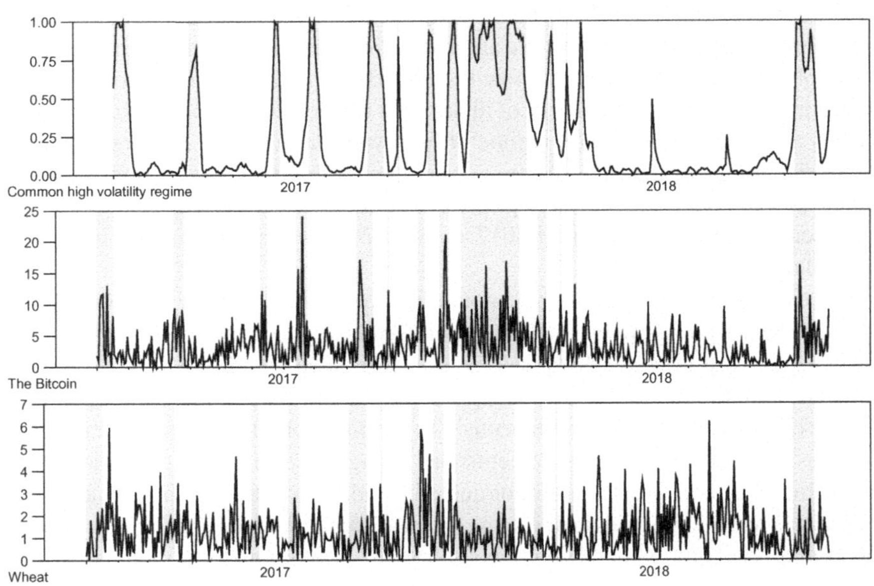

Fig. 2 Wheat, Bitcoin, and common regimes

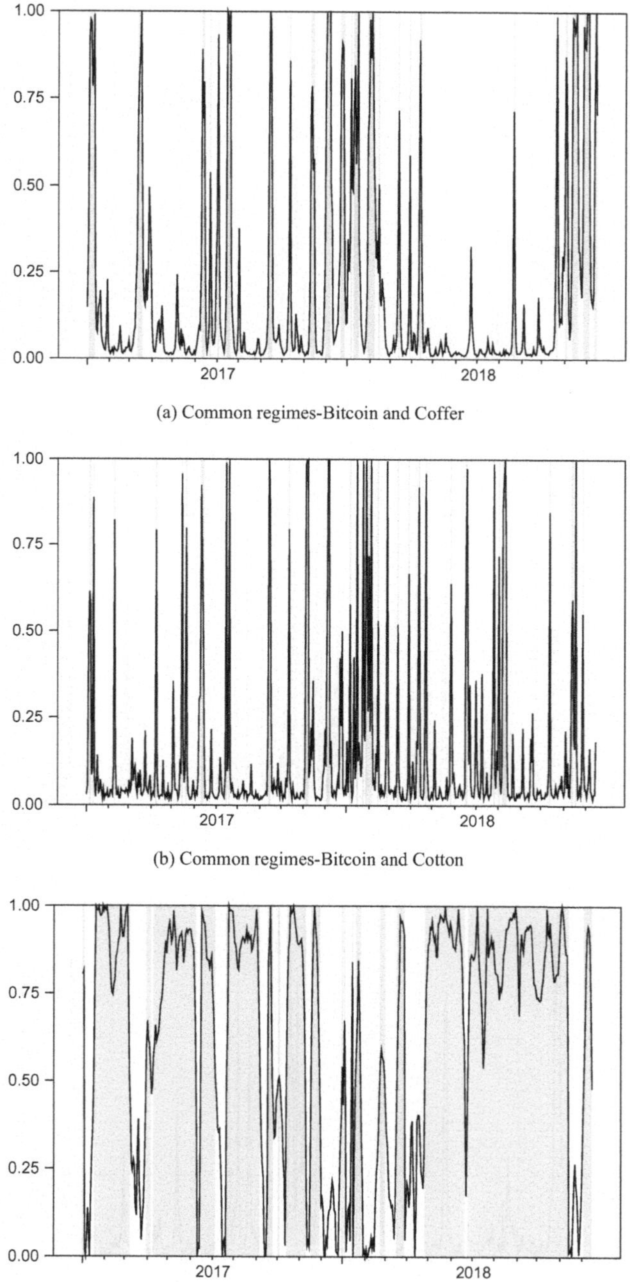

(a) Common regimes-Bitcoin and Coffer

(b) Common regimes-Bitcoin and Cotton

(c) Common regimes-Bitcoin and Silver

Fig. 3 Regime probabilities

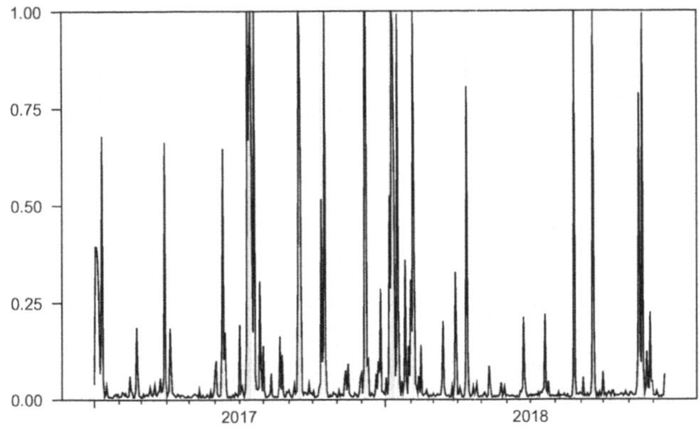

(a) Common regimes - Bitcoin and Copper

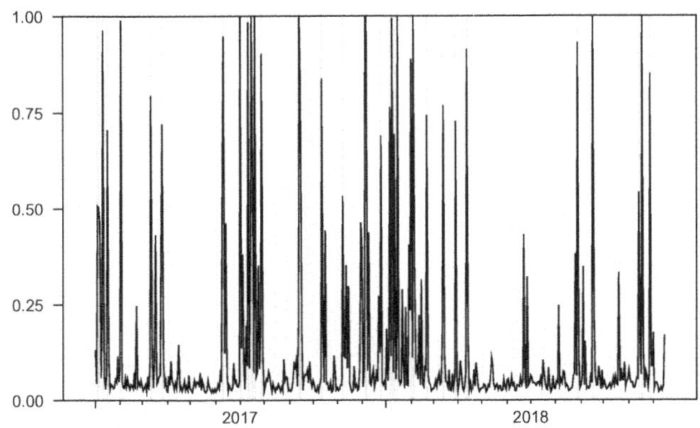

(b) Common regimes - Bitcoin and Lead

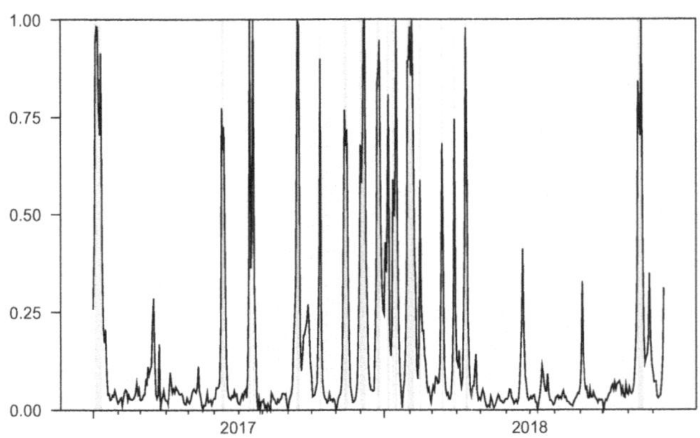

(c) Common regimes - Bitcoin and Ethanol

Fig. 4 Regime probabilities

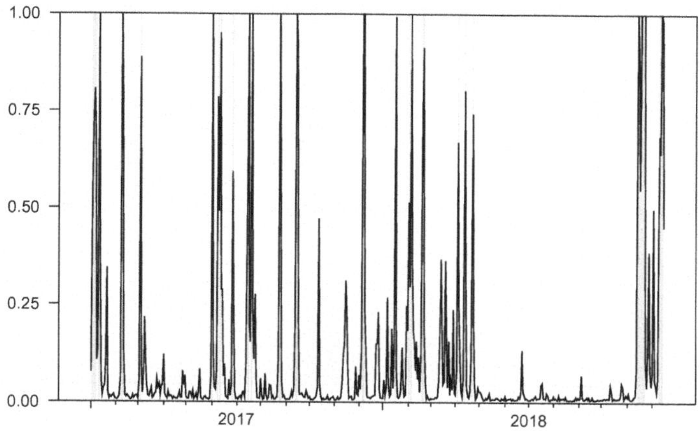

(a) Common regimes - Bitcoin and Gas

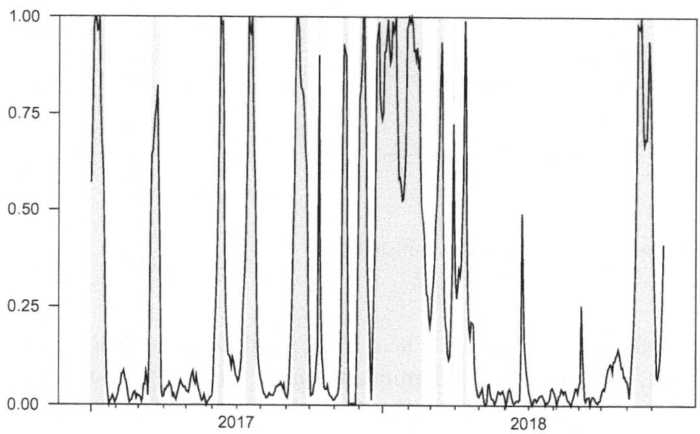

(b) Common regimes - Bitcoin and Wheat

Fig. 5 Regime probabilities

more important in the high volatility regime for all commodities except Cotton. For instance, we observe that absolute returns of the Bitcoin significantly increase by 1% following a 1% shock in Coffee futures after three days, while the response decreases gradually and vanish after ten days. However, the responses are insignificant in the first regime. We find a similar pattern for Lead that reacts by 1.5% after four days. We also observe that commodities do not respond similarly to Bitcoin fluctuations. For Ethanol, Zinc, Gas, and Silver, we find a initial negative response before turning positive for Zinc only. As for Coffee, the Bitcoin response is non-significant or very close to zero in the first regime. Note that the negative impact is particularly obvious for Ethanol with a decrease in volatility of 0.8% after two days. For Cotton, we find a significant response in the low volatility regime, with a moderate negative impact

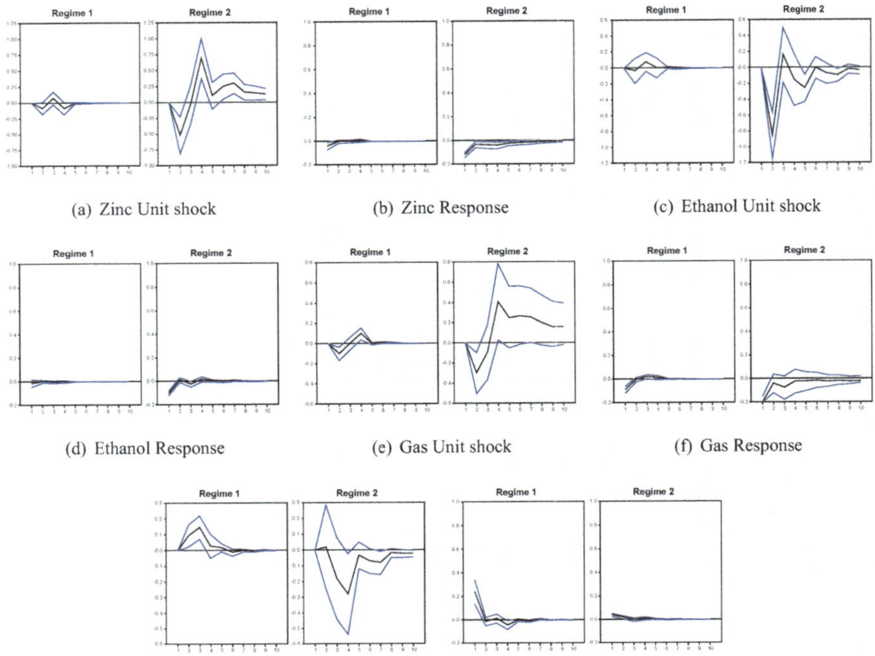

(a) Zinc Unit shock (b) Zinc Response (c) Ethanol Unit shock

(d) Ethanol Response (e) Gas Unit shock (f) Gas Response

(g) Wheat Unit shock (h) Wheat Response

Fig. 6 Bitcoin responses to commodity unit shocks

of 0.25% after three days. Finally, Wheat is not sensitive to the Bitcoin as the IRF is non-significant, but the impact is significant during tranquil periods with a positive response around 0.15% after three days. As a whole, we find that the Bitcoin exerts a significant impact on many commodities with a shape in the responses that are clearly heterogeneous.

The second striking feature is the very moderate impact of Bitcoin fluctuations on commodities, although the situation seems to differ according to the commodity. Accordingly, we clearly find asymmetric spillovers between both assets, with a stronger impact running from the commodities to the Bitcoin. This should indicate that the cryptocurrency market has only a very limited effect on other assets such as commodities. We even find that a volatility shock on the the Bitcoin reduces commodities volatility such as Coffee, Cotton, Copper, Lead, Zinc, and Ethanol. But the spillovers are very low, while statistically significant, ranging from 0.2 to 0.1%. Noting that we observe a transmission mechanism only in the second regime as for volatility shocks on commodities.

However, the Bitcoin seems to play a role for investor as it is highly connected especially during episodes of financial stress. This result has strong implication in terms of portfolio management and diversification for investors engaged in trading activity combining both assets. Besides, the fact that causality is running only from

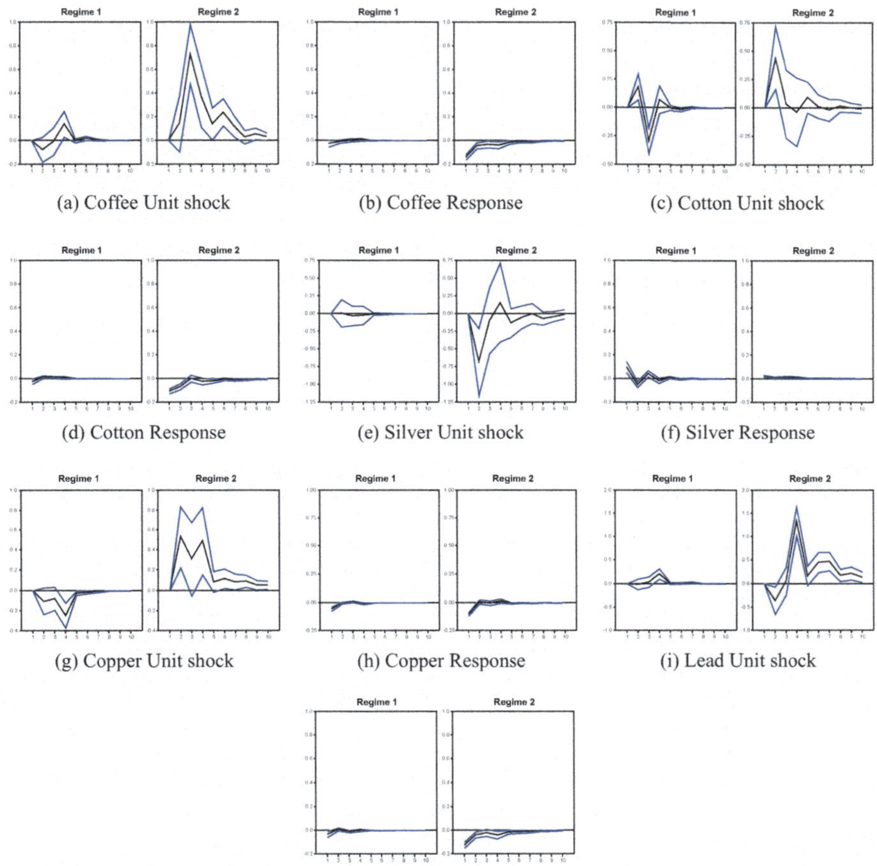

(a) Coffee Unit shock (b) Coffee Response (c) Cotton Unit shock

(d) Cotton Response (e) Silver Unit shock (f) Silver Response

(g) Copper Unit shock (h) Copper Response (i) Lead Unit shock

(j) Lead Response

Fig. 7 Bitcoin responses to commodity unit shocks

commodity to cryptocurrency the market reveals that the Bitcoin does not have any particular destabilizing effect on financial markets. However, investors with cryptocurrencies in their portfolio should be particularly sensitive to events affecting commodities volatility.

4 Conclusion

In this paper, we have investigated how commodities and the Bitcoin are linked together. More precisely, we have tried to detect common volatility regimes between both assets with the aim to measure bidirectional spillovers during episodes of financial stress and tranquil periods. To the best of our knowledge, there are only few studies that studied the connectedness of these two markets. In this paper, we have proposed a nonlinear approach using daily-frequency data on the Bitcoin and a set of commodities ranging from Gas to Cotton or Copper. First, we find that the Bitcoin impacts significantly many commodities with a shape in the responses that are clearly heterogeneous. More importantly, this impact is clearly stronger during episodes of financial stress. Finally, we have observed that spillovers between both assets are asymmetric, with a stronger impact running from the commodities to the Bitcoin. These results have strong implications for investors who are engaged in portfolio activity and search for hedge during episodes of financial stress.

References

Alvarez-Ramirez, J., Rodriguez, E., & Ibarra-Valdez, C. (2018). Long-range correlations and asymmetry in the Bitcoin market. *Physica A: Statistical Mechanics and its Applications*, *492*, 948–955.

Aslanidis, N., Bariviera, A. F., & Martinez-Ibanez, O. (2019). An analysis of cryptocurrencies conditional cross correlations. *Finance Research Letters*, *31*, 130–137.

Cheah, E.-T., Mishra, T., Parhi, M., & Zhang, Z. (2018). Long memory interdependency and inefficiency in Bitcoin markets. *Economic Letters*, *167*, 18–25.

Chevallier, J., Goutte, S., Guesmi, K., & Saadi, S., (2019). *On the Bitcoin price dynamics: An augmented Markov-Switching model with Lévy jumps*. Working Paper lshs-02120636.

Ehrmann, M., et al. (2003). Regime-dependent impulse response functions in a Markov-switching vector autoregression model. *Economic Letters*, *78*, 295–299.

G7 Working Group on Stablecoins. *Investigating the impact of global stablecoins*. Committee on Payments and Market Infrastructures. Bank for International Settlements.

Gillaizeau, M., Jayasereka, R., Maaitah, A., Mishra, T., Parhi, M., & Volokitina, E. (2019). Giver and the receiver: Understanding spillover effects and predictive power in cross-market Bitcoin prices. *International Review of Financial Analysis*, *63*, 86–104.

Kurka, J. (2019). Do cryptocurrencies and traditional asset classes influence each other? *Finance Research Letters*, *31*, 38–46.

Mensi, W., Al-Yahyaee, K. H., & Kang, S. H. (2019). Structural breaks and double long memory of cryptocurrency prices: A comparative analysis from Bitcoin and Ethereum. *Finance Research Letters*, *29*, 222–230.

Philippas, D., Rjiba, H., Guesmi, K., & Goutte, S. (2019). Media attention and Bitcoin prices. *Finance Research Letters*, *30*, 37–43.

Phillip, A., Chan, J., & Peiris, S. (2019). On long memory effects in the volatility measure of cryptocurrencies. *Finance Research Letters*, *28*, 95–100.

Zargar, F. N., & Kumar, D. (2019). Informational inefficiency of Bitcoin: A study based on high-frequency data. *Research in International Business and Finance*, *47*, 344–353.

Typology of Nonlinear Time Series Models

Aditi Chaubal

1 Introduction

> ...in many areas of applied mathematics and physical sciences, we have had to be content with simplifications of nonlinear problems which result in linearized versions. ... We must face up to nonlinear problems with their special demands and peculiar difficulties.
>
> –Preface in 'Nonlinear Mathematics' (Saaty and Bram 1960)

Time series models provide effective tools to understand the dynamics underlying economic data. Most economic data series seem to exhibit nonlinear structures.[1] In practice, linear models continue to be used extensively as they can be treated as approximations locally for explaining various complicated phenomena. The methodology of linear time series models to explain dynamics of economic data has been developed and implemented extensively. The need for nonlinear techniques stems from linear models being unable to explain certain phenomena (e.g. asymmetries, periodicities in the data, regime switching behaviour, asymmetric adjustments to deviations in data, etc.) in the data.

An early example of nonlinear time series modelling in economics is that of Hansen (1997) who models the United States (US) unemployment rate using nonlinear time series modeling techniques. Hansen (1997, pp. 10, Fig. 2) studied the monthly unemployment rate data for men over 20 years' age from January 1959–July 1997. Figure 1 indicates the monthly US unemployment rate for men over 20 years'

[1] **Disclaimer**: The views expressed here do not reflect the views of the Indian Institute of Technology Bombay, Mumbai. The author is extremely grateful to Prof. Gilles Dufrenot and Prof. Takashi Matsuki for accepting this paper as a chapter. This paper was part of my PhD thesis submitted and defended at the Indira Gandhi Institute of Development Research, Mumbai. Responsibility for any remaining shortcomings and errors rests solely with the author.

A. Chaubal (✉)
Indian Institute of Technology Bombay, Mumbai, India
e-mail: aditichaubal@gmail.com; aditichaubal@iitb.ac.in

G. Dufrénot and T. Matsuki (eds.), *Recent Econometric Techniques for Macroeconomic and Financial Data*, Dynamic Modeling and Econometrics in Economics and Finance 27, https://doi.org/10.1007/978-3-030-54252-8_13

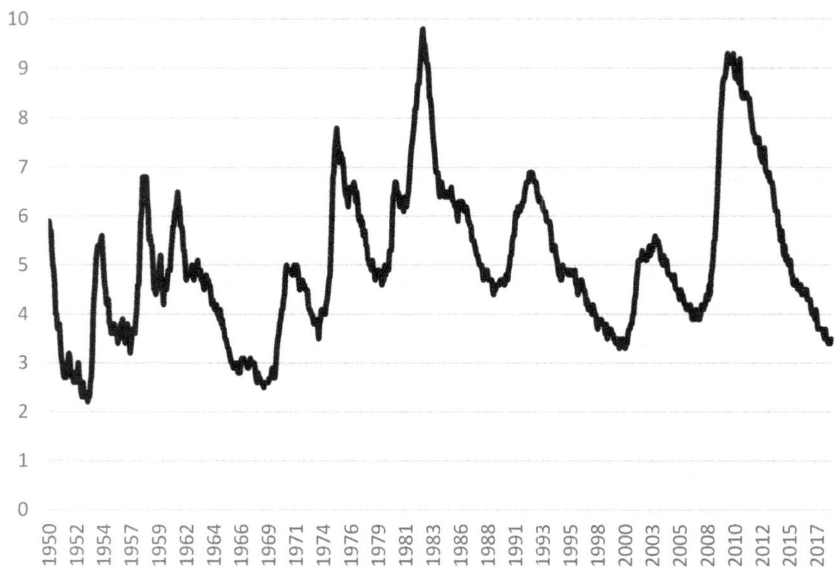

Source: U.S. Bureau of Labour Statistics (2020), Labor Force Statistics, Current Population Survey

Fig. 1 US monthly unemployment rate (Men 20+years) (Jan 1950–Dec 2018)

age from January 1950–December 2018 (The data has been updated up to December 2018 using data from the US Bureau of Labor Statistics).

It may appear that a linear time series model (e.g., an autoregressive moving average (ARMA) model) would help explain the various crests and troughs in this data. However, using a model to explain the ups and downs in the data (instead of the other way round) may result in neglecting any major events that could have led to the crests/troughs and thus may have permanently altered the behavior of data. The argument that the important events can be captured with the help of dummy variable in the linear model also has its drawbacks. It does not account for the fact that the break in data could also lead to a change in the structure (parameters) of the model. Figure 2 highlights the two regimes as indicated in Hansen (1997, pp. 12, Fig. 4) through which the US unemployment rate transits along with an uncertain phase that cannot be explained. Hansen (1997) found that the recoveries and boom phases fell in one regime (indicated by blue in Fig. 2) while the downturns fell in the second regime (indicated by red in Fig. 2).

1.1 Economic Examples

The importance of complicated nonlinear methods has evolved over time in order to address the need for models which can capture certain features of data inexplicable

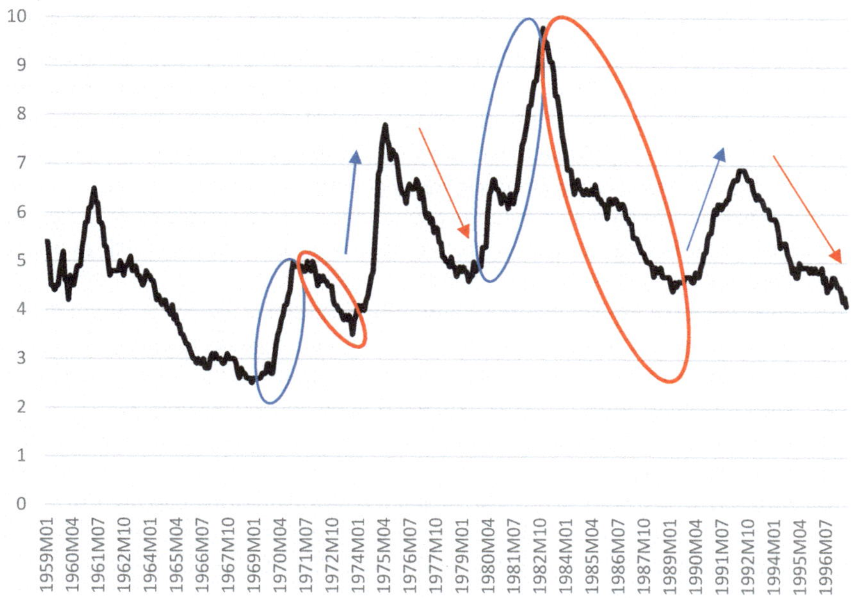

Source: U.S. Bureau of Labor Statistics, Labor Force Statistics, Current Population Survey

Fig. 2 Different regimes through which the unemployment rate passes as found in Hansen (1997, pp. 12) (1959: 01–1997: 07)

by linear models. Among many such features in economic time series data, mention may be made of the following:

- *Asymmetries in the cyclical behaviour* of data [e.g., asymmetric business cycles (Morley and Piger 2010)],
- *Limit cycles*[2] if any in the data [limit cycles in models of growth which may be due to physical characteristics of the data or may be merely endogenous due to the nonlinear dynamics involved (Zhang 1988)],
- *Regime switching behaviour* [e.g., may be caused due to economic, social, or political shocks or changes in the economy (Teraesvirta 1994)],
- Other features arefat tail behaviour i.e. anomalously large fourth moments (kurtosis) (Thurner et al. 2012), volatility clustering (refers to "… *large changes tend to be followed by large changes of either sign and small changes tend to be followed by small changes*" (Mandelbrot 1963; Kirchler and Huber 2007), and persistence i.e. long memory in a process or interdependence of the process across time (Carnero et al. 2004).

In order to understand asymmetric cyclical phenomena, consider the phases that an economy goes through viz. the boom periods and the recessionary phases. It has

[2]Refer to Appendix 1 for definition of a limit cycle.

been observed (Montgomery et al. 1998; Morley and Piger 2010) that important macroeconomic indicators which define an economy's health (e.g., unemployment rate, output gap levels, etc.) exhibit highly significant variations between recessionary phases and expansionary times. Morley and Piger (2010) reiterates that recessionary phases are marked by significant transitory variation in output whereas changes in output during the expansionary phases are in accordance with the trend. This type of cyclical behavior which also exhibits asymmetries in data in terms of the period under consideration cannot be captured effectively using linear models. The nonlinear models can explain such phenomena better.

The phenomena of limit cycles are commonly observed in electrical circuit data as well as in economic growth models. However, economic cycles resulting from such complex nonlinear dynamics are considered endogenous and less intuitive (Zhang 1988). The concept of limit cycles is explained in detail in Appendix 1.

These are some of the many attributes that can be analyzed using the advanced nonlinear models which are not captured by their simple linear counterparts. However, the reasons as to why these characteristics cannot be captured by linear models have been elaborated in the next section along with a formal introduction to various concepts necessary to understand the nonlinear time series models.

1.2 Time Series Models—A Mathematical Introduction

1.2.1 Time Series Models

The linear time series models form the foundation for the understanding of the nonlinear time series models. A limitation of this methodology is that the analysis pertains to stationary time series. According to the Box-Jenkins methodology (Box and Jenkins 1970; Hamilton 1994) if a time series exhibits non-stationarity, the series is transformed to a stationary series [depending on the type of nonstationarity (discussed in the next section)] and modelled.

A linear model, thus, provides a simple additive linear relationship between the variables. This may be inadequate in many real world economic applications where data assume a functional form manifestation which is nonlinear. These may be explained by a power relation or a trigonometric function[3] or a step function, etc. These nonlinear manifestations are also not exact representations of the data but hopefully provide better approximations in trying to explain additional dynamics than those captured by their linear counterparts.

[3]It is not necessary that the presence of such a function indicates a nonlinear relationship with certainty. Such relations can also be analyzed under the linear time series modeling framework by transforming the relation into a linear one.

1.2.2 Nonlinear and Nonstationary Time Series

A nonlinear time series model is one that does not satisfy Eq. (1). Nachane (2011), Tsay (2002) define a nonlinear time series as follows.

A time series $\{X_t\}$ is said to be ***nonlinear*** if it can be expressed as:

$$X_t = f\left(X_{t-1}, \ldots, X_{t-p}, \varepsilon_t, \varepsilon_{t-1}, \ldots, \varepsilon_{t-q}\right) \tag{1}$$

where $f(.)$ is a nonlinear function that describes the behaviour of $\{X_t\}$, and $\varepsilon_t \sim N\left(0, \sigma^2\right)$ is a white noise process and time is discrete.

A limitation of the Box-Jenkins methodology is that it deals with stationary time series. Stationarity is an inherent assumption underlying the analysis of linear time series models using the Box-Jenkins methodology. This is another aspect that economic data may not exhibit owing to the presence of trends, cycles, or random walk behavior. Moreover, persistence, if any, in these characteristics would affect the statistical properties of the series over time. Such characteristics are usually taken into account by differencing (to make a series stationary), or de-trending (removal of trend from series) etc. before further analyzing the data.

In our further discussion, the term '*stationarity*' will refer to covariance stationary or weak stationary. There are two common types of non-stationarity that are encountered while dealing with economic time series data (Hamilton 1994) viz. the processes that exhibit a *deterministic* time trend or a *stochastic* time trend[4].

Thus, both the above aspects viz. functional form manifestation (nonlinear form) and non-stationarity cannot be explained simultaneously using the linear time series model setup. This leads to the significance of developing an alternative framework for analyzing nonlinear time series models.

2 Issues with Nonlinearity and Non-stationarity Testing

Any time series which needs to be analyzed can be partitioned into the following categories (Nachane 2011, 2006):

(i) Linear and stationary
(ii) Linear and non-stationary
(iii) Nonlinear and stationary
(iv) Nonlinear and non-stationary.

The term 'non-stationary' in this context focuses on only one type of non-stationarity viz. unit roots unless mentioned otherwise. The testing of (ii) against (i) thus constitutes the familiar domain of unit root testing which includes the Dickey-Fuller (DF) test, the Phillips-Perron (PP) test, and the augmented DF test. If (i)

[4]Deterministic time trend implies the trend in the time series is a deterministic function of time; stochastic time trend implies that the trend is not predictable (Gujarati and Porter 2008, pp. 745).

is tested against (ii), then it results in the Kwiatowski, Phillips, Schmidt and Shin (KPSS) unit root test.

In the case of nonlinear models, the tests of linearity will need to be carried out to ascertain whether the model is linear or nonlinear under the assumption of stationarity [testing (i) against (iii)]. If nonlinearity is accepted, then the next step is the testing of stationarity of the nonlinear model [testing (iii) against (iv)].

The testing for non-stationarity and nonlinearity has been a much-debated topic in recent literature. Nachane and Clavel (2008) discuss this issue over the existing literature and further discuss an alternative of second-generation nonlinear models. It is also difficult to discriminate among alternative nonlinear models as several provide plausible estimates and forecasts. Most existing and frequently used approaches are based on the assumption of stationarity and focus on the nonlinear aspects. They have the following two-step procedure:

- Remove any non-stationarity, if it exists: First, assuming linearity, check for stationarity of the data and remove any non-stationarity, if it exists [Testing (i) vs. (ii)].
- The transformed series is modeled by alternative nonlinear methods: Assuming stationarity (this holds from the first step), check for nonlinearity and fit the necessary linear or nonlinear model. [Testing (i) vs. (iii)]

Thus, in practice, the difficult task of disentangling of non-stationarity and nonlinearity of the data is a formidable one. They are intertwined within a series and the above two-step approach is based on the assumption that one holds (or is ignored completely) when testing for the other. This could lead to erroneous results in the following manner: if testing for stationarity is carried out (under the assumption of or by ignoring linearity), then it may lead to accepting of stationarity (when the data is actually non-stationary given it is nonlinear); or rejecting stationarity (when the data is actually stationary given it is nonlinear). A similar argument holds when testing for linearity first assuming stationarity, and then testing for stationarity; and fitting a model accordingly.

However, due to the lack of development of appropriate methods, the above two-step procedure has been commonly implemented in literature, though we must recognize that the rejection of the null of stationarity and linearity (i) does not help in determining whether the rejection is in the direction of linearity or stationarity or both.

The efforts in the literature to account for these issues can be categorized into the following two types:

- There are parametric and non-parametric linearity tests [testing (i) vs. (iii)] which are based on the assumption of stationarity. The initial developments in this regard were the non-parametric tests which include the RESET test (1969), the bispectral tests (due to Hinich 1982; Subba Rao and Gabr 1980, etc.) where the null is tested against a general form of nonlinearity.
- These were then followed by developments with respect to testing against specified forms of nonlinearities; parametric tests include Teräsvirta's test (1994) etc.

The data, if found nonlinear, is then subjected to unit root tests which differ from those conducted for linear time series. These include the range ADF (RADF) test (1991), modified ADF (MADF) test (1998), range unit root (RUR) test (2006), etc. which are based on the assumption that the data is nonlinear and then test for the presence of unit roots.

This approach has however been criticized due to the following reasons:

- The tests may have low power (as in case of the Hinich bispectrum test, for instance), **or**
- The assumptions may be restrictive (the assumption that the test statistic converges to a χ-square distribution is violated depending on the bispectrum estimate used in case of the bispectrum test).

Thus, an alternative approach has emerged wherein the non-stationarity and nonlinearity are modeled simultaneously (Nachane and Clavel 2008). This is done by either making the coefficients time varying in a deterministic or random fashion or by attempting to approximate the non-stationarity by locally stationary structures. This genre of methods is known as the ***second-generation nonlinear methods***. These methods include wavelet based neural networks (WANN), mixed spectrum (MS) analysis, analysis based on Fourier coefficients of nonlinear ARMA models (FNLARMA), etc. These three methods are discussed by Nachane and Clavel (2008) in detail who illustrate their uses with a model for interest rates.

The next section discusses some of the stationarity and nonlinearity tests mentioned above, though it does not explain the second-generation methods in detail.

3 Tests of Linearity

The testing of linearity in nonlinear time series is the first step in the specification strategy. Before directly going for a nonlinear model, it is important to test for linearity because:

(1) A nonlinear model may be erroneously fit to data from a linear process leading to wrong results and conclusions. If the time series is short, then the erroneous nonlinear fit may be successful in situations where the nonlinear model is not identified.
(2) The type of nonlinearity most appropriate for describing the data may not be clear at the outset.
(3) More statistical theory is available for linear models and parameter estimation than for nonlinear models.
(4) Multi-step forecasting with nonlinear models is more complicated than with linear models.

If nonlinearity is accepted, then the next step is the testing of stationarity of the nonlinear model [testing (iv) against (iii)]. The various *linearity tests* involved can be categorized into two types: tests against a specific alternative and those without a specific alternative.

3.1 Tests Against a Specific Alternative

In these tests, the null of a linear model is tested against an alternative hypothesis of a specific type of nonlinear model.

Teräsvirta's test (1994): Teräsvirta (1994) proposed a test based on the *score* or LM *principle* which tests the null of linearity against a particular nonlinear model specification (viz. the logistic or exponential smooth transition AR (LSTAR and ESTAR) model) as the alternative.

The model specification of a general LSTAR and ESTAR model of order p for a time series y_t are given below but models are discussed in detail in the typology Sect. 5:

LSTAR =>

$$y_t = \pi_{10} + \boldsymbol{\pi}_1' \boldsymbol{w}_t + \left(\pi_{20} + \boldsymbol{\pi}_2' \boldsymbol{w}_t\right)[(1 + exp\{-\gamma(y_{t-d} - c)\})^{-1} - (1/2)] + \varepsilon_t \tag{2}$$

where $\varepsilon_t \sim n\text{id}(0, \sigma_\varepsilon^2)$, $\gamma > 0$, $\boldsymbol{w}_t = \left(y_{t-1}, \ldots, y_{t-p}\right)'$, d is the delay parameter and $\boldsymbol{\pi}_j = \left(\pi_{j1}, \ldots, \pi_{jp}\right)'(j = 1, 2)$. The constant (1/2) is added to this specification as it is useful in the derivation of the linearity test. It is not a part of the general specification or used while estimation (Teräsvirta 1994).

ESTAR =>

$$y_t = \theta_{10} + \theta_1' w_t + \left(\theta_{20} + \theta_2' w_t\right)(1 - \exp\{-\gamma^*(y_{t-d} - c^*)^2\}) + \upsilon_t \tag{3}$$

where $\varepsilon_t \sim nid(0, \sigma_\upsilon^2)$, $\gamma^* > 0$, $\boldsymbol{w}_t = \left(y_{t-1}, \ldots, y_{t-p}\right)'$, d is the delay parameter and $\boldsymbol{\theta}_j = \left(\theta_{j1}, \ldots, \theta_{jp}\right)'(j = 1, 2)$. The test for linearity for the above model specifications would entail testing the null of $\gamma = 0$ (in case of LSTAR) or $\gamma^* = 0$ (in case of ESTAR). As models (2) and (3) are not identified under the null, Teräsvirta (1994) initially uses the suggestion of Davies (1977) to derive the LM test by keeping the unidentified values fixed.

LM-test for LSTAR and ESTAR effects (Notations have been referred to from Teräsvirta 1994):

Define the following under the null of linearity: $\boldsymbol{\tau} = \left(\boldsymbol{\tau}_1', \tau_2'\right)$ where $\boldsymbol{\tau}_1 = \left(\pi_{10}, \boldsymbol{\pi}_1'\right)'$ and $\tau_2 = \gamma$; define $\hat{\tau}_1$ to be the least squares estimator of $\boldsymbol{\tau}_1$ (under the null) and let $\hat{\tau} = \left(\hat{\tau}_1', 0\right)'$. Define $z_t = z_t(\boldsymbol{\tau}) = \partial\varepsilon_t/\partial\tau$ and $\hat{z}_t = z_t\left(\hat{\boldsymbol{\tau}}\right) = \left(\hat{z}_{1t}', \hat{z}_{2t}'\right)'$. The general form of the LM statistic is:

$$\mathrm{LM} = \sigma^{-2}\left(\sum_{t=1}^{T}\hat{\varepsilon}_t\hat{z}_{2t}\right)^2\left\{\sum_{t=1}^{T}\hat{z}_{2t}^2 - \sum_{t=1}^{T}\hat{z}_{2t}\hat{z}_{1t}'\left(\sum_{t=1}^{T}\hat{z}_{1t}\hat{z}_{1t}'\right)^{-1}\sum_{t=1}^{T}\hat{z}_{1t}\hat{z}_{2t}\right\}^{-1} \quad (4)$$

where $\sigma^2 = (1/T)\sum_{t=1}^{T}\hat{\varepsilon}_t^2$, $\hat{\varepsilon}_t = y_t - \hat{\boldsymbol{\tau}}_1'\bar{\boldsymbol{w}}_t$, $\bar{\boldsymbol{w}}_t = \left(1, \boldsymbol{w}_t'\right)'$, $(t = 1, \ldots, T)$.

In the LSTAR(p) model, $\hat{\boldsymbol{z}}_{1t} = -\bar{\boldsymbol{w}}_t$, $\boldsymbol{\pi} = \left(\pi_{20}, \boldsymbol{\pi}_2', c\right)'$ and $\hat{z}_{2t}(\boldsymbol{\pi}) = \hat{z}_{2t} = -(1/4)\left\{\pi_{20}(y_{t-d} - c) - c\boldsymbol{\pi}_2'\boldsymbol{w}_t + \boldsymbol{\pi}_2'\boldsymbol{w}_t y_{t-d}\right\}$. The auxiliary regressions for the tests are then constructed depending on the alternative (LSTAR or ESTAR)[5]. The LM test statistic for testing against the alternative of LSTAR, $\mathrm{LM}(\pi) = (\mathrm{SSR}_0 - \mathrm{SSR}(\pi))/\hat{\sigma}^2 \sim \chi^2(1)$ under the null of linearity given by $\tilde{\beta}_2 = 0$ i.e. $\gamma = 0$ (in Eq. 2); where SSR_0 is the sum of squared residuals of the auxiliary regression and $\mathrm{SSR}(\boldsymbol{\pi})$ is the sum of squared residuals of the auxiliary regression under the null.

Issues and extensions: In order to overcome the dependence of the distribution of the above statistic for LSTAR on $\boldsymbol{\pi}$, Davies (1977) suggested considering the infimum of the $\mathrm{SSR}(\boldsymbol{\pi})$ to construct the test statistic. The conservative statistic proposed by Davies (1977) is formulated as:

$$\mathrm{LM}_1 = \sup_{\pi}\mathrm{LM}(\pi) = \left(\mathrm{SSR}_0 - \inf_{\pi}\mathrm{SSR}(\pi)\right)/\hat{\sigma}^2$$

The distribution of this test statistic is generally unknown. However, Teräsvirta (1994, pp. 209) derives the asymptotic distribution of this statistic as $\chi^2(p)$ if $E\left(u_t^4\right) < \infty$.

Testing against the ESTAR alternative implies testing the null of linearity with respect to Eq. (3) i.e. testing whether $\theta_{20} = c^* = 0$ (in Eq. 3). The general LM statistic can be modified for testing against the ESTAR alternative.[6] The LM statistic is constructed the same as for the LSTAR alternative but follows a $\chi^2(2p)$ distribution.

A **likelihood ratio test for threshold nonlinearity** was formulated by Chan (1990), which tests the null of linearity against a threshold AR model (with a single threshold) (parametric test). The test is detailed in Chan (1990). Thus, these are some of the tests which test the null of linearity with respect to alternatives of parametric nonlinear models.

---

[5]The auxiliary regression is (Teräsvirta et al. 1994,Teräsvirta 1994):

$\hat{\varepsilon}_t = \hat{z}_{1t}'\tilde{\beta}_1 + \hat{z}_{2t}(\boldsymbol{\pi})\tilde{\beta}_2 + u_t(\boldsymbol{\pi}), t = 1, \ldots, T$ where $\tilde{\beta}_1 = \left(\tilde{\beta}_{11}, \ldots, \tilde{\beta}_{1,p+1}\right)'$ and $u_t(\boldsymbol{\pi})$ is the error term.

[6]The auxiliary regression is then formulated as:

$\hat{\upsilon}_t = \tilde{\beta}_1'\tilde{z}_{1t} + \tilde{\beta}_2'\boldsymbol{w}_t y_{t-d} + \tilde{\beta}_3'\boldsymbol{w}_t y_{t-d}^2 + e_t', t = 1, \ldots, T$ where $\tilde{\beta}_1 = (\tilde{\beta}_{10}, \tilde{\beta}_1')'$, $\tilde{\beta}_{10} = \theta_{10} - (c^*)^2\theta_{20}$, e_t' is the error term and $\beta_2 = 2c^*\boldsymbol{\theta}_2 - \theta_{20}\boldsymbol{e}_d$ and $\beta_3 = -\boldsymbol{\theta}_2$.

3.2 Tests Without a Specific Alternative

These tests do not have any particular nonlinear model against which they are tested (in a sense, non-parametric). These tests test the linearity hypothesis against any general form of nonlinearity. Some of the major tests in this regard are explained briefly in this section, in the order of evolution:

Regression Error Specification Test (RESET): This test was first proposed by Ramsey (1969). It is an F-test conducted on a constructed auxiliary regression of the estimated residuals of a linear model against a general model misspecification. The test procedure is detailed in Granger and Teräsvirta (1993), with the test statistic following the F-distribution if the lags of the endogenous and exogenous variables, are uncorrelated with its lagged innovations; and following an asymptotic-χ^2 distribution under the null if the lags of the exogenous variables and its lagged innovations are uncorrelated but the regressors include lags of the endogenous variable.

Keenan's test (1985): Developed a test to examine the linearity of a model against an unspecified nonlinear alternative form (under the assumption of stationarity). Keenan's test is a special case of the RESET test wherein the estimated residuals are fitted on M lags (M is a large fixed value which may be chosen in the "… *range of 4–8 though larger values seem to do as well, if not better.*" (Keenan 1985, pp. 43) of the time series of interest $\{Y_t\}$, (with sample size, n) a constant and the current fitted value Y_t. The test procedure detailed in Keenan (1985) indicates that the test statistic follows the F-distribution and tests the null of linearity against that of nonlinearity. It is a parametric test as the value of M is pre-specified, with the low power of the test being a major drawback.

Tsay's test (1986): This has proved to be a powerful test and works against an unspecified nonlinear alternative (non-parametric). It is a generalization of the test proposed by Keenan (1985) and tests for the quadratic serial dependence in the data. It retains the simplicity of Keenan's test but is considerably more powerful. The Tsay test statistic is the usual F-statistic for testing the null of linearity against the alternative of whether the residuals of the linear model are correlated with quadratic linear terms. The test detailed in Tsay (1986), however, cannot handle models with concurrent nonlinear terms such as $e_t e_{t-2}$ (where e denotes the error in the model) and exhibits low power for the same.

Spectral analysis tests: These tests, based on the frequency domain, rely on the concept of the polyspectrum (defined in Brillinger 1965). The various tests under this category are those proposed by Subba Rao and Gabr (1984), Ashley et al. (1986), Hinich (1982, 1996), etc. These tests are based on the *underlying principle* that the normalized bispectrum is constant over all frequencies if the series is linear. The normalized bispectrum is defined as:

$$\bar{f}(\omega_1, \omega_2) = \frac{f(\omega_1, \omega_2)}{s(\omega_1)s(\omega_2)s(\omega_1 + \omega_2)} \tag{5}$$

where $s(\omega)$ is the power spectrum.[7]

$f(\omega_1, \omega_2)$ is the bispectrum defined as:

$$f(\omega_1, \omega_2) = s(\omega_1)s(\omega_2)s(\omega_1 + \omega_2)\lambda_3 \quad (6)$$

where $\lambda_3 = E\left[\varepsilon_t^3\right]$, ε_t is i.i.d. with zero mean.

The disadvantage of this approach is that the interpretation of the diagram of the spectrum is difficult as they are often jagged and have an unclear shape. Even for simple nonlinear models such as the bilinear or nonlinear (nonlinear) AR (of order 1) model, there are no distinctive bispectral shapes and some nonlinear models may have $f(\omega_1, \omega_2) = 0$, which will make the normalized bispectrum constant but not indicative as a property of nonlinearity. This method also has poor estimation properties which does not make it a recommended tool for analysis.

Brock-Dechert-Scheinkmen (BDS) test (Brock et al. 1987): It is a powerful test based on the correlation integral. It tests for the null that the elements of a time series are i.i.d with no alternative specified. It is useful for detecting deterministic chaos and also helps examine the goodness of fit of a model. The BDS test statistic is commonly used and follows a standard normal distribution. It is discussed in detail in Brock et al. (1987).

4 Testing for Stationarity in Nonlinear Models

The traditional Dickey-Fuller (DF) or augmented DF (ADF) tests are misleading in order to test for the joint presence of nonlinearity and non-stationarity as:

- they reject the null of unit root incorrectly for long memory[8] processes and,
- other transformations of the integrated processes are also rejected too often.

Thus, new tests have been devised in the nonlinear context some of which are: *Rank ADF test* (Granger and Hallman 1991), *modified ADF test* (Franses, McAleer 1998), and the *Range Unit Root test (RUR)* (Aparicio et al. 2006).

RADF unit root test tests the stationarity of nonlinear transformations of integrated processes. It is invariant to monotone transformations. It tests the null of a

[7] The power spectrum is defined as:

$$s(\omega) = \frac{1}{2\pi} \sum_{k=-\infty}^{\infty} R_k e^{-ik_\infty}$$

where R_k are the k autocovariances of x_t which is a zero mean linear stationary process.

[8] **Mixing processes** (Dufrénot and Mignon 2002): Mixing is a concept used to measure the degree of dependence in the memory of a time series. Strong mixing can be understood as short-range dependence. Mixing implies that as the time span between two events increases, the dependence between past and future events becomes negligible.

Refer to **Footnote 38** for formal definition of memory in time series.

unit root against that of no unit root (and also indicates the linearity of the process). Granger and Hallman (1991) based this test on the rank ordering of a series which is unaffected by strictly monotone transformations as tests based on these notions have distributions that are unaffected by monotone transformations of the data. The RADF test can be carried out by calculating the ADF test statistic of the ranks of the ordered observations (of the time series under consideration, say $\{y_t\}$), and comparing it with the ADF statistic of the series. If the two are significantly different, then the montonic transformation of the time series, say f (.) is linear and nonlinear otherwise. In other words, if the ADF test rejects the null of unit root and RADF does not, then the series has a unit root *and* is nonlinear. However, the distribution of the RADF statistic (under the null) is not specified analytically as the rank distribution varies for every sample. Thus, there is no well-defined distribution to which the test statistic converges under the null. This test is applied when it is known that the normalized sample ranks converge to the population distribution function. The test has relatively good power but poor size.

MADF test: Franses and McAleer (1998) first proposed this test for logarithmic transformations (nonlinear) of a time series $\{y_t\}$. The ADF (J) *auxiliary regression* for a log-transformed $\{y_t\}$ is given by its Box-Cox logarithmic transformation.[9] The modified ADF auxiliary regressions are obtained by using the Taylor series expansion on the auxiliary regressions around a known specified λ^*(Franses and McAleer 1998, pp. 154–155). The OLS fitted values (of the log-transformed ADF regression) are then substituted in the Taylor expansion after setting $\lambda^* = 0$. This equation along with a modified version wherein the deterministic trend is excluded, together give the MADF (J) auxiliary regressions. The steps proposed to test the null of a single unit root against no unit root under a nonlinear transformation (assumed to be known) are carried out as given in Franses and McAleer (1998). The test is based on testing the significance of the appropriate added variable in the MADF(J) regression (where the lag length J of the auxiliary regression is determined by estimating the ADF for a sufficiently long lag-length; and, by presuming the correct transformation of $\{y_t\}$ and examining the residuals for serial correlation). If the additional variable is statistically significant, then the ADF(J) regression has been inappropriately transformed and therefore there is no valid inference for testing for a single unit root in $y_t(\lambda)$. If the added variable is not significant, then the ADF(J) regression is used to test for a single unit root in $y_t(\lambda)$. The drawback of this test is the issue of determining the appropriate nonlinear transformation in empirical applications for this test to be applied.

[9]The Box-Cox transformation is given as:

$$\begin{aligned} y_t(\lambda) &= \frac{y_t^\lambda - 1}{\lambda}, \lambda \neq 0, y_t \geq 0 \\ &= \log y_t, \lambda = 0, y_t > 0 \end{aligned}$$

where t represents the inclusion of a time trend and λ denotes the set of parameters that enter in the nonlinear model.

Range Unit Root Test

The RUR test was first proposed by Aparicio et al. (2003)[10] highlighting the fact that most stationarity tests are not robust to error distributions, outliers, structural breaks and nonlinearities. Some of the unit root tests which account for nonlinearities in the data, e.g., the RADF test, MADF test, etc. have been described in detail above. The advantage of these tests is that they assume nonlinearity of the data structures. However, the RUR test tackled some of the drawbacks (in terms of invariance to nonlinear monotonic transformations, robust to additive outliers and level shifts and has an error-model free asymptotic distribution) of the tests described above and is thus, an optimal test to implement while examining the stationarity properties of a time series. It tests the null of a random walk with i.i.d. $(0, \sigma_\epsilon^2)$ errors against a stationary alternative. The test statistic is based on the *running ranges* of the series, lending the test its many advantages over its counterparts. Given a time series x_t, $x_{1,i}(= \min\{x_1, \ldots, x_i\})$ and $x_{i,i}(= \max\{x_1, \ldots, x_i\})$ ($\forall i = 1, \ldots$ n) are termed the ith extremes. The two test statistics that are then implemented under this test are.

RUR test statistic:

$$J_0^{(n)} = n^{-1/2} \sum_1^n \mathbf{1}(\Delta R_t^{(x)} > 0) \tag{7}$$

and the Forward-Backward (FB) RUR test statistic:

$$J_*^{(n)} = \frac{1}{\sqrt{2n}} \sum_1^n \left\{ \mathbf{1}(\Delta R_t^{(x)} > 0) + \mathbf{1}(\Delta R_t^{(x')} > 0) \right\} \tag{8}$$

where $I(.)$ is the indicator function, $R_i^{(x)} = x_{i,i} - x_{1,i}$ and $x_t'(= x_{n-t+1}, t = 1, 2, \ldots, n)$ is the time reversed series. The test being a one-tailed test, the test statistics take larger values for an $I(1)$ time series and smaller values for an $I(0)$ time series. The critical values for both the RUR and FB-RUR are provided by Aparicio et al. (2003). The FB-RUR statistic has improved size and power over its simpler version (especially in cases where additive outliers occur at the beginning of the sample).

5 Typology of Nonlinear Models

The tree-diagram (Fig. 3) gives a brief (but not exhaustive, as indicated by the models indicated alongside the bilinear models) structure of different types of nonlinear time series models that exist in the literature. The Figure does not include the commonly used heteroscedasticity models, which are also nonlinear in nature

[10]Proposed by Aparicio et al. (2003, 2006).

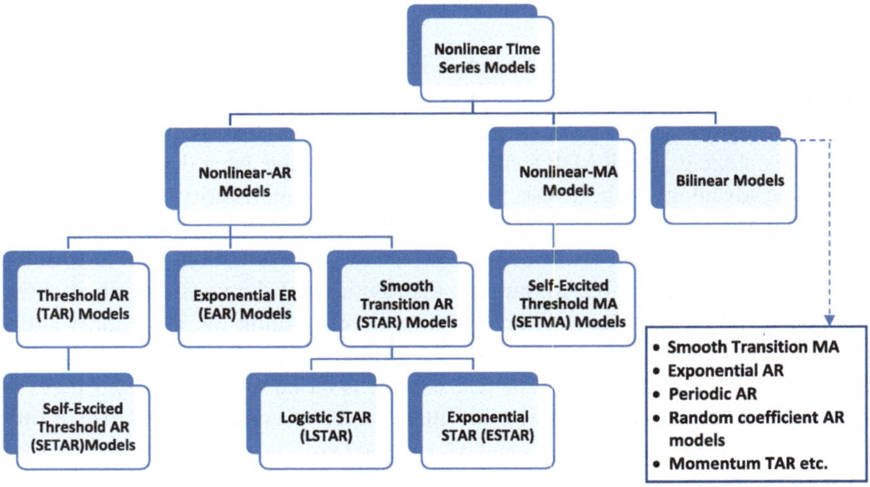

Fig. 3 Typology of nonlinear models

and used frequently in the analysis of financial time series. A brief description of these models in given following the Figure to maintain completeness. This section discusses some of these models relevant to economic analyses in detail. The Figure mentions a few models which have not been discussed in the text viz. the smooth transition MA, exponential AR etc. which have found fewer economic application and hence not been included (the periodic AR and bilinear models have been explained). Their applications do however range from mathematics and applied statistics to computer science, electrical engineering, river flow analysis etc.

5.1 Nonlinear Models Frequently Used in Economic Theory

(i) Autoregressive Conditional Heteroscedastic (ARCH) and Generalized ARCH (GARCH) Models

This class of models deals with the time varying volatility (heteroscedasticity) that may exist in the time series data. The concept of conditional heteroscedasticity[11] was introduced in economics by Engle (1982). Engle (1982) formulated the ARCH model wherein the conditional variance[12] of a time series is a function of past shocks.

[11]Econometricians refer to conditional variance while dealing with the volatility of the time series and the time varying volatility is referred to as conditional heteroscedasticity (Harris and Sollis 2006).

[12]The conditional mean (and variance) of a time series are the mean (and variance) conditional on the information set available at time t (Harris and Sollis 2006).

An AR(p)-ARCH(q) model is given by[13]:

$$y_t = \mu + \sum_{i=1}^{p} \rho_i y_{t-i} + u_t, t = 1, \ldots, T, \quad \text{(AR}(p)\text{ process) (Mean equation)} \quad (9)$$

$$u_t = \varepsilon_t (h_t)^{1/2} \text{ where } h_t = \alpha_0 + \sum_{i=1}^{q} \alpha_i u_{t-i}^2 \text{ (ARCH } (q) \text{ process)}$$
$$\text{(Conditional variance equation) where } \varepsilon_t \sim \text{IID}(0, 1). \quad (10)$$

ARCH model estimation involves formulating the log-likelihood function, the details of which are provided in Harris and Sollis (2006). The ARCH LM-test (Engle 1982, 1984; Harris and Sollis 2006) tests for the presence of heteroscedasticity in the data based on a χ^2 distributed Lagrange Multiplier (LM) test statistic. This test also has power against general GARCH alternatives and can also be used as a specification test for GARCH effects. The drawbacks of the ARCH models are that large number of lagged squared error terms may be required to explain the time-varying heteroscedasticity of the time series; and the values of the ARCH parameters are constrained by the conditions imposed on them.

The ***generalized*** **ARCH (GARCH) model** (developed by Bollerslev (1986)) accounts for these drawbacks by augmenting the conditional variance model with lags of the conditional variance as regressors. The model specification for a GARCH (p, q) (where the mean equation is given by Eq. 9: AR(k) process => $y_t = \mu + \sum_{i=1}^{k} \rho_i y_{t-i} + u_t, t = 1, \ldots, T$) model is given as:

$$u_t = \varepsilon_t (h_t)^{1/2} \quad (11)$$

where $\varepsilon_t \sim \text{NID}(0, 1)$ (here NID refers to normal and identically distributed), $p \geq 0$, $q > 0$, $\alpha_0 > 0$, $\alpha_i \geq 0$, $i = 1, \ldots, q$, and $\beta_i \geq 0$, $i = 1, \ldots, p$. As derived for the ARCH model in Eq. (10), the conditional variance of u_t is given by h_t by:

$$h_t = \alpha_0 + \sum_{i=1}^{q} \alpha_i u_{t-i}^2 + \sum_{j=1}^{p} \beta_j h_{t-j}^2 = \alpha_0 + A(L) u_t^2 + B(L) h_t \quad (12)$$

where $A(L) = \alpha_1 L + \alpha_2 L^2 + \ldots + \alpha_q L^q$ and $B(L) = \beta_1 L + \beta_2 L^2 + \ldots + \beta_p L^p$ where L denotes the lag operator ($L u_t = u_{t-1}$). Bollerslev (1986) derives the condition for the weak stationarity of u_t in a GARCH(p, q) model as: $A(1) + B(1) < 1$. The GARCH model estimation is similar to that of the ARCH models but with different log-likelihood function specification (Zivot 2008, pp. 9–12). A limitation is that the squared errors u_t^2 are functions of u_t^2 and h_t. This formulation implies that the conditional variance is invariant to the algebraic signs of the innovations and changes

[13]The AR(p) process can also be replace by series of exogenous variables which include lagged dependent values of the dependent variable as well.

in the signs. Thus, only the magnitudes affect and factor into the conditional variance. The parameter restrictions on α and β imply that the conditional variance is an increasing function and cannot account for oscillatory behaviour of the conditional variance. The non-negativity restrictions also result in estimation issues because in practice, the coefficients can be negative. In order to address some of these limitations, various extensions of the GARCH model have been developed in literature such as the exponential GARCH (EGARCH) model, the integrated GARCH (IGARCH), etc. These models are discussed in detail in Zivot (2008).

(ii) Nonlinear AR (NLAR) Models

There are various types of nonlinear models which are explored below. The most basic univariate nonlinear model is given as follows:

$$y_t = f(y_{t-1}) + \varepsilon_t \tag{13}$$

This is a *NLAR model* in the case when errors are i.i.d. with zero mean. An alternative form is:

$$y_t = g(y_{t-1}) \cdot y_{t-1} + \varepsilon_t \tag{14}$$

Equation (14) can be generalized to p lags as follows:

$$y_t = f\left(y_{t-j}, j = 1, \ldots, p\right) + \varepsilon_t \tag{15}$$

These models are used through their various forms such as the threshold AR (TAR), the smooth transition AR (STAR) etc.

(iii) Threshold AR (TAR) Models

The TAR models, in particular, are mainly useful for modeling phenomena which follow different regimes depending on the values of the endogenous variable. The basic idea is to consider a linear model for a series $\{X_t\}$ and then allow the parameters of the model to vary according to the values of a finite number of past values of X_t or a finite number of past values of an associated series $\{Y_t\}$. (Note the reference for the notations in this section is Tong 1983) For example a first order TAR model [TAR(1)] is of the form

$$\left.\begin{aligned} X_t &= a^{(1)} X_{t-1} + e_t^{(1)} \quad \text{if } X_{t-1} < \delta \\ &= a^{(2)} X_{t-1} + e_t^{(2)} \quad \text{if } X_{t-1} \geq \delta \end{aligned}\right\} \tag{16}$$

where $a^{(1)} \neq a^{(2)}$, $a^{(1)}$ and $a^{(2)}$ are constants, $\{e_t^{(1)}\}$, $\{e_t^{(2)}\}$ are white noise processes, and δ is the 'threshold' or transition parameter. A higher order (kth order or k lags) TAR model [TAR (k)] (with d as the threshold as in Eq. 16) can be constructed similarly as,

$$X_t = a_0^{(i)} + a_1^{(i)} X_{t-1} + \ldots + a_k^{(i)} X_{t-k} + e_t^{(i)} \text{ if}(X_{t-1}, X_{t-2}, \ldots, X_{t-k}) \in \boldsymbol{R}^{(i)} \tag{17}$$

where $\boldsymbol{R}^{(i)}$, $i = 1, 2, \ldots, l$ is a given region of the k-dimensional Euclidean space $\boldsymbol{R}^k$. In practice, however, it is not feasible to fit a k-order model to data as the determination of the threshold regions would involve a search over a k-dimensional space. Thus, in most cases, only those models are considered where the various sets of parameter values are determined by just a single past value, X_{t-d}, say (where d is called the threshold delay parameter). Equation (17) then takes the form,

$$X_t = a_0^{(j)} + \sum_{i=1}^{k} a_i^{(j)} X_{t-i} + e_t^{(j)}, \quad \text{if } X_{t-d} \in R^{(j)}, \quad j = 1, 2, \ldots, 1 \tag{18}$$

where $\boldsymbol{R}^{(j)}$ is a given subset of the real line $\boldsymbol{R}^1$ and a common order k is considered for each of the autoregressions. In practice, however, different threshold regions give rise to models of different orders; which is also captured by Eq. (18) as k can be assigned to be the largest order involved and all the redundant coefficients may be set to 0. Equation (18) can then be rewritten as

$$X_t = a_0^{(j)} + \sum_{i=1}^{k_j} a_i^{(j)} X_{t-i} + e_t^{(j)}, \quad \text{if } X_{t-d} \in R^{(j)}, \quad j = 1, 2, \ldots, 1 \tag{19}$$

where $k_1, k_2, \ldots, k_l$ denote the orders of the autoregressions in the l threshold regions. The class of threshold models can be further extended to include cases where the switching between the different sets of parametric values is ***determined by a past value of an associated process***, $\{Y_t\}$, rather than a past value of $\{\boldsymbol{X}_t\}$. These models are of the form,

$$X_t = a_0^{(j)} + \sum_{i=1}^{m_j} a_i^{(j)} \boldsymbol{X}_{t-i} + \sum_{i=0}^{m'_j} b_i^{(j)} Y_{t-i} + e_t^{(j)}, \quad if\ Y_{t-d} \in R^{(j)}, \quad j = 1, 2, \ldots, l \tag{20}$$

This model is called the **TARSO**$\{\boldsymbol{l}, (\boldsymbol{m}_1, \boldsymbol{m}'_1), \ldots, (\boldsymbol{m}_l, \boldsymbol{m}'_l)\}$ model (an **O**pen loop TAR **S**ystem).

In order to deal with a bivariate series (X_t, Y_t) for which X_t satisfies an equation of the form (20), and Y_t satisfies:

$$Y_t = \alpha_0^{(j)} + \sum_{i=1}^{m_j} \alpha_i^{(j)} X_{t-i} + \sum_{i=0}^{m'_j} \beta_i^{(j)} Y_{t-i} + \eta_t^{(j)} \quad \text{if } X_{t-d} \in \boldsymbol{R}'_j \quad \text{, say } (j = 1, \ldots, l'), \tag{21}$$

with $\left\{\eta_t^{(j)}\right\}$ being strict white noise process, then Eqs. (19) and (20) together constitute the **TARSC model** (a **C**losed loop TAR **S**ystem). A canonical form for the threshold AR models is:

$X_t = a_0^{(J_t)} + \sum_{i=1}^{k} a_i^{(J_t)} X_{t-i} + h^{(J_t)} e_t$ where $\{J_t\}$ is a general "indicator process" taking integer values, $J_t = 1, 2, \ldots, l$, and conditional on $J_t = j$, $a_i^{(J_t)} = a_i^{(j)}$, etc. The TAR(k) model corresponds to the special case where $\{J_t\}$ is a function of $\{X_{t-d}\}$ i.e. a function of the lagged values of the given series itself. It is hence, called a '*self-exciting TAR model*' [**SETAR (*l*, *k*₁, *k*₂, ..., *k*ₗ)**] model where ***l*** denotes number of regimes and k_i denote the number of lags in each regime.

Structural properties of threshold models

It is difficult to obtain necessary and sufficient conditions for a nonlinear class of models, and in case of threshold models, there are no general conditions available for the ergodicity[14] of a higher order model. These conditions have been derived for some special cases of models such as for a self-excited TAR(1) [SETAR (2, 1, 1)] model (Petrucelli and Woolford 1984), general SETAR (2, 1, 1) model with constant coefficients (Chen and Tsay 1991), and a general SETAR (l, 1, 1, ..., 1) model (Tong 1983). Tong and Chan (1986) also studied the stationary marginal distributions of the first-order TAR models. They showed how the moment generating function (MGF) of the distribution of $\{e_t\}$ may be obtained, given the MGF of the stationary distribution of $\{X_t\}$. However, the dual of the above problem, i.e. determining the distribution of $\{X_t\}$ from a distribution of $\{e_t\}$, is difficult as the solution (if exists) takes the form of an integral equation which has analytic solutions only under special conditions. The theoretical autocovariance and spectral density functions of threshold models do not take any general form. Hence, when fitting threshold models to the data, the sample autocovariance function of the data is compared with that generated with the fitted model. This is done by using the model to generate artificial data and then estimating the autocovariance function of the model from the artificial data.

Estimation of parameters in threshold models

Given N observations $(X_1, X_2, \ldots, X_N)$ from a series $\{X_t\}$, a SETAR (l) model of the form given by Eq. (19) can be fit to the data by fitting each of the l models separately to the appropriate subset of observations. Though the estimation of coefficients is difficult, the determination of the structural parameters, namely the delay parameter d, the threshold regions $\{R^{(j)}\}$ and the individual model orders $k_1, k_2, \ldots, k_l$, pose a bigger problem. Tong (1983) proposed an algorithm based on the AIC criterion for the estimation of the structural parameters, viz. the maximum lag value and the threshold parameter δ (assuming δ is the only single threshold in the model). However, these may not determine the exact choice of the model and are indicative of a relatively small subclass of plausible models. These may further be narrowed down by examining certain special properties which could be desirable according to certain

[14]**Ergodicity**: It is an attribute of stochastic systems; generally, a system that tends in probability to a limiting form that is independent of the initial conditions.

considerations. A set of diagnostic checks aimed at assessing whether the fitted model shares the main characteristics of the data helps in this process and the checks are detailed in Tong (1983). The variance covariance matrix of the estimated coefficients (as an approximation), can be computed for each jth component model by applying the standard least squares theory to that subset of the observations appropriate to the jth model. Due to the extremely complicated computation of the variance-covariance matrix for most models except for some simple ones, Tong (1983) has suggested that in cases where it is easy to derive the asymptotic variance-covariance matrix, simulation results indicate that the two methods are in close agreement with the former being commonly used.

STAR Models

Since the switching across regimes is rather abrupt and thus unrealistic in the above TAR models, hence a smoother switching model is proposed (called the ***STAR model***). The simplest STAR model has the following form:

$$y_t = g(y_{t-2}) \cdot y_{t-1} + \varepsilon_t \tag{22}$$

where $g(y)$ is a smooth non-decreasing function with $g(\underline{y}) = \alpha_1$ and $g(\bar{y}) = \alpha_2$ and $\alpha_1 < g(y) < \alpha_2$ for all other y. The function $g(y)$ is thus like a cumulative density function (c.d.f). The switching process across regimes is smoothened by means of this function $g(.)$ which can take various functional forms as long as the above conditions are satisfied. The most commonly used functional forms are the logistic and exponential functions. These result in the logistic STAR (**LSTAR**) and the exponential STAR (**ESTAR**) models.

The model specification of a **general LSTAR** model of order p for a time series y_t is given as (Teräsvirta 1994):

$$y_t = \pi_{10} + \boldsymbol{\pi}'_1 \boldsymbol{w}_t + \left(\pi_{20} + \boldsymbol{\pi}'_2 \boldsymbol{w}_t\right)[(1 + exp\{-\gamma(y_{t-d} - c)\})^{-1}] + \varepsilon_t \tag{23}$$

where $\varepsilon_t \sim \text{nid}(0, \sigma_\varepsilon^2)$, $\gamma > 0$, $\boldsymbol{w}_t = \left(y_{t-1}, \ldots, y_{t-p}\right)'$, d is the delay parameter and $\boldsymbol{\pi}_j = \left(\pi_{j1}, \ldots, \pi_{jp}\right)'$ $(j = 1, 2)$.

The **ESTAR model** is specified as follows (Teräsvirta 1994):

$$y_t = \theta_{10} + \boldsymbol{\theta}'_1 \boldsymbol{w}_t + \left(\theta_{20} + \boldsymbol{\theta}'_2 \boldsymbol{w}_t\right)(1 - exp\{-\gamma^*(y_{t-d} - c^*)^2\}) + \upsilon_t \tag{24}$$

where $\varepsilon_t \sim \text{nid}(0, \sigma_\upsilon^2)$, $\gamma^* > 0$, $\boldsymbol{w}_t = \left(y_{t-1}, \ldots, y_{t-p}\right)'$, d is the delay parameter and $\boldsymbol{\theta}_j = \left(\theta_{j1}, \ldots, \theta_{jp}\right)'$ $(j = 1, 2)$. The specification, estimation and selection procedure for the appropriate STAR model is elaborated in Teräsvirta (1994).

(iv) Random Coefficient AR Models

The models discussed above consider nonlinearities in variables. The nonlinearities in errors are accounted for in the class of '*nonlinear moving average models*'. The most common type in the latter category is the conditional heteroscedasticity model.

These models, however, do not allow for time-varying coefficients. In order to provide a more general framework, the random coefficient AR (RCA) models allow for both, nonlinearity in variables in the form of time varying parameters and, multiplicative noise (nonlinearity in the errors).

A time series $\{X_t\}$ is said to be generated by a p-variate RCA model of order k (RCA (k)) (Nicholls and Quinn 1980) if X_t satisfies an equation of the form

$$X_t = \sum_{i=1}^{k} [\beta_i + B_i(t)] X_{t-i} + \varepsilon_t \tag{25}$$

where the following conditions hold: $\{\varepsilon_t\}$ is a sequence of i.i.d. random variables with zero mean and common non-negative definite covariance matrix G; β_i are constants $\forall i = 1, 2, \ldots, k$ where β_i are $p \times p$ matrices; if $B(t) = [B_k(t), \ldots, B_1(t)]$, then $\{B(\mathrm{t})\}$ is a sequence of independent $p \times kp$ vectors with zero mean and $E[B(t) \otimes B(t)] = C$ where $\otimes$ denotes the Kronecker product; $\{B(\mathrm{t})\}$ is independent of $\{\varepsilon_t\}$; X_t is generated with the initial values $\{x_{1-n}, x_{2-n}, \ldots, x_0\}$ for the variables $\{X_{1-n}, X_{2-n}, \ldots, X_0\}$.

The conditional variance is modeled as

$$h_t^* = \sigma_e^2 + \sum_{i=1}^{p} \sum_{j=1}^{p} \sigma_{ij} X_i X_j \tag{26}$$

The class of RCA models may appear to be complicated in their structure, properties, and estimation. However, their potential cannot be undermined. They have found certain applications in stock market transaction volume data (Wang and Ghosh 2004), interest rate volatility (Miller 2006), etc. These models allow for the conditional variance to evolve with previous observations (Tsay 1987) whereas in the ARCH models, it uses the past innovations (refer Eq. 10). Tsay (1987) also proves as to how ARCH models can be regarded as special cases of RCA models under conditions of stationarity and equivalence. The ARCH models can be considered as special cases of RCA models if they have the same conditional expectation and variance (Tsay 1987). Both these models handle the same phenomenon of varying conditional variance but use different representations. The most important drawback of these models is that they are not parsimonious. These models provide a generalization for models such as the volatility models and thus help in better understanding them.

(v) Periodic AR Models

The nonlinear time series models discussed above do not take into account an important aspect of the inherent nature of time series—periodicity in the data. In several economic situations, agents are expected to behave differently in different seasons. For example, trend in the sales of seasonal fruits or vegetables, effect of tax measures on seasonal variation in inflation. The periodic AR (PAR) model accounts for this effect by assuming that observations in each of the seasons can be described using a different model. The most commonly used type of models is the linear periodic models—linear with respect to the individual models specified across each season. These models, however, fall in the nonlinear category owing to their structural similarity with the TAR models discussed above.

PAR (p) model

Consider a univariate time series, y_t which is observed *quarterly* (an assumption which can be extended to monthly or daily frequencies) for N years ($t = 1, 2, \ldots, n$). Assuming without loss of generality that $n = 4\ N$, a PAR model of order p is given as

$$y_t = \mu_s + \varphi_{1s} y_{t-1} + \cdots + \varphi_{ps} y_{t-p} + \varepsilon_t \tag{27}$$

or

$$\varphi_{p,s}(B) y_t = \mu_s + \varepsilon_t \tag{28}$$

where

$$\varphi_{p,s}(B) = 1 - \varphi_{1s} B - \cdots - \varphi_{ps} B^p \tag{29}$$

μ_s is the seasonally varying intercept term; $\varphi_{1s}, \ldots, \varphi_{ps}$ are AR parameters with order p which vary with season s ($s = 1, 2, 3, 4$) and assume $\varepsilon_t \sim WN\left(0, \sigma^2\right)$. A practical reason for analyzing this set of models is that the empirical experience with *economic* time series shows that low-order PAR models often fit periodic time series data rather well; in case of higher frequency data (e.g., monthly, daily etc.), the appropriate PAR model may be of higher order (Kar 2010). One of the drawbacks of the PAR models is the large number of parameters to be estimated given the limited number of observations. This drawback was overcome by Gersovitz and MacKinnon (1978) by imposing smoothness restrictions on the AR parameters in case of quarterly data. However, monthly data applications of PAR models indicate that such restrictions are not strictly necessary to obtain useful estimation results. It is important to note that data generated from PAR models cannot be decomposed into seasonal and non-seasonal components as the AR components are seasonally varying and cannot be separated from the y_t series. Thus, seasonal adjustment in PAR models does not lead to non-periodic models. A detailed exposition of the PAR models (structural properties, estimation, testing, and order selection) is discussed in Kar (2010).

(vi) Bilinear Models (BM)

The above models deal with the nonlinearity in variables and errors individually. In order to deal with both these nonlinearities simultaneously, a third class of models called *bilinear models* exist.

They are natural nonlinear extensions of ARMA models as they can approximate any well-behaved Volterra series relationship over a finite time horizon to an arbitrary degree of accuracy (ARMA models can approximate any general linear relationship between $\{X_t-\}$ and $\{e_t\}$ to an arbitrary degree of accuracy). This helps capture some types of nonlinear features such as a series with a number of 'bursts' of large amplitudes, etc. This result, however, does not hold over an infinite time horizon which limits its degree of generality i.e. a BM cannot capture certain nonlinear characteristics such as limit cycle behavior. This chapter does not discuss these models in detail as they appear to have found limited applicability in economic applications and thus to maintain brevity of the chapter. The bilinear model is discussed in detail (multivariate representations, stationarity and invertibility conditions, estimation, etc.) in Priestley (1988), Nachane (2011), etc.

(vii) Markov Switching Regressions

We briefly describe the Markov switching models as they are non-parametric nonlinear models unlike the models described so far. These models were first considered by Goldfeld and Quandt (1973) as a tool to model finite number of parameter changes where each possible state of the parameter vector signifies a regime. Kuan (2002) provides a comprehensive analysis of the Markov switching model which encompasses the model specification (including generalisations and special cases), estimation and hypothesis testing. The Markov switching model involves multiple structures (equations) which characterize time series behaviour in different regimes (Kuan 2002). The switching mechanism across the regimes is controlled by an unobservable state variable that satisfies the Markovian property.[15] This section details the simple Markov switching model of order one (for the context of two regimes) [MS(1)] followed by the generalisation of the same. The estimation procedure for the MS(k) model (with two regimes) (Kuan 2002) and the hypothesis tests for the existence of these models are mentioned but not detailed to maintain brevity of the chapter. They are discussed in detail in Hamilton (2005), Kuan (2002), Hansen (1992) etc.). Hamilton (1989, 1994) considers the Markov switching autoregressive (MSA) model where the transition between the regimes is driven by an unobservable first-order Markov chain. A time series $\{y_t\}_{t=1}^{T}$ follows a MSA model if it satisfies (Tsay 2002; Kuan 2002):

$$\begin{aligned} y_t &= \alpha_1 + \sum_{i=1}^{p} \beta_{1,i} y_{t-i} + \varepsilon_{1t} \quad , s_t = 1 \qquad \text{(MSA } (p) \text{ model)} \\ &= \alpha_2 + \sum_{i=1}^{p} \beta_{2,i} y_{t-i} + \varepsilon_{2t} \quad , s_t = 2 \end{aligned} \tag{30}$$

where ε_{1t} and ε_{2t} are i.i.d random variables with mean 0 and variances $\sigma^2_{\varepsilon_1}$ and $\sigma^2_{\varepsilon_2}$ respectively and; s_t can take values in $\{1, 2\}$ denoting the regimes, and is a first-order Markov chain i.e.

---

[15]Markovian property implies that the current value of the state variable depends on its immediate past value.

$P\{s_t = j|s_{t-1}, s_{t-2}, \ldots, y_{t-1}, y_{t-2}, \ldots\} = P\{s_t = j|s_{t-1}\}$, where $j = 1$ or 2. Define the transition matrix as the matrix containing the conditional probabilities of the current state s_t given the state s_{t-1} as:

$$\mathrm{P} = \begin{bmatrix} P(s_t = 1|s_{t-1} = 1) & P(s_t = 2|s_{t-1} = 1) \\ P(s_t = 1|s_{t-1} = 2) & P(s_t = 2|s_{t-1} = 2) \end{bmatrix} = \begin{bmatrix} p_{11} & p_{12} \\ p_{21} & p_{22} \end{bmatrix} \quad (31)$$

Thus, p_{ij} $(i, j = 1, 2)$ denote the transition probabilities of current state $s_t = j$ given that $s_{t-1} = i$. These transition probabilities satisfy the sum $p_{i1} + p_{i2} = 1 (i = 1, 2)$ and determine the persistence of each regime. Equations (30) and (31) constitute an MSA model of order p (MSA(p)) (Mizrach and Watkins 1999). This model can be generalised to account for multiple regimes and exogenous variables.

There are various methods of estimation that have been developed to estimate the Markov-switching model viz. the quasi-maximum likelihood estimation[16] (QMLE), the Gibbs sampling estimation method etc. (Kuan 2002; Hamilton 1989, 1994, 2005; Kim et al. 2008; Tsay 2002 etc.). The QMLE involves estimating the conditional expectations based on the information sets (prediction probabilities, filtering probabilities and smoothing probabilities) which are used to formulate the likelihood function and the quasi-likelihood function; which are in turn used to compute the QML estimates (by maximizing the quasi-log-likelihood function subject to constraints) (Kuan 2002; Hamilton 2005, pp. 5–6). The Gibbs method, on the other hand, is a Bayesian approach based on the Markov Chain Monte Carlo (MCMC) simulation method (Kuan 2002, pp. 9–11; Hamilton 2005, pp. 6–7).

The hypothesis tests to test the appropriateness of a Markov switching model can be categorized into three types viz. (i) test for the existence of regimes (test for switching parameters), (ii) test for the Markovian structure of the model, and (iii) test for optimal number of regimes. A test developed by Hansen (1992) tests for the first type and using a standardized supremum test statistic. The critical values are generated by a simulation approach (Hansen 1992) and implementation of the test is computationally quite intensive (Kuan 2002). Hamilton (1996) also derived a conditional score statistic[17] and formulated various specification tests for the Markov-switching models to test for misspecification of Markov dynamics, omitted autocorrelation and omitted ARCH effects.

Thus, this section has described the nuances of the nonlinear time series models in the context of univariate as well as some multivariate models (such as the PAR and RCA models). However, the multivariate models did not account for any long-run equilibrium relationship which may exist between the variables. The next step is to thus analyze the dynamics of nonlinear models in a multivariate cointegration

[16] Quasi-maximum likelihood estimators refer to the maximum likelihood estimators obtained when normality is assumed but the true conditional distribution is non-normal (Harris and Solis 2006).

[17] Hamilton (1996) defines the conditional score statistic as the derivative of the conditional log-likelihood of the tth observation with respect to the parameter vector. This score can be calculated using the procedure for smoothed probabilities; thus, it does not require estimating additional parameters by maximum likelihood.

framework. This is done by analyzing the nonlinear cointegration framework in the succeeding section.

6 Linear Versus. Nonlinear Cointegration: Structural Properties of Nonlinear Cointegration

Nonlinear cointegration (NLCI) has not been explored as extensively as its linear counterpart. It has developed in three different directions—equilibration, cointegration and co-trending. The first is based on the generalization of linear cointegration. It introduces nonlinear concepts such as attractors and Lyapunov stability, in order to capture richer dynamics than are possible via its linear counterpart. The second is based on the reasoning that cointegration may exist if the two series have the same order of integration and their combination (linear or nonlinear) is *mixing*.[18] This technique has the advantage of being applicable to linear and nonlinear cointegration relations. However, the property of mixing is not enough as the "*dependency between variables changes over time and is highly conditioned by the shape of the nonlinear function used in the adjustment mechanism*" (Dufrénot and Mignon 2002, pp. 194). Thus, imposing bound conditions on the nonlinear functions is necessary as they help in determining whether the adjustment mechanism is correcting or explosive. In the explosive case, the presence of a short-term disequilibrium indicates perpetual persistence. The last strand of literature is based on the concept of nonlinear trending. It is based on the property of cointegration of deterministic trends. The term co-trending is used to indicate that combination of nonlinear trends provides linear trends. The nonlinear trends are modeled as Chebyshev polynomials, which allow multiple representations of nonlinear trends. This co-trending can be detected by tests based on the eigen value problem.

The concept of cointegration helps in establishing the existence of a long run equilibrium or stationary relationship between two or more time series, each of which is individually nonstationary. The standard definitions of long memory (persistence) and cointegration are inadequate when dealing with non-Gaussian and nonlinear time series as the autocorrelation function (ACF) fails to capture the higher-order dependencies in the data. The ACF is usually used to capture the dependencies in data in the linear case. A similar problem also occurs when two or more time series are nonlinearly related. In such a case, the series may not appear to be linearly cointegrated, but may be cointegrated after a nonlinear transformation. Thus, these two problems give rise to the need for a nonlinear measure of serial dependence, and reformulation of the cointegration concept in nonlinear terms. There are two approaches proposed to model the concept of nonlinear cointegration: an

[18]**Mixing processes** (Dufrénot and Mignon 2002): Mixing is a concept used to measure the degree of dependence in the memory of a time series. Mixing implies that as the time span between two events increases, the dependence between past and future events becomes negligible.

information theory-based approach (Aparicio and Escribano 1998) and the conventional Granger-Hallman based approach (Dufrénot and Mignon 2002). Dufrénot and Mignon (2002) provide one of the most comprehensive accounts of nonlinear cointegration (including on nonlinear equilibration and co-trending) starting with testing procedures, identification and model specification and estimation; and finally applying it to empirical applications.

Aparicio and Escribano (1998) were among the first to explain nonlinear cointegration using an information-theoretic approach. They provided a general characterization of memory and cointegration. The generalization of the concept of memory requires a generalization of the ACF measure (need to capture higher order dependencies). They suggested a non-negative measure of serial dependence which captures the higher order dependency structure. This measure is denoted by $i_x(\tau, t)$ based on which memory is generalized.[19]

Dufrénot and Mignon (2002) explain that nonlinear cointegration has two conceptions, one attributed to Granger and Hallman (1991) (based on the nonlinearity of the long term attractor) and the other based on characterizing the nonlinear adjustment mechanism assuming a linear attractor (proposed by Dufrénot and Mignon 2002). The two definitions are reproduced here from Dufrénot and Mignon (2002).

Granger and Hallman (1991): "*A pair of series* $\{X_t\}_{t=1}^{\infty}$ and $\{Y_t\}_{t=1}^{\infty}$ *are said to have a cointegrating nonlinear attractor if there are nonlinear measurable functions* $f(.)$ *and* $g(.)$ *such that* $f(x_t)$ *and* $g(y_t)$ *are both* $I(d)$ *and* $z_t = g(y_t) - f(x_t) \sim I(d_z)$ *where* $d_z < d$."

Dufrénot and Mignon (2002, pp. 219): "*Consider two processes* $\{X_t\}_{t=1}^{\infty}$ and $\{Y_t\}_{t=1}^{\infty}$ *that are I(1) or non-mixing. Suppose there exists a measurable nonlinear function* $g(X_t, Y_t, \theta)$ such that the sequence $\{g(X_t, Y_t, \theta)\}_{t=1}^{\infty}$ *is SMM for* $\theta = \theta^*$ *and EMM for* $\theta \neq \theta^*$; *then* X_t *and* Y_t *are said to be nonlinearly cointegrated (co-mixing) with cointegrating function g*".

The cointegrating function is not unique as any measurable function of a mixing process is also mixing. The restrictions that are imposed by this definition on the definition of the cointegrating function regarding the mixing condition necessarily holding for $\theta = \theta^*$, imply that nonlinear functions that always yield a short memory process are avoided.

An important lemma formulated by Granger and Hallman (1991) helps understand whether this definition of cointegration holds if $\{X_t\}_{t=1}^{\infty}$ and $\{Y_t\}_{t=1}^{\infty}$ are nonlinear. It states that if there exists a nonlinear cointegrating function for X_t and Y_t, then there exists a nonlinear cointegration function for $\{\tilde{X}_t\}_{t=1}^{\infty}$ and $\{\tilde{Y}_t\}_{t=1}^{\infty}$ and vice versa (where $\{\tilde{X}_t\}_{t=1}^{\infty}$ and $\{\tilde{Y}_t\}_{t=1}^{\infty}$ are $I(1)$ series which result from invertible transformations $h_X(X)$ and $h_Y(Y)$).

[19]Aparicio and Escribano (1998), pp. 121 suggest the general characterization of mean reversion, long- and short- memory and integrated of order d where $i_x(\tau, t)$ is considered to be a non-negative measure of serial dependence which captures higher-order dependency structure in the series.

6.1 Detection of Nonlinear Cointegration

Based on the first definition (*nonlinearity of the long term attractor*), Aparicio and Escribano (1998) suggested a test to detect nonlinear cointegration which tests the null of nonlinear cointegration against the alternative of no nonlinear cointegration. The test statistic is given by:

$S_{m,q}(X, Y) = \frac{1}{T}\sum_{t=1}^{T}\sum_{\tau=m}^{m+q}\left(1 - \frac{\hat{i}_{X,Y}(\tau)}{\hat{i}_X(\tau)}\right)$ where $i_{X,Y}(\tau)$ is the nonlinear measure for *cross*-serial dependence; m captures long-term dependence; and q is chosen such that $m + q < T$. The decision criteria is that if X_t and Y_t are cointegrated then for large values of the delay parameter τ, $\hat{i}_{X,Y}(\tau)$ and $\hat{i}_X(\tau)$ must be of the same order; a similar argument for *no* cointegration implies that $S_{m,q}(X, Y)$ should tend to one and $\hat{i}_{X,Y}(\tau) < \hat{i}_X(\tau)$.

Dufrénot and Mignon (2002) use the above test to detect nonlinear cointegration and also suggest an alternative test based on *third-order cumulants*. In order to test the effectiveness of the proposed test, they successfully apply it to Monte Carlo simulations and an empirical application with promising results. The nonlinear error correction (NEC) models that they analyze characterize the nonlinear adjustment mechanism under the assumption of a linear attractor. Dufrénot and Mignon (2002, pp. 220) also provide conditions for existence of a nonlinear error correction relationship which need to hold before the estimation procedures are conducted. The assumptions include mixing error processes with finite second-order moments and cross-moments.

6.2 Conditions for Nonlinear Cointegration and Model Specification

Consider two sets of $I(1)$ variables, an endogenous variable y_t and vector X_t of N explanatory variables. Dufrenot and Mignon (2002, pp. 220) postulate a set of assumptions which include assumptions regarding the error distribution and moments, and regarding the continuity and differentiability of the nonlinear function used to capture the short-run dynamics of the system. The nonlinear error correction (NEC) model is then formulated as (notations from Dufrenot and Mignon 2002, pp. 220):

$$\Delta y_t = \sum_{i=1}^{q} \gamma_i' \Delta X_{t-i} + \sum_{j=1}^{p} \delta_i' \Delta y_{t-j} + \lambda z_{t-1} + f(z_{t-1}, \theta) + u_t \tag{32}$$

$$\Delta X_t = v_t \tag{33}$$

$$z_t = y_t - \beta X_t \tag{34}$$

where $\gamma_i's$ and $\delta_i's$ are vectors of parameters. Under the set of assumptions postulated (Dufrénot and Mignon 2002) regarding the errors and function f, y_t and X_t are cointegrated with the cointegrating vector $(1, -\beta)$.

6.3 Nonlinear Error Correction Model (NECM) Estimation

The parameter θ can be estimated using a 4-step procedure (Dufrénot and Mignon 2002):

1. Estimate β using OLS estimation. Let $\hat{z}_{t-1} = y_{t-1} - \hat{\beta}' X_{t-1}$ denote the lagged estimated error term.
2. $\hat{z}_{t-1}$ is substituted for z_{t-1} in $f\,(z_{t-1}, \theta)$.
3. Estimate θ using nonlinear least squares (NLLS).
4. Estimate remaining coefficients of the model using an OLS regression.

Assuming that the assumptions hold, the OLS estimator in step 1 is super-consistent and has an asymptotic non-standard distribution. The steps involved in the entire modeling of nonlinear ECM have been summarized effectively in Dufrénot and Mignon (2002, pp. 223). These steps include starting with the unit root and nonlinear cointegration and causality tests, followed by the estimation of the model and residual diagnostics. Escribano and Mira (2002) address some of the issues of the NLECM and provide a nonlinear extension to the linear Granger Representation theorem.

6.4 Threshold Vector Error Correction Model (TVECM)

In the NECM, we first established the existence of a long-run relationship between the variables and then attempted to explain the dynamics between them using a nonlinear function form. This, however, does not account for changes in the long-run relationship itself across time. Moreover, in linear cointegration, the implicit assumption made when dealing with cointegrating variables is that any deviation from equilibrium triggers an instantaneous error correction mechanism which is '*equilibrium-reverting*'. The implicit assumptions of symmetric and instantaneous adjustment to deviations from equilibrium may not always apply to practical applications. The TVECM or threshold cointegration model addresses these limitations and provides a framework to allow for asymmetric adjustment to deviations from equilibrium.

Hansen and Seo (2002) and Stigler (2011) provide a comprehensive analysis regarding the threshold vector error correction model (TVECM) (first proposed by Balke and Fomby (1997)) along with a comparative view with respect to its linear counterparts as well. This section discusses the concept of TVECM wherein the nonlinear adjustment (to the long-run equilibrium) is dependent on the values of a transition variable.

6.4.1 Model Specification

Let y_t be a p-dimensional nonstationary time series which is cointegrated with (say) one cointegrating vector β. Let ECT denote the stationary error correction term. A two-regime threshold cointegration (TVECM) model can be specified as (Note the notations are referred to from Hansen and Seo 2002; Stigler 2011):

$$\Delta y_t = \begin{cases} A_1' X_{t-1}(\beta) + u_t, \; ECT_{t-1}(\beta) \leq \gamma \\ A_2' X_{t-1}(\beta) + u_t, \; ECT_{t-1}(\beta) > \gamma \end{cases} \tag{35}$$

where $X_{t-1}(\beta) = (1, ect_{t-1}, \Delta y_{t-1}, \Delta y_{t-2}, \ldots, \Delta y_{t-l})'$ is a k x 1 regressor vector[20]; A_i are $k \times p$ coefficient matrices and $k = pl + 2$; u_t is a vector martingale difference series with a finite covariance matrix $\sum = E\left(u_t u_t'\right)$, γ denotes the threshold parameter. The matrices A_1 and A_2 determine the dynamics in each of the regimes. In this model specification, all the coefficients switch between the regimes with the cointegrating vector remaining constant. This assumption can be relaxed by imposing restrictions on A_1 and A_2. Note that the time delay of the transition variable (*ECT*) is assumed to be one in the specification above but it may vary depending on the data under consideration. Another important aspect to note is that the threshold effects are valid only if the $0 < P(ECT_{t-l}(\beta) \leq \gamma) < 1$ (specifically less than one). If this condition does not hold, then the model (Eq. 35) reduces to that of linear cointegration. This condition is thus imposed by introducing a *trimming parameter* ($\lambda_0 > 0$) such that

$$\lambda_0 \leq P(\text{ECT}_{t-1}(\beta) \leq \gamma) \leq (1 - \lambda_0) \tag{36}$$

The trimming parameter can be interpreted as providing the minimum percentage of observations that should exist within a regime.

6.4.2 Estimation

The estimation of a threshold cointegration (TVECM) model is carried out using the method of concentrated log-likelihood maximization (Hansen and Seo 2002). The maximum likelihood function is formulated by Hansen and Seo (2002, pp. 296) as:

---

[20]Hansen and Seo (2002): The notation of $X_{t-l}(\beta)$ implies that the variables are evaluated at generic values and not the true values of β. The variables evaluated at the true values are denoted by X_{t-l}. A similar argument holds for the *ECT* term.

$$L_n\left(A_1, A_2, \sum, \beta, \gamma\right) = -\frac{n}{2}\log|\sum| - \frac{1}{2}\sum_{t=1}^{n} u_t\left(A_1, A_2, \sum, \beta, \gamma\right)' \sum^{-1} u_t\left(A_1, A_2, \sum, \beta, \gamma\right) \quad (37)$$

where

$$u_t\left(A_1, A_2, \sum, \beta, \gamma\right) = \Delta y_t - A_1' X_{t-1}(\beta) d_{1t}(\beta, \gamma) - A_2' X_{t-1}(\beta) d_{2t}(\beta, \gamma). \quad (38)$$

In order to compute the concentrated log-likelihood function, the estimates of $\left(A_1, A_2, \sum\right)$ are computed by maximizing a constrained MLE while holding (β, γ) fixed (Hansen and Seo 2002, pp. 296–97). In order to compute the optimal (β, γ), Hansen and Seo (2002) suggest a two-dimensional grid search over the (β, γ) space in smaller dimension ($p \leq 2$) models which can be implemented to find the maximizing log-likelihood estimates for $\left(A_1, A_2, \sum\right)$. The grid-search algorithm is summarized as follows (Hansen and Seo 2002, pp. 299). These estimates of the threshold and cointegrating values are super-convergent (Seo 2009), though their consistency is conjectured in Hansen and Seo (2002).

6.4.3 Hypothesis Tests

There are two tests that have been proposed to establish the presence of threshold cointegration: (i) test of linear cointegration against threshold cointegration and, (ii) test of no cointegration against threshold cointegration. The first helps establish the presence of threshold cointegration whereas the second ensures that rejection of the null in the former is not due to the absence of cointegration (linear and threshold, both). The two tests proposed to test these hypotheses are Hansen and Seo (2002) (**HS**) **test** for the first hypothesis and the Seo (2006) test for the second hypothesis. The HS test examines the null of linear cointegration ($A_1 = A_2$) against threshold cointegration ($A_1 \neq A_2$). The test is based on the supremum of the '*LM-like*' (Hansen and Seo 2002) (supLM) test statistic.[21] The test statistic is derived for the case of known and unknown cointegrating vectors, and is robust to heteroscedasticity. Under the null, if the test cointegrating vector is unknown, then the test statistic is evaluated at the point estimates obtained under the null. Thus, there is an estimate of β i.e.$\tilde{\beta}$, but there is no estimate for γ under the null (i.e. γ is not identified under the null). Thus, in order to find the supremum of the LM-test statistic, the grid evaluation over the grid $\left[\gamma_L, \gamma_U\right]$ is necessary. This estimate of γ is not the same as the ones which maximize the log-likelihood in the previous section. This is because the test statistic

[21] The sup-LM value is the maximal value for which the test is most favourably rejected. A supremum statistic is an aggregation possibility in case of an unknown threshold parameter (which would result in a non-standard distribution and the threshold parameter thus being unidentified under the null).

is formulated based on parameter estimates under the null; and the test statistics are constructed using heteroscedasticity-consistent covariance matrix estimates. The asymptotic critical values and the *p*-values are calculated using the fixed regressor bootstrap. The test is derived in detail in Hansen and Seo (2002, pp. 299–305).

The Seo (2006) **test** examines the null of no cointegration ($A_1 = A_2 = 0$) against threshold cointegration ($A_1 \neq A_2$) by implementing a supremum of the Wald (supWald) test statistic. The basis for a Wald test-based statistic is that it is also used to test for the null of no cointegration in the case of linear cointegration (when a single cointegrating relation is known under the alternative). The supremum of the Wald test statistic is chosen to avoid the issue of arbitrariness caused due to the selection of the weighting function for averaging. The supWald statistic is chosen as it is asymptotically equivalent to the likelihood ratio statistic under normality (Seo 2006). Since the threshold γ is not identified under the null (referred to as the Davies problem), the asymptotic critical values and the *p*-values are calculated using the residual-based bootstrap.[22] The drawback of this test, however, is that it assumes a known cointegrating vector. The test is derived in detail in Seo (2006, pp. 131–135).

7 Economic Intuition Behind the Different Models

This section highlights a few economic studies where some of the models discussed earlier have been implemented. There are various models which have been proposed in the nonlinear context, though not all have been applied to economic theories and results due to either the "*curse of dimensionality*" or being too complex. The fact that a variable can be explained using a particular model for a certain period of time and some other model for a different period of time, [depending on the range of values it takes (or any other influential variable)] cannot be captured using a simple dummy variable or structural break model. Such data which indicate change in the inherent structure of the model (which may be due to major policy revision, calamity, etc.) can be better modeled using a family of models namely the *threshold models*. The threshold AR models (TAR) have been widely implemented to explain various economic data series (Mohanty et al. 2011; Hansen 2011; Sarno 2003; Kilian and Taylor 2003). Hansen (2011) provides a comprehensive account of different applications of the TAR models in economics. Kilian and Taylor (2003) investigated the PPP puzzle and found evidence that the deviation of exchange rates from macroeconomic fundamentals can be captured by nonlinear adjustment. The theoretical findings indicate long run stability of exchange rates i.e. deviations from PPP are stationary. However, the data suggests otherwise when modeled using linear models; exchange rates have been found to exhibit persistent behavior (presence of unit roots). Thus,

[22]**Residual-based bootstrap**: The time series under consideration, y_t are nonstationary and cannot be resampled directly. Given the assumption that u_t are iid (pp. 72) and unobservable, the least-squares residuals of the TVECM are resampled independently with replacement. This is called the residual-based bootstrap (Seo 2006).

when exchange rates are found to have unit roots which are contradictory to the long run stability expected according to PPP, it is possibly due to the existence of transaction costs (Kilian and Taylor 2003). Ahmad and Glosser (2007) which gives rise to some of the persistent misalignments or large movements. Exchange rates, however, exhibit asymmetric adjustment above and below equilibrium. Thus, a model which can capture nonlinear deviations[23] as well as asymmetric adjustments is required. The exponential smooth transition AR (ESTAR) model provides such a framework (Kilian and Taylor 2003; Lahtinen 2006). A similar intuitive explanation can be provided for the nonlinear study of unemployment rates. Montgomery et al. (1998) observed that the US unemployment rates exhibited sharp peaks with gradual and longer declines ending in mild troughs[24] (cyclical asymmetry). However, the business cycle patterns indicated counter-cyclical behavior of the unemployment rates versus the movements in the former. Since unemployment would be greater during the contractionary periods, it is of great social (and political) concern (Montgomery et al. 1998) to be able to correctly predict the duration of these cycles especially during the recessions. Such asymmetric cycle features in the data are difficult to capture using linear models. Montgomery et al. (1998) explain the data using various linear and nonlinear models, the latter including TAR models and Markov-switching AR models. They found that the nonlinear models scored over their linear counterparts in terms of forecasting accuracy in this regard. The idea behind the regime switching models is that they allow for the threshold in a TAR model to be stochastic. Basu (2008) tries to capture the persistence and long-run stability in the US and Japanese exchange rates using this model. There have been studies analyzing the periodic and other seasonal properties of various quarterly series (Dua and Kumawat 2005; Krishnan 2007). Kar (2010) has applied advanced seasonal modelling techniques and estimated a PAR model for the Indian monthly wholesale price index series. Ters and Urban (2018) establish the existence of a long-run equilibrium in the spot and derivative markets which accounts for nonlinear adjustments of prices towards the equilibrium using a TVECM model. The paper finds that the adjustment results in 3 regimes which are driven by the disequilibrium in the spot and derivative prices, resulting in a no-arbitrage middle regime and higher and lower arbitrage regimes.

8 Empirical Application to Examine the Money Multiplier in India

As an example of an empirical application of the univariate nonlinear time series models, we examine the Indian money multiplier and its components viz. the broad money $M3$ and reserve money H. The money multiplier is an important indicator which gives the proportion of total money supply in the system when compared to the

[23]Transactions costs lead to large (more than proportional) changes in real exchange rates which is captured using an exponential functional form in the STAR model.

[24]Thus, resulting in a complete peak-trough cycle.

actual amount injected by the monetary authority (in the Indian context, the Reserve Bank of India (RBI)). The multiplier is an indicator of the liquidity in the economy as it highlights the changes in the money supply (which is a measure of liquidity in the economy) with respect to the monetary base (which is the liability of the RBI). It is a ratio of the total stock of money to the stock of high-powered money in an economy. The change in its value is a measure of whether the monetary authority has increased or decreased the money supply in the economy, thus capturing the role of money creation by the central bank. It is defined as $m = M/H$ where M is the stock of money i.e. the money supply in the economy (which is one of the measures of money) and H is the high-powered money or the monetary base. In the Indian context, the broad money measure, termed $M3$ consists of the currency with the public, demand, time and other deposits with the banking system. Thus, the measure we use is $m = M3/H$. It is the most commonly used measure of money supply in the economy. Several studies have examined and analysed the multiplier in the Indian context (Darbha 2002; Jha and Rath 2001; Nachane 1992; Nachane and Ray 1997; Chitre 1986; Rangarajan and Singh 1984). There have been different approaches ranging from a threshold cointegrating relation between the money stock measure and reserve money (Darbha 2002) to the use of cointegration and time domain techniques to examine its stability (Nachane 1992). This chapter thus concludes with a simple univariate threshold model for the Indian broad money multiplier for the period January 1951–September 2012.

8.1 Analysis of the Indian Multiplier

The data[25] for the Indian multiplier is log transformed (natural logarithms) (Fig. 4), and the preliminary tests include testing for stationarity while accounting for nonlinearity (using the range unit root test), and testing for nonlinearity while accounting for stationarity (using the Keenan test, Tsay's test and Teraesvirta's neural network test).[26] The results indicated the presence nonlinearity as well as nonstationarity. The optimal lag of each of the regimes was set to 12 lags based on the AIC criteria. The optimal threshold delay in turn was found to be one and the optimal number of thresholds were found to be 2 based on the AIC criterion.[27] The optimal estimated SETAR (3, 12, 12, 12) model is given in Table 1 after the model specification.

[25]Data source—Reserve Bank of India (2013): Handbook of Statistics on the Indian Economy, Reserve Bank of India, Mumbai.

[26]The results of the stationarity and nonlinearity tests are available on request. The augmented Dickey Fuller test for unit roots was also conducted for the sake of completeness.

[27]The results, code are available on request from the author.

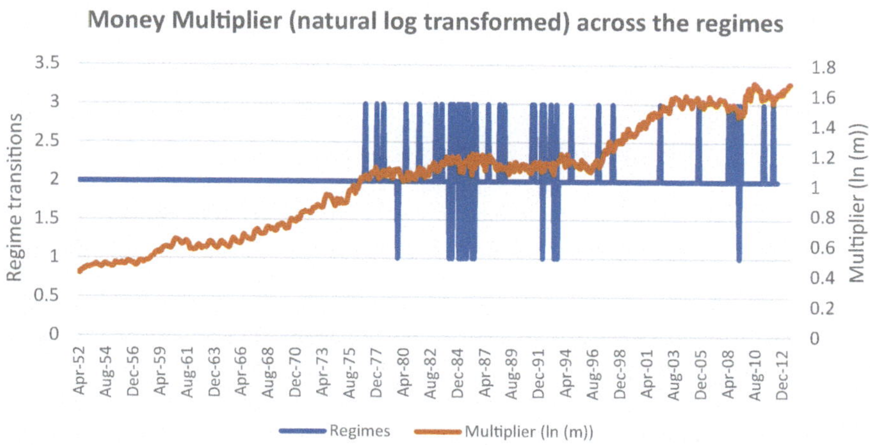

Fig. 4 Money multiplier [natural log transformed where $m = (M3/H)$] across the estimated regimes

Table 1 Results of the SETAR (2, 12, 12)

Lower regime		Middle regime		Higher regime	
Coefficient	Estimate	Coefficient	Estimate	Coefficient	Estimate
$c^{(1)}$	-0.995^{a}	$c^{(2)}$	0.0023^{b}	$c^{(3)}$	-0.014
ΔX_{t-1}	-4.900^{a}	ΔX_{t-1}	-0.252^{a}	ΔX_{t-1}	-0.848^{a}
ΔX_{t-2}	-18.111^{a}	ΔX_{t-2}	-0.065	ΔX_{t-2}	-0.137
ΔX_{t-3}	-5.876^{a}	ΔX_{t-3}	-0.028	ΔX_{t-3}	-0.217^{d}
ΔX_{t-4}	-5.483^{a}	ΔX_{t-4}	-0.155^{a}	ΔX_{t-4}	-0.005
ΔX_{t-5}	-7.733^{a}	ΔX_{t-5}	0.0012	ΔX_{t-5}	0.0138
ΔX_{t-6}	0.1931	ΔX_{t-6}	-0.055	ΔX_{t-6}	0.1215
ΔX_{t-7}	0.422^{c}	ΔX_{t-7}	-0.100^{b}	ΔX_{t-7}	-0.187
ΔX_{t-8}	-3.896^{a}	ΔX_{t-8}	-0.075^{c}	ΔX_{t-8}	0.0752
ΔX_{t-9}	-6.063^{a}	ΔX_{t-9}	-0.087^{c}	ΔX_{t-9}	0.1849
ΔX_{t-10}	-3.937^{a}	ΔX_{t-10}	-0.008	ΔX_{t-10}	-0.231
ΔX_{t-11}	-5.715^{a}	ΔX_{t-11}	0.0220	ΔX_{t-11}	-0.572^{a}
ΔX_{t-12}	-1.981^{a}	ΔX_{t-12}	0.3534^{a}	ΔX_{t-12}	0.2005

Significance codes key: ^{a} 0.001/ ^{b} 0.01/ ^{c}0.05/ ^{d} 0.1
MSE = 0.0003373, AIC = −5810
Proportion of points in low regime: 1.79%, Middle regime: 93.1%, High regime: 5.1%
Threshold values: $d1 = -0.05718$, $d2 = 0.03741$

The estimated model is of the form:

$$\Delta X_t = c^{(1)} + a_1^{(1)}\Delta X_{t-1} + a_2^{(1)}\Delta X_{t-2} + a_3^{(1)}\Delta X_{t-3} + \ldots + a_{12}^{(1)}\Delta X_{t-12} + e_t^{(1)},\quad \text{if } X_{t-11} < d1$$

$$
\begin{aligned}
&= c^{(2)} + a_1^{(2)} \Delta X_{t-1} + a_2^{(2)} \Delta X_{t-2} + a_3^{(2)} \Delta X_{t-3} + \ldots + a_{12}^{(2)} \Delta X_{t-12} + e_t^{(2)}, \\
&\qquad \text{if } d1 \leq X_{t-11} < d2 \\
&= c^{(3)} + a_1^{(3)} \Delta X_{t-1} + a_2^{(3)} \Delta X_{t-2} + a_3^{(3)} \Delta X_{t-3} + \ldots + a_{12}^{(3)} \Delta X_{t-12} + e_t^{(3)}, \\
&\qquad \text{if } X_{t-11} \geq d2
\end{aligned}
$$

The coefficient values and their significance levels are given in Table 1.

The regime analysis indicates that the middle regime, which also contains most of the data points, is mostly during the early decades post-Independence. The results indicate a significant impact of the multiplier just over half a year after on the current value in this regime. The transition between the three regimes is relatively volatile in the period 1984–1989 and then again from 1991 to 1995. The former phase refers to the period when the RBI undertook a review of the Indian monetary system under the Chakravarty Committee headed by Dr. Sukhamoy Chakravarty, when price stability was recommended as a "dominant" objective of monetary policy alongside a commitment to fiscal discipline (RBI 2004). The latter phase (1991–1995) refers to the period of India's balance of payments crisis.

This empirical application and brief analysis merits a more detailed exposition and analysis which provides scope for further research and, it has not been analysed in greater detail to maintain the brevity of the chapter. It can further be analysed using the smooth transition models and can include other exogenous determinants to help explain the Indian money multiplier.

9 Conclusion

This chapter has provided the essence of the nonlinear time series models. It has attempted to explain the various nonlinear models which are economically relevant and the need for such complexity in order to explain economic theory better. It has also highlighted issues dealing with the nonlinearity and non-stationarity testing in these models which is of prime importance. The chapter has provided an overview into the field of nonlinear time series modeling in the univariate as well as multivariate context. As an illustrative example, the paper further estimates a SETAR model for the Indian money multiplier to examine the dynamics of the liquidity measure. It finds phases of relatively low volatility as well as some volatile phases which have not been analysed in greater detail. The application can be further improved to include other exogenous determinants as well as a more parsimonious nonlinear model, which may also account for the different long-run and short-run dynamics in the multiplier.

However, the paper does not provide an in-depth analysis of the wide plethora of models that exist in literature. Such models include the threshold moving average models, exponential autoregressive models, etc. The models that have been studied in the chapter have economic interpretations and implications and have found applications in economics. The proofs to all the estimation methods and tests have also not been discussed in detail. The issues regarding the testing of non-stationarity and nonlinearity in a time series that have been discussed do not provide detailed alternative solutions to deal with them. The second generation methods (Nachane and Clavel 2008) which have been mentioned above have not been dealt with in detail. These are some of the shortcomings which have not been dealt with in this paper as it aims to provide a brief but detailed overview of the field of nonlinear time series models and has therefore focused on nonlinear models which have been used in economics or which have potential to be so used.

Appendix 1

Limit cycles

Threshold models have the distinction of being able to demonstrate limit cycle behavior (under suitable conditions), encountered in the study of nonlinear differential equations. If e_t is set to zero for $t > t_0$ (t_0 is a threshold beyond which noise e_t is 'switched off'), then Eq. (24) may possess a solution $\tilde{X}_t$, which has an asymptotic periodic form. This solution is closely related to the forecast function of the model.

Note: Though $\tilde{X}_t = E[X_{t_0+1}|X_{t_0}, X_{t_0-1}, \ldots]$ but $\tilde{X}_t \neq E[X_t|X_{t_0}, X_{t_0+1}, \ldots], t > t_0 + 1$.

This assumption clears the fact that the solution would then closely help in describing the cyclical phenomena.

Thus, such models help describe cyclical phenomena better than the conventional models based on superposition of harmonic components on a linear stationary residual i.e. a nonlinear model may provide satisfactory description of the data in case of a series exhibiting cyclical form with asymmetrical cycles.

Definition of a limit cycle: (for **discrete time** system) (Tong 1983): Let $\boldsymbol{x_t} \in \Re^k$ denote a k-dimensional vector satisfying the recurrence relation

$$x_t = f(x_{t-1})$$

where $f(.)$ is a $[k]$ vector-valued function. Let $f^{(j)}$ denote the jth iterate of f, i.e.

$$f^{(j)}(x) = f(f(\ldots(f(x))\ldots)$$

Then a [k] vector $\mathbf{c_1}$ **is** called a ***stable periodic point of period T*** w.r.t a domain $D \subset \Re^k$ if

$$\forall x_0 \in D,\ f^{(jT)}(x_0) \to c_1 \ as\ j \to \infty,$$

T being the smallest integer for which such convergence holds. In this case, $c_1, f^{(1)}(c_1), f^{(2)}(c_1), \ldots, f^{(T-1)}(c_1)$ are all distinct stable limit points of period T. Then rewriting these distinct limit points in a recursive relation as

$$c_{i+1} = f^{(i)}(c_1), \quad i = 1, 2, \ldots, T - 1,$$

the set of vectors ($\boldsymbol{c}_1$, $\boldsymbol{c}_2$, ..., $\boldsymbol{c}_T$) is called a *stable* ***limit cycle*** *of period* T (w.r.t. D).

References

Ahmad, Y., & Glosser, S. (2007). *Searching for nonlinearities in real exchange rates, Manuscript*. Whitewater, WI: University of Wisconsin.

Aparicio, F., & Escribano, A. (1998). Information-theoretic analysis of serial dependence and cointegration. *Studies in Nonlinear Dynamics and Econometrics, 3*(3), 119–140.

Aparicio, F., Escribano, A., & García, A. (2003). Range unit root (RUR) tests. *Working paper 03–11 Statistics and Econometrics Series 26*, Universidad Carlos III de Madrid.

Aparicio, F., Escribano, A., & Siplos, A. (2006). Range unit root (RUR) tests: robust against nonlinearities, error distributions, structural breaks and outliers. *Journal of Time Series Analysis, 27*, 545–576.

Ashley, R., Patterson, D. M., & Hinich, M. J. (1986). A diagnostic test for nonlinear serial dependence in time series fitting errors. *Journal of Time Series Analysis, 7*, 165–178.

Balke, N. S., & Fomby, T. B. (1997). Threshold cointegration. *International Economic Review, 38*(3), 627–645.

Basu, D. (2008). Essays on dynamic nonlinear time series models and on gender inequality. *PhD Dissertation*. The Ohio State University.

Bollerslev, T. (1986). Generalized autoregressive conditional heteroskedasticity. *Journal of Econometrics, 31*, 307–327.

Box, G., & Jenkins, GM. (1970). *Time series analysis: forecasting and control*. San Francisco: Wiley.

Brillinger, D. (1965). An Introduction to polyspectra. *Annals of Mathematical Statistics, 36*(5), 1351–1374.

Brock , W., Dechert, W., & Scheinkman, J. (1987). *A test for independence based on the correlation dimension*. Madison, Mimeo: Department of Economics, University of Wisconsin.

Bureau of Labor Statistics. (2020). Labour force statistics, current population survey. US Department of Labor, on the Internet at https://www.bls.gov/cps/data.htm.

Carnero, M., Peña, D., & Ruiz, E. (2004). Persistence and kurtosis in GARCH and stochastic volatility models. *Journal of Financial Econometrics, 2*(2), 319–342.

Chan, K., & Tong, H. (1986). A note on certain integral equations associated with nonlinear time series analysis. *Probability Theory and Related Fields, 73*, 153–158.

Chan, K. S. (1990), Testing for Threshold Autoregression, The Annals of Statistics, Vol. 18, No. 4, pp. 1886-1894.

Chen, R., & Tsay, R. S. (1991). On the ergodicity of TAR(1) processes. *The Annals of Applied Probability, 1*(4), 613–634.

Chitre, V. (1986). Quarterly predictions of reserve money multiplier and money stock in India. *Arthavijnana, 28*, 1–119.

Darbha, G. (2002). Testing for long-run stability—an application to money multiplier in India. *Applied Economics Letters, 9*, 33–37.

Davies, R. (1977). Hypothesis testing when a nuisance parameter is present only under the alternative. *Biometrika, 74*(1), 33–43 1987. https://doi.org/10.1093/biomet/74.1.33.

Gersovitz, M., & MacKinnon, J. M. (1978). Seasonality in regression: An application of smoothness priors, *Journal of the American Statistical Association, 73*(362), 264–73.

Dufrénot, G., & Mignon, V. (2002). *Recent developments in nonlinear cointegration with applications to macroeconomics and finance*. Dordrecht, Netherlands: Kluwer Academic Press.

Dua, P., & Kumawat, L. (2005). *Modelling and forecasting seasonality in Indian macroeconomic time series*, Centre for Development Economics, Working Paper 136.
Engle, R. (1982). Autoregressive conditional heteroscedasticity with estimates of the variance of United Kingdom inflation. *Econometrica, 50,* 987–1007.
Engle, R. (1984). Wald, likelihood ratio, and lagrange multiplier tests in econometrics. In Z. Griliches, & M. Intriligator (Eds.), *Chapter 13*: *Handbook of Econometrics*, (Vol. 2, pp. 775–826). Amsterdam: Elsevier.
Escribano and Mira. (2002). Nonlinear error correction models. *Journal of Time Series Analysis, 23*(5), 509–522.
Franses, P. H., & McAleer, M. (1998). Testing for unit roots and nonlinear transformations. *Journal of Time Series Analysis, 19*(2), 147–164.
Goldfeld, S. M., & Quandt, R. (1973). A Markov model for switching regressions. *Journal of Econometrics, 1*(1), 3–16.
Granger, C. W., & Teräsvirta, T. (1993). *Modelling nonlinear economic relationships*. Oxford: Oxford University Press.
Granger, C. W. J., & Hallman, J. (1991). Nonlinear transformations of integrated time series. *Journal of Time Series Analysis, 12*(3), 207–218.
Gujarati, D., & Porter, D. C. (2008). *Basic econometrics* (5th ed.). New York: McGraw-Hill.
Hamilton, J. D. (1989). A new approach to the economic analysis of nonstationary time series and the business cycle. *Econometrica, 57*(2), 357–384.
Hamilton, J. D. (1994). *Time series analysis*. Princeton, NJ: Princeton University Press.
Hamilton, J. D. (1996). Specification testing in Markov-switching time series models. *Journal of Econometrics, 70*(1), 127–157.
Hamilton, J. D. (2005). Regime-switching models. In S. Durlauf, & L. Blume (Eds.), *New palgrave dictionary of economics* (2nd ed) London: Palgrave McMillan.
Hansen, B. E. (1992). The likelihood ratio test under nonstandard conditions: testing the Markov switching model of GNP. *Journal of Applied Econometrics, 7*(S1), S61–S82.
Hansen, B. (1997). Inference in TAR models. *Studies in Nonlinear Dynamics and Econometrics, 2*(1), 1–14.
Hansen, B. (2011). Threshold autoregression in economics. *Statistics and Its Interface, 4,* 123–127.
Hansen, B., & Seo, B. (2002). Testing for two-regime threshold cointegration in vector error-correction models. *Journal of Econometrics, 110*(2), 293–318.
Harris, R., & Sollis, R. (2006). *Applied time series modelling and forecasting*. Singapore: Wiley.
Hinich, M. J. (1982). Testing for gaussianity and linearity of a stationarv time series. *Journal of Time Series Analysis, 3,* 169–176.
Hinich, M. J. (1996). Testing for dependence in the input to a linear time series model. *Nonparametric Statistics, 6*, 205–221.
Jha, R., & Rath, D. P. (2001). *On the endogeneity of the money multiplier in India*, ASARC Working Papers 2001-12, Australia South Asia Research Centre, The Australian National University, available at: https://taxpolicy.crawford.anu.edu.au/acde/asarc/pdf/papers/conference/CONF2001_06.pdf.
Kar, S. (2010). A periodic autoregressive model of Indian WPI inflation, *Margin—The Journal of Applied Economic Research, 4*(3), 279–292.
Keenan, D. M. (1985). A Tukey nonadditivity-type test for time series nonlinearity. *Biometrika, 72*(1), 39–44.
Kilian, L., & Taylor, M. (2003). Why is it so difficult to beat the random walk forecast of exchange rates? *Journal of International Economics, 60*(1), 85–107.
Kim, C. J., Piger, J., & Startz, R. (2008). Estimation of Markov regime-switching regression models with endogenous switching. *Journal of Econometrics, 143*(2), 263–273. https://doi.org/10.1016/j.jeconom.2007.10.002.
Kirchler, M., & Huber, J. (2007). Fat tails and volatility clustering in experimental asset markets. *Journal of Economic Dynamics and Control, 31*(6), 1844–1874.

Krishnan, R. (2007). Seasonal characteristics of Indian time series. *Indian Economic Review, 42*(2) (July-December), 191–210.
Kuan, Chung-Ming. (2002). *Lecture on the Markov switching model.* Academia Sinica, Taipei: Institute of Economics.
Lahtinen, M. (2006). The purchasing power parity puzzle: a sudden nonlinear perspective. *Applied Financial Economics, 16*(1), 119–125.
Mandelbrot, B. (1963). The variation of certain speculative prices. *Journal of Business, XXXVI*, 392–417.
Miller, J. (2006). A random coefficients autoregressive model with exogenously-driven stochastic unit roots. In *2006 International Symposium on Econometric Theory and Applications*, Xiamen, China.
Mizrach, B., & Watkins, J. (1999). A Markov switching cookbook. In P. Rothman (Ed.), *Nonlinear time series analysis of economic and financial data.* New York: Springer.
Mohanty, D., Chakraborty, A. B., Das, A. & John, J. (2011). Inflation threshold in India: an empirical investigation. *RBI Working Paper Series (DEPR)*: 18/2011, Department of Economic and Policy Research. http://www.rbi.org.in/scripts/PublicationsView.aspx?id=13838.
Montgomery, A., Zarnowitz, V., Tsay, R., & Tiao, G. (1998, June). Forecasting the US unemployment rate. *Journal of the American Statistical Association, 93*(442), 478–493.
Morley, J., & Piger, J. (2010). The asymmetric business cycle. http://research.economics.unsw.edu.au/jmorley/abc.pdf .
Nachane, D. (1992). The money multiplier in India: short-run and long-run aspects. *Journal of Quantitative Economics, 8*(1), 51–66.
Nachane, D., & Ray, D. (1997). Non-linear dynamics of the money multiplier—selected case studies. *Indian Economic Journal, 45*(1), 36–53.
Nachane, D. (2006). *Econometrics: theoretical foundations and empirical perspective.* New Delhi: Oxford University Press.
Nachane, D. (2011). Selected problems in the analysis of nonstationary and nonlinear time series. *Journal of Quantitative Economics, 9*(1), 1–17.
Nachane, D., & Clavel, J. (2008). Forecasting interest rates: a comparative assessment of some second generation nonlinear models. *Journal of Applied Statistics, 35*(5), 493–514.
Nicholls, D. & Quinn, B. (1980). *Random coefficient autoregressive models: an introduction.* Berlin: Springer.
Petruccelli, J. D., & Woolford, S. W. (1984). A threshold AR(1) model. *Journal of Applied Probability, 21*(2), 270–286.
Priestley, M. B. (1988). *Nonlinear and nonstationary time series analysis.* New York, NY: Academic Press.
Ramsey, J. (1969). Tests for specification errors in classical linear least-squares regression analysis. *Journal of the Royal Statistical Society Series B (Methodological), 31*(2), 350–371.
Rangarajan, C., & Singh, A.. (1984, June). *Reserve money: Concepts and Implications for India.* Reserve Bank of India Occasional Papers.
Reserve Bank of India. (2004). *Report on currency and finance.* Mumbai: Reserve Bank of India.
Reserve Bank of India. (2013). *Handbook of statistics of the Indian economy.* Mumbai: Reserve Bank of India.
Saaty, T., & Bram, J. (1960). *Nonlinear mathematics.* Mineola: Courier Dover Publications.
Sarno, L. (2003). Nonlinear exchange rate models: a selective overview. *IMF Working Paper* WP/03/111.
Seo, M. (2006). Bootstrap testing for the null of no cointegration in a threshold vector error correction model. *Journal of Econometrics, 127*(1), 129–150.
Seo, M. (2009). Estimation of nonlinear error-correction models. *Discussion paper.* London School of Economics, http://personal.lse.ac.uk/seo/PDF/SEO-NECM.pdf.
Stigler, M. (2011). Threshold cointegration: overview and implementation. *Working paper.* http://cran.r-project.org/web/packages/tsDyn/vignettes/ThCointOverview.pdf.

Subba Rao, T., & Gabr, M. M. (1984). *An introduction to bispectral analysis and bilinear time series models, Vol. 24 of lecture notes in statistics*. New York: Springer.
Subba Rao, T., & Gabr, M. M. (1980). A test for linearity of stationary time series. *Journal of Time Series Analysis, 1*(2), 145–158.
Teräsvirta, T. (1994). Specification, estimation, and evaluation of smooth transition autoregressive models. *Journal of the American Statistical Association, 89*(425), 208–218.
Teräsvirta, T., Tjøstheim, D., & Granger, C. W. J. (1994). Aspects of modelling nonlinear time series. In R. F. Engle & D. McFadden (Eds.), *Handbook of Econometrics* (Vol. 4). Amsterdam: Elsevier.
Ters, K., & Urban, J. (2018). *Estimating unknown arbitrage costs: Evidence from a 3-regime threshold vector error correction model*, BIS Working Papers No. 689, Monetary and Economic Department, Bank for International Settlements.
Thurner, S., Farmer, J., & Geanakoplos, J. (2012). Leverage causes fat tails and clustered volatility. *Quantitative Finance, 12*(5), 695–707.
Tong, H. (1983). *Threshold models in nonlinear time series analysis*. Berlin: Springer.
Tsay, R. (1986). Nonlinearity tests for time series. *Biometrika, 73*(2), 461–466.
Tsay, R. (1987). Conditional heteroscedastic time series models. *Journal of the American Statistical Association, 82,* 590–604.
Tsay, R. (2002). *Analysis of financial time series* (1st ed.). New York: Wiley.
Wang, D., & Ghosh, S. (2004). *Bayesian analysis of random coefficient autoregressive models*, Institute of Statistics Mimeo Series, NC State University Libraries, available at: https://repository.lib.ncsu.edu/bitstream/handle/1840.4/184/ISMS2566.pdf?sequence=1&isAllowed=y.
Zhang, W.-B. (1988). Limit cycles in van der ploeg's model of economic growth and conflict over the distribution of Income. *Journal of Economics, 48*(2), 159–173
Zivot, E. (2008). Practical issues in the analysis of univariate GARCH models. In T. G. Andersen, R. A. Davis, J. P. Kreiss, & T. V. Mikosch (Eds.), *Handbook of financial time series*. Berlin: Springer. http://faculty.washington.edu/ezivot/research/practicalgarchfinal.pdf.

Pareto Models for Risk Management

Arthur Charpentier and Emmanuel Flachaire

JEL Classification C13 · C18 · C46 · G22 · G32

1 Introduction

The Gaussian distribution is one of the most popular distributions used initially to model observation errors. It can be used to model individual measurements that are (somehow) centered, such as the height or the weight of individuals. But not all data exhibit such a property, where the distribution has a peak around a typical value (the median or the mean), such as the distribution of the population among cities or the distribution of wealth among individuals. Those distribution are known to be right-skewed, where most observations are bulked with small values, a small proportion can reach much higher values than the median (or the mean), leading to long tail on the right of the distribution. For instance, Pareto (1895) introduced the following model for the distribution of income: let $p(x)\mathrm{d}x$ denote the proportion of individuals with income between x and $x + \mathrm{d}x$; if the histogram is a straight line on the log–log scales, then $\log p(x) = a - b\log(x)$ (since the slope is clearly always negative), or equivalently, $p(x) = cx^{-b}$. Such distributions are said to have a *power law* or a Pareto distribution since it was introduced by the economist Vilfredo Pareto (see Charpentier and Flachaire 2019 for applications to model income distributions). The

A. Charpentier (✉)
Université du Québec à Montréal (UQAM), 201, avenue du Président-Kennedy, Montréal (Québec) H2X 3Y7, Canada
e-mail: arthur.charpentier@uqam.ca

E. Flachaire
Aix-Marseille Université AMSE, CNRS and EHESS, 5 bd Maurice Bourdet, 13001 Marseille, France
e-mail: emmanuel.flachaire@univ-amu.fr

G. Dufrénot and T. Matsuki (eds.), *Recent Econometric Techniques for Macroeconomic and Financial Data*, Dynamic Modeling and Econometrics in Economics and Finance 27, https://doi.org/10.1007/978-3-030-54252-8_14

constant b might be called the exponent of the power law (especially in the context of networks or in physical applications), but in economic application, the exponent will be the one of the survival distributions $S(x) = \int_x^\infty p(y)dy$, which is also a power function, with index $1 + b$.

The Gaussian distribution is a natural candidate when focusing on the center of the distribution, on average values, because of the central limit theorem: the Gaussian distribution is stable by summing and averaging. If $\{X_1, \ldots, X_n\}$ are independent Gaussian variables, so is $X_1 + \cdots + X_n$ (and any linear transformation). As mentioned in Jessen and Mikosch (2006) and Gabaix (2009), the power distribution satisfies similar interesting properties: if X_1 is a power distribution with exponent b, independent of X_2, another power distribution with the same exponent b_1, then $X_1 + X_2, X_1 \cdot X_2$, or $\max\{X_1, X_2\}$ are also power distributed, with exponent b_1.[1] And those properties of invariance and stability are essential in stochastic modeling. As mentioned in Schumpeter (1949), in an obituary article about Vilfredo Pareto, "*few if any economists seem to have realized the possibilities that such invariants hold for the future of our science. In particular, nobody seems to have realized that the hunt for, and the interpretation of, invariants of this type might lay the foundations for an entirely novel type of theory.*" If several applications can be found on proportional (random) growth theory (as discussed in Gabaix (2009), with connections to recursive equations), or on matching and networks, an important application of this Pareto model is risk management.

Insurers and actuaries have used Pareto distribution to model large losses since Hagstroem (1925, 1960). Balkema and de Haan (1974) suggested to use such a distribution in a life insurance context (to model the remaining lifetimes for very old people), but the Pareto distribution has mainly been used to model (large) insurance losses since reinsurance premiums have simple and analytical expression, as in Vajda (1951), even for more complex treaties than standard stop-loss or excess-of-loss treaties as in Kremer (1984). Beirlant and Teugels (1992), McNeil (1997) (discussed in Resnick 1997) and Embrechts et al. (1997) discussed more applications in insurance, with connections to extreme value theory.

The value-at-risk has been the finance benchmark risk measure for the past 20 years, estimating a quantile for small probabilities, such as the 1%-quantile of the profit and loss distribution for the next 10 days. Historically, practitioners used a standard Gaussian model and then multiply by a factor of 3 (as explained in Klüppelberg 2004, this factor 3 is supposed to account for certain observed effects, also due to the model risk; it is based on backtesting procedures and can be increased by the regulatory authorities, if the backtesting proves the factor 3 to be insufficient). A less ad hoc strategy is to use a distribution that fits better tails of profit and loss distribution, such as the Pareto one.

Nevertheless, if the Pareto distribution remains extremely popular because several quantities can easily be derived (risk measures or insurance premium), in practice, distributions are only Pareto in very high tails (say above the 99.9% quantile). So

[1] Gabaix (2009) claimed that similar results can be obtained when exponents are different, unfortunately, this yields only asymptotic power tails, which will be discussed in this chapter.

it becomes necessary to take into account second-order approximation, to provide a better fit. Thus, we will see in this article how to derive risk measures and insurance premiums when distribution are Pareto-type above a (high) threshold u.

In Sect. 2, we will define the two popular Pareto models, the strict Pareto distribution and the generalized Pareto distribution (GPD), emphasizing differences between the two. A natural extension will be related to the concept of Pareto-type distribution, using regular variation function. In Sect. 3, we will get back on the concepts of regular variation and present the extended Pareto distribution (EPD), introduced in Beirlant et al. (2009). The inference of Pareto models will be discussed in Sect. 4, with a discussion about the use of Hill's estimator, still very popular in risk management, and maximum likelihood estimation. In Sect. 5, we will define classical measures and indices, and discuss expression and properties when underlying distribution are either strict Pareto, or Pareto-type, with high quantiles (Q), expected shortfall (ES), and large claim or top share indices (TS). And as Pareto models are usually assumed above a high threshold u, we will see, at the end of that section, how analytical expressions of those measures can be derived when Pareto-type distributions are considered only above threshold u (and define Q_u, ES_u, etc). Finally, applications on real data will be considered, with insurance and reinsurance pricing in Sect. 6 and log-returns and classical financial risk measures in Sect. 7.

2 Strict Pareto Models

Pareto models are obtained when the loss distribution $\mathbb{P}[X > x]$ exhibits a power decay $x^{-\alpha}$. From a mathematical perspective, such a power function is interesting because it is the general solution of Cauchy's functional equation (its multiplicative version), $h(x \cdot y) = h(x) \cdot h(y)$, $\forall x, y$. A probabilistic interpretation is the stability of the conditional distribution X given $X > u$ for any threshold u: the conditional distribution still exhibits a power decay, with the same index. But if we try to formalize more, two distributions will be obtained: a Pareto I distribution will be obtained when modeling the distribution of *relative excesses* X/u given $X > u$—see Eq. (1)—while the generalized Pareto distribution (GPD) will be obtained when modeling the distribution of *absolute excesses* $X - u$ given $X > u$—see Eq. (6).

2.1 Pareto I Distribution

A Pareto type I distribution, bounded from below by $u > 0$, with tail parameter α, has probability density function and cumulative density function (CDF) equal to

$$f(x) = \frac{\alpha u^{\alpha}}{x^{\alpha+1}} \quad \text{and} \quad F(x) = 1 - \left(\frac{x}{u}\right)^{-\alpha}, \quad \text{for } x \geq u \tag{1}$$

If a random variable X has distribution (1), we will write $X \sim \mathcal{P}_1(u, \alpha)$. This distribution has an attractive property: the average above a threshold is proportional to the threshold, and it does not depend on the scale parameter u,

$$\mathbb{E}(X|X > u') = \frac{\alpha u'}{\alpha - 1}, \qquad \alpha > 1 \tag{2}$$

where $u' \geq u$. Such a function is related to the *mean excess function*—or *expected remaining lifetime* at age u' when X denotes the (random) life length—defined as

$$e(u') = \mathbb{E}(X - u'|X > u'). \tag{3}$$

In the case of a Pareto distribution, it is also a linear function

$$e(u') = \frac{u'}{\alpha - 1}, \qquad \alpha > 1 \tag{4}$$

Remark 1 In several textbooks and articles, Pareto models are defined with tail index $\xi = 1/\alpha$, so that the survival function is proportional to the power function $x^{-1/\xi}$, instead of $x^{-\alpha}$. In that case, we have the following expressions

$$\mathbb{E}(X|X > u') = \frac{u'}{1-\xi} \quad \text{and} \quad e(u') = \frac{\xi u'}{1-\xi}, \qquad \xi < 1. \tag{5}$$

2.2 Generalized Pareto Distribution

A generalized Pareto distribution (GPD), bounded from below by $u \geq 0$, with scale parameter σ and tail parameter α, has cumulative density function (CDF)

$$F(x) = 1 - \left[1 + \left(\frac{x-u}{\sigma}\right)\right]^{-\alpha} \qquad \text{for } x \geq u \tag{6}$$

where $\sigma > 0$ and $\alpha \in (0, \infty]$. If a random variable X has distribution (6), we will write $X \sim \mathcal{GPD}(u, \sigma, \alpha)$. The $\mathcal{GPD}(0, \sigma, \alpha)$ distribution is also called "Lomax" in the literature, see Lomax (1954). The GPD is "general" in the sense that Pareto I distribution is a special case, when $\sigma = u$:

$$\mathcal{GPD}(u, u, \alpha) = \mathcal{P}_1(u, \alpha) \tag{7}$$

Scollnik (2007) suggested an alternative expression, instead of (6),

$$F(x) = 1 - \left[\frac{u+\lambda}{x+\lambda}\right]^{\alpha} \qquad \text{for } x \geq u \tag{8}$$

where $\lambda = \sigma - u$. Note that this function is used in Rigby and Stasinopoulos (2005) to introduce a regression type model, where both λ and α will be the exponential of linear combinations of some covariates, as in the `gamlss` package, in `R`.

Remark 2 In several textbooks and articles, not only the generalized Pareto has tail index $\xi = 1/\alpha$, but ξ appears also in the fraction term

$$F(x) = 1 - \left[1 + \xi\left(\frac{x-u}{\sigma}\right)\right]^{-1/\xi} \qquad \text{for } x \geq u, \tag{9}$$

which could be seen as a simple alternative expression of the shape parameter σ. This expression is interesting since the limit when ξ tends to 0 can easily be defined, as well as the case where $\xi < 0$, which could be interesting in extreme value theory.

The average above a higher threshold, $u' \geq u$, depends now on all parameters of the distribution,

$$\mathbb{E}(X|X > u') = \frac{\sigma - u}{\alpha - 1} + \frac{\alpha}{\alpha - 1}u'. \tag{10}$$

The linearity of this function (as well as the mean excess function) characterizes the GPD class (see Guess and Proschan 1988; Ghosh and Resnick 2010). A sketch of the proof is that we want

$$e(u') = \mathbb{E}(X|X > u') = \frac{1}{\overline{F}(u')}\int_{u'}^{\infty} x\mathrm{d}F(x) = A + Bu', \tag{11}$$

or, when differentiating with respect to u', we obtain

$$B\overline{F}(u')\mathrm{d}u' = s - \left[A + (B-1)u'\right]\mathrm{d}\overline{F}(u'), \tag{12}$$

thus, up to some change of parameters

$$\frac{\mathrm{d}\overline{F}}{\overline{F}} = -\alpha\frac{\mathrm{d}u}{u+c} \quad \text{with } \alpha = \frac{B}{B-1} \tag{13}$$

which is a GPD distribution, with tail index α.

One of the most important results in the extreme value theory states that, for most heavy-tailed distributions, the conditional excess distribution function, above a threshold u, converges toward a GPD distribution as u goes to infinity (Pickands 1975; Balkema and de Haan 1974), for some parameters α and σ,

$$F_u(x) \longrightarrow \text{GPD (or Pareto II)} \quad \text{as } u \to +\infty \tag{14}$$

where $F_u(x) = P(X - u \leq x|X > u)$. This result is known as the Pickands–Balkema–de Haan theorem, also called the second theorem in extreme value theory (as discussed in Footnote 3, page 7). It provides strong theoretical support for modeling the upper tail of heavy-tailed distributions with GPD, also known as the Pareto type II distribution. In a very general setting, it means that there are α and σ such that F_u can be approximated by the CDF of a $\mathcal{GPD}(u, \sigma, \alpha)$, see Embrechts et al. (1997, Theorem 3.4.13, p. 165).

2.3 Threshold Selection

Whether to use a Pareto I or GPD model to fit the upper tail is related to the choice of the threshold. From (1) and (6), we have seen that Pareto I is a special case of GPD, when $\sigma = u$. They differ by an affine transformation when $\sigma \neq u$. A key property of Pareto distributions is that, if a distribution is Pareto for a fixed threshold u, it is also Pareto with the same tail parameter α for a higher threshold $u' \geq u$. For GPD, we have

$$\overline{F}_u = \mathcal{GPD}(u, \sigma, \alpha) \quad \Rightarrow \quad \overline{F}_{u'} = \mathcal{GPD}(u', \sigma + u' - u, \alpha). \tag{15}$$

where $\overline{F}_u$ is the survival excess function above u.[2] Thus, Pareto I and GPD are the same for all $u' \geq u$, if $\sigma = u$. Otherwise, we have $\sigma + u' - u \approx u'$ for very large values of u' only. It follows that a GPD above a threshold will behave approximately as a Pareto I above a *higher* threshold, much higher as σ differs from u. This point will be discussed further in Sect. 5.5.

3 Pareto-Type Models

Since our interesting in risk management is usually motivated by the description of so-called *downside risks*, or (important) losses, as described in Roy (1952). Thus, we do not need that have a (strict) Pareto distribution for all x's, in $\mathbb{P}[X > x]$, but possibly only when x's are large enough. In this section, we will introduce some regular variation concepts, used to model that asymptotic property, and exhibit a distribution that is not strictly Pareto, by introducing a second-order effect: the *extended Pareto distribution*.

[2] $\overline{F}_{u'}$ is a truncated Pareto distribution, with density equals to $f(x)/(1 - F(u'))$. This property can be observed directly using Eq. (8), where both α and λ remain unchanged.

Note that this property is quite intuitive, since the GPD distribution appears as a limit for exceeding distributions, and limit in asymptotic results are always fixed points: the Gaussian family is stable by addition (and appears in the Central Limit Theorem) while Fréchet distribution is max-stable (and appears in the first theorem in extreme value theory).

3.1 First- and Second-Order Regular Variation

The tail index α is related to the max-domain of attraction of the underlying distribution, while parameter σ is simply a scaling parameter.[3] The shape of the conditional excess cumulative distribution function is a power function (the Pareto distribution) if the threshold is large enough. Tails are then said to be Pareto-type and can be described using so-called *regularly varying* functions (see Bingham et al. 1987).

First- and second-order regular variations were originally used in extreme value theory, to study, respectively, the tail behavior of a distribution and the speed of convergence of the extreme value condition (see Bingham et al. 1987; de Haan and Stadtmüller 1996; Peng and Qi 2004, or Sect. 2 in de Haan and Ferreira 2006 for a complete survey). A function H is said to be regularly varying (at infinity) with index $\gamma \in \mathbb{R}$ if

$$\lim_{t\to\infty} \frac{H(tx)}{H(t)} = x^{\gamma} \text{ or } \lim_{t\to\infty} x^{-\gamma}\frac{H(tx)}{H(t)} = 1. \tag{16}$$

A function regularly varying with index $\gamma = 0$ is said to be slowly varying. Observe that any regularly varying function of index $-\gamma$ can be written $H(x) = x^{-\gamma}\ell(x)$ where ℓ is some slowly varying function.

Consider a random variable X, its distribution is regularly varying with index $-\gamma$ if, up-to some affine transformation, its survival function is regularly varying. Hence,

$$\lim_{t\to\infty} x^{-\gamma}\frac{\overline{F}(tx)}{\overline{F}(t)} = 1 \tag{17}$$

or

$$\overline{F}(x) = x^{-\gamma}\ell(x), \tag{18}$$

where $\overline{F}(x) = 1 - F(x)$. A regularly varying survival function is then a function that behaves like a power law function near infinity. Distributions with survival function as defined in (18) are called *Pareto-type distributions*. It means that the survival function tends to zero at polynomial (or power) speed as $x \to \infty$, that is, as $x^{-\gamma}$. For instance, a Pareto I distribution, with survival function $\overline{F}(x) = x^{-\alpha}u^{\alpha}$, is regularly varying with index $-\alpha$, and the associated slowly varying function is the constant u^{α}. And a GPD or Pareto II distribution, with survival function $\overline{F}(x) = \left(1 + \sigma^{-1}x\right)^{-\alpha}$, is

[3]Historically, extremes were studied through block-maximum—yearly maximum, or maximum of a subgroup of observations. Following Fisher and Tippett (1928), up to some affine transformation, the limiting distribution of the maximum over n i.i.d observations is either Weibull (observations with a bounded support), Gumbel (infinite support, but light tails, like the exponential distribution) or Fréchet (unbounded, with heavy tails, like Pareto distribution). Pickands (1975) and Balkema and de Haan (1974) obtained further that not only the only possible limiting conditional excess distribution is GPD, but also that the distribution of the maximum on subsamples (of same size) should be Fréchet distributed, with the same tail index γ, if $\gamma > 0$. For instance in the USA, if the distribution of maximum income per county is Fréchet with parameter γ (and if county had identical sizes), then the conditional excess distribution function of incomes above a high threshold is a GPD distribution with the same tail index γ.

regularly varying with index $-\alpha$, for some slowly varying function. But in a general setting, if the distribution is not *strictly* Pareto, ℓ will not be constant, and it will impact the speed of convergence.

In de Haan and Stadtmüller (1996), a concept of second-order regular variation function is introduced, that can be used to derive a probabilistic property using the quantile function,[4] as in Beirlant et al. (2004). Following Beirlant et al. (2009), we will consider distributions such that an extended version of Eq. (17) is satisfied,

$$\lim_{t\to\infty} x^{-\gamma}\frac{\overline{F}(tx)}{\overline{F}(t)} = 1 + \frac{x^{\rho}-1}{\rho}, \text{ for some } \rho \leq 0, \tag{19}$$

that we can write, up to some affine transformation,

$$\overline{F}(x) = cx^{-\gamma}[1 - x^{\rho}\ell(x)], \tag{20}$$

for some slowly varying function ℓ and some second-order tail coefficient $\rho \leq 0$. The corresponding class of Pareto-type distributions defined in (20) is often named the Hall class of distributions, referring to Hall (1982). It includes the Singh–Maddala (Burr), Student, Fréchet and Cauchy distributions. A mixture of two strict Pareto-I distributions will also belong to this class.

Since $\rho \leq 0, x \mapsto 1 - x^{\rho}\ell(x)$ is slowly varying, and therefore, a distribution $\overline{F}$ that satisfies (20) also satisfies (18). More specifically, in (20), the parameter ρ captures the rate of convergence of the survival function to a strict Pareto distribution. Smaller is ρ, faster the upper tail behaves like a Pareto, as x increases. Overall, we can see that

- γ is the first order of the regular variation, and it measures the tail parameter of the Pareto distribution,
- ρ is the second order of the regular variation, and it measures how much the upper tail deviates from a strictly Pareto distribution.

In the following, we will write $\text{RV}(-\gamma, \rho)$. There are connections between tail properties of the survival function $\overline{F}$ and the density f (see also Karamata theory for first-order regular variation). More specifically, if $\overline{F}$ is $\text{RV}(-\gamma, \rho)$, with $\gamma > 1$, then f is $\text{RV}(-\gamma - 1, \rho)$.

For instance, consider a Singh–Maddala (Burr) distribution, with survival distribution $\overline{F}(x) = [1 + x^a]^{-q}$, then a second-order expansion yields

$$\overline{F}(x) = x^{-aq}[1 - qx^{-a} + o(x^{-a})] \text{ as } x \to \infty \tag{21}$$

which is regularly varying of order $-aq$ and with second-order regular variation $-a$, that is $\text{RV}(-aq, -a)$.

---

[4]The quantile function U is defined as $U(x) = F^{-1}(1 - 1/x)$.

3.2 Extended Pareto Distribution

Beirlant et al. (2009) show that Eq. (20) can be approximated by

$$\overline{F}(x) = \left[x\left(1 + \delta - \delta x^{\tau}\right)\right]^{-\alpha} \qquad \text{for } x \geq 1 \tag{22}$$

where $\tau \leq 0$ and $\delta > \max(-1, 1/\tau)$.[5] The main feature of this function is that it captures the second-order regular variation of the Hall class of distributions, that is, deviation to a strictly Pareto tail, as defined in Hall (1982),

$$\overline{F}(x) = ax^{-\alpha}\left[1 + bx^{-\beta} + o(x^{-\beta})\right] \tag{23}$$

or, using Taylor's expansion,

$$\overline{F}(x) = ax^{-\alpha}\left[1 + b_1 x^{-\alpha} + \cdots + b_k x^{-k\alpha} + o(x^{-k\alpha})\right] \tag{24}$$

with general notations. For more details, see also Albrecher et al. (2017, Sect. 4.2.1).

From (22), we can define the extended Pareto distribution (EPD), proposed by Beirlant et al. (2009), as follows:

$$F(x) = 1 - \left[\frac{x}{u}\left(1 + \delta - \delta\left(\frac{x}{u}\right)^{\tau}\right)\right]^{-\alpha} \qquad \text{for } x \geq u \tag{25}$$

where $\tau \leq 0$ and $\delta > \max(-1, 1/\tau)$. If a random variable X has (25) as its CDF, we will write $X \sim \mathcal{EPD}(u, \delta, \tau, \alpha)$.

Pareto I is a special case when $\delta = 0$ and GPD is a special case when $\tau = -1$:

$$\mathcal{EPD}(u, 0, \tau, \alpha) = \mathcal{P}_1(u, \alpha) \tag{26}$$

$$\mathcal{EPD}(u, \delta, -1, \alpha) = \mathcal{GPD}(1, u/(1 + \delta), \alpha) \tag{27}$$

The mean over a threshold for the EPD distribution has no closed form expression.[6] Numerical methods can be used to calculate it. Since $u' \geq u > 0$, X given $X > u'$ is a positive random variable and

$$\mathbb{E}[X|X > u'] = \int_0^{\infty} \overline{F}_{u'}(x)\mathrm{d}x \tag{28}$$

---

[5]Using the expansion $(1 + y^a)^b \approx 1 + by^a$, for small y^a, in (22) yields (20).

[6]Albrecher et al. (2017, Sect. 4.6) give an approximation, based on $(1 + \delta - \delta y^{\tau})^{-\alpha} \approx 1 - \alpha\delta + \alpha\delta y^{\tau}$, which can be very poor. Thus, we do not recommend to use it.

where $\overline{F}_{u'}(x) = \mathbb{P}[X > x|X > u']$ for $x > u$, i.e.,

$$\overline{F}_{u'}(x) = \frac{\overline{F}(x)}{\overline{F}(u')} \quad \text{where} \quad \overline{F} \text{ is the s.d.f. of } X \tag{29}$$

Thus

$$\mathbb{E}[X|X > u'] = u' + \frac{1}{\overline{F}(u')} \int_{u'}^{\infty} \overline{F}(x) \mathrm{d}x \tag{30}$$

The integral in Eq. (30) can be computed numerically. Since numerical integration over a finite segment could be easier, we can make a change of variable $(1/x)$ to obtain an integral over a finite interval:

$$E_{u'} = \int_{u'}^{\infty} \overline{F}(x) \mathrm{d}x = \int_{0}^{1/u'} \frac{1}{x^2} \overline{F}\left(\frac{1}{x}\right) \mathrm{d}x \tag{31}$$

The extended Pareto distribution has a stable tail property: if a distribution is EPD for a fixed threshold u, it is also EPD for a higher threshold $u' \geq u$, with the same tail parameter α. Indeed, deriving a truncated EPD distribution, we find

$$\overline{F}_u = \mathcal{EPD}(u, \delta, \tau, \alpha) \quad \Rightarrow \quad \overline{F}_{u'} = \mathcal{EPD}(u', \delta', \tau, \alpha). \tag{32}$$

where $\delta' = \delta(u'/u)^\tau/[1 + \delta - \delta(u'/u)^\tau]$. A plot of estimates of the tail index α for several thresholds would then be useful. If the distribution is extended Pareto, a stable horizontal straight line should be plotted. This plot is similar to Hill plot for Hill estimates of α from Pareto I distribution. It is expected to be more stable if the distribution is not strictly Pareto. Indeed, Embrechts et al. (1997, p. 194) and Resnick (2007, p. 87) illustrate that the Hill's estimator can perform very poorly if the slowly varying function is not constant in (18). It can lead to very volatile Hill plots, also known as Hill *horror* plots. This will be discussed in Sect. 4.1.

4 Inference Based on Pareto-Type Distributions

In this section, we briefly recall two general techniques used to estimate the tail index, and additional parameters: Hill estimator and maximum likelihood techniques.[7] For further details, see Embrechts et al. (1997), de Haan and Ferreira (2006) or Beirlant et al. (2004), and references therein.

[7]Even if Hill estimator can be can be seen as a Maximum Likehood estimator, for some properly chosen distribution.

4.1 Hill's Estimator

In the case where the distribution above a threshold u is supposed to be $\mathcal{P}_1(u,\alpha)$, Hill's estimator of α is

$$\widehat{\alpha}_u = \left(\frac{1}{n_u}\sum_{i=1}^{n_u}\log(x_{n+1-i:n}) - \log(u)\right)^{-1} \quad (33)$$

This estimator is the maximum likelihood estimator of α.

From Theorem 6.4.6 in Embrechts et al. (1997) and Sect. 4.2 in Beirlant et al. (2004), if we assume that the distribution of X is Pareto-type, with index α, $\overline{F}(x) = x^{-\alpha}\ell(x)$, if $u \to \infty$, $n \to \infty$ and $n_u/n \to 0$, then $\widehat{\alpha}_u$ is a (strongly) consistent estimator of α when observations are i.i.d., and further, under some (standard) technical assumptions,

$$\sqrt{n_u}\big(\widehat{\alpha}_u - \alpha\big) \xrightarrow{\mathcal{L}} \mathcal{N}(0,\alpha^2). \quad (34)$$

This expression can be used to derive confidence interval for α, but also any index mentioned in the previous section (quantile, expected shortfall, top shares, etc.) using the Δ-method. For instance

$$\frac{\sqrt{n_u}Q(1-p)}{\sqrt{1+(\log[n_u/n]-\log[p])^2}}\big(\widehat{Q}_u(1-p) - Q(1-p)\big) \xrightarrow{\mathcal{L}} \mathcal{N}(0,\alpha^{-2}), \quad (35)$$

as shown in Sect. 4.6 of Beirlant et al. (2004).

Nevertheless, as discussed in Embrechts et al. (1997) and Beirlant et al. (2004), this asymptotic normality is obtained under some appropriate choice of u (as a function of n) and ℓ: u should tend to infinity sufficiently slowly, otherwise, $\widehat{\alpha}_u$ could be a biased estimator, even asymptotically. More specifically, in Sect. 3.1, we introduced second-order regular variation: in Eq. (20), assume that $\ell(x) = -1$, so that

$$\overline{F}(x) = cx^{-\alpha}[1 + x^{\rho}] \quad (36)$$

From Theorem 6.4.9 in Embrechts et al. (1997), if $n_u = o(n^{(2\rho)/(2\rho+\alpha)})$, then the asymptotic convergence of Eq. (34) is valid. But if $n^{-(2\rho)/(2\rho+\alpha)}n_u$ tends to λ as $n \to \infty$, then

$$\sqrt{n_u}\big(\widehat{\alpha}_u - \alpha\big) \xrightarrow{\mathcal{L}} \mathcal{N}\left(\frac{\alpha^3\lambda}{\rho-\alpha},\alpha^2\right), \quad (37)$$

has an asymptotic bias. Such a property makes Hill's estimator dangerous to use for Pareto-type distributions.

4.2 Maximum Likelihood Estimator

For the GPD, a popular approach is to use the maximum likelihood estimator. When threshold u is given, the idea is to fit a $\mathcal{GPD}(0,\sigma,\alpha)$ distribution on the sample $\{y_1,\dots,y_{n_u}\}=\{x_{n+1-n_u:n}-u,\dots,x_{n:n}-u\}$. The density being

$$f(y;\sigma,\alpha)=\frac{\alpha}{\sigma}\left(1+\frac{y}{\sigma}\right)^{-\alpha-1} \tag{38}$$

the log-likelihood is here

$$\log\mathcal{L}(\sigma,\alpha;\boldsymbol{y})=-n\log\frac{\alpha}{\sigma}-(\alpha+1)\sum_{i=1}^{n_u}\left(1+\frac{y_i}{\sigma}\right). \tag{39}$$

As expected, the maximum likelihood estimator of (σ,α) is asymptotically Gaussian, as shown in Sect. 6.5 of Embrechts et al. (1997) (with a different parameterization of the GPD). In the case of the extended Pareto distribution, there might be numerical complications, but Beirlant et al. (2009) provided theoretical properties, and Reynkens (2018) provided R codes, used in Albrecher et al. (2017).

Note that since α is usually seen as the most important parameter (and σ is more a nuisance parameter), it can be interesting to use profile likelihood techniques to derive some confidence interval, as discussed in Sect. 4.5.2 in Davison (2003). Consider some Pareto-type model, with parameter $(\alpha,\boldsymbol{\theta})$. Let

$$(\widehat{\alpha},\widehat{\boldsymbol{\theta}})=\underset{\alpha,\boldsymbol{\theta}}{\text{argmax}}\,\{\log\mathcal{L}(\alpha,\boldsymbol{\theta})\} \tag{40}$$

denote the maximum likelihood estimator. From the likelihood ratio test, under technical assumptions,

$$2\big(\log\mathcal{L}(\widehat{\alpha},\widehat{\boldsymbol{\theta}})-\log\mathcal{L}(\alpha,\boldsymbol{\theta})\big)\to\chi^2\big(\dim(\alpha,\boldsymbol{\theta})\big)$$

where $I_n(\alpha,\boldsymbol{\theta})$ is Fisher information. The idea of the profile likelihood estimator is to define, given α

$$\widehat{\boldsymbol{\theta}}_\alpha=\underset{\boldsymbol{\theta}}{\text{argmax}}\,\{\log\mathcal{L}(\alpha,\boldsymbol{\theta})\} \tag{41}$$

and then

$$\widehat{\alpha}_p=\underset{\alpha}{\text{argmax}}\,\{\log\mathcal{L}(\alpha,\boldsymbol{\theta}_\alpha)\} \tag{42}$$

Then

$$2\big(\log\mathcal{L}(\widehat{\alpha}_p,\widehat{\boldsymbol{\theta}}_{\widehat{\alpha}_p})-\log\mathcal{L}(\alpha,\boldsymbol{\theta})\big)\to\chi^2(1)\big).$$

Thus, if $\log \mathcal{L}_p(\alpha)$ denotes the profile likelihood, defined as $\log \mathcal{L}_p(\alpha) = \log \mathcal{L}(\alpha, \widehat{\boldsymbol{\theta}}_\alpha)$, then a 95% confidence interval can be obtained,

$$\left\{\alpha\colon \log \mathcal{L}_p(\alpha) \geq \log \mathcal{L}_p(\widehat{\alpha}_p) - \frac{1}{2} q_{\chi^2(1)}(95\%)\right\} \tag{43}$$

where $q_{\chi^2(1)}(95\%)$ is the 95% quantile of the $\chi^2(1)$ distribution.

4.3 Application on Simulated Data

In this section, three distributions were simulated: a strict Pareto on Fig. 1 with tail index $\alpha = 1.5$, a generalized Pareto on Fig. 2 with tail index $\alpha = 1.5$, and an extended Pareto on Fig. 3 with tail index $\alpha = 1.5$. Each time, on the left, the Hill plot is plotted, i.e., the plot $u \mapsto \widehat{\alpha}_u$. The dotted lines are bounds of the confidence interval (at level 95%). The vertical segment (—) is the confidence interval for α when u is the 80%-quantile of the sample. On the right, the profile likelihood of a GPD distribution is plotted, including the horizontal line (—) that defines the 95%-confidence interval.

In the case of the EPD, the smaller the value of τ, the more bias Hill's estimator has (see Charpentier and Flachaire 2019 for a detailed discussion). When generating a Pareto-type distribution, using the top 80% observations does not guarantee anymore that Hill's estimator could be a relevant estimator. On Fig. 4, we used simulated samples of size $n = 1000$ with the same Monte Carlo seed (on the left), and another

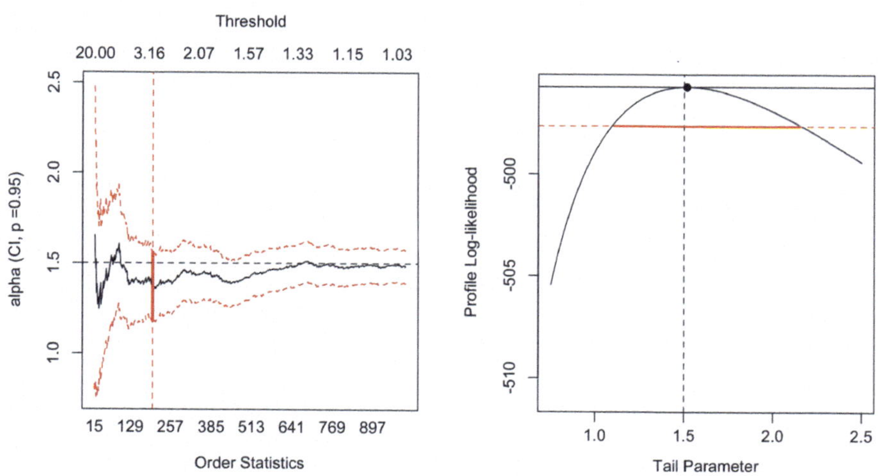

Fig. 1 Hill's estimator (on the left) for α and profile likelihood for α from a GPD model (on the right) on the top 20% of the observations, when $n = 1000$ observations were generated from a (strict) Pareto distribution, with $\alpha = 1.5$

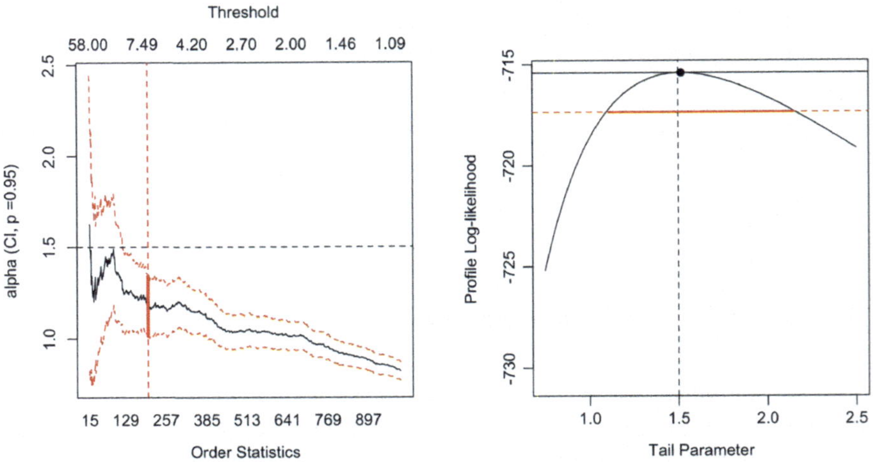

Fig. 2 Hill's estimator (on the left) for α and profile likelihood for α from a GPD model (on the right) on the top 20% of the observations, when $n = 1000$ observations were generated from a generalized Pareto distribution, with $\alpha = 1.5$, and $\sigma = 1$

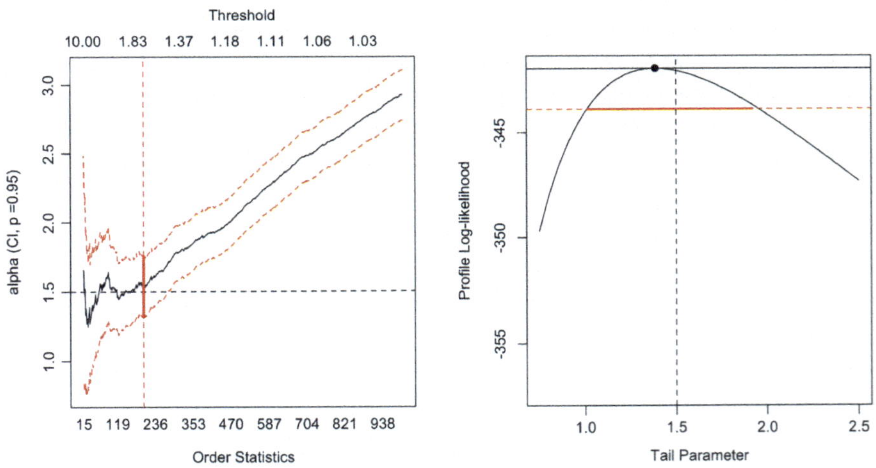

Fig. 3 Hill's estimator (on the left) for α and profile likelihood for α from a GPD model (on the right) on the top 20% of the observations, when $n = 1000$ observations were generated from an extended Pareto distribution, with $\alpha = 1.5$, and $\tau = -2$

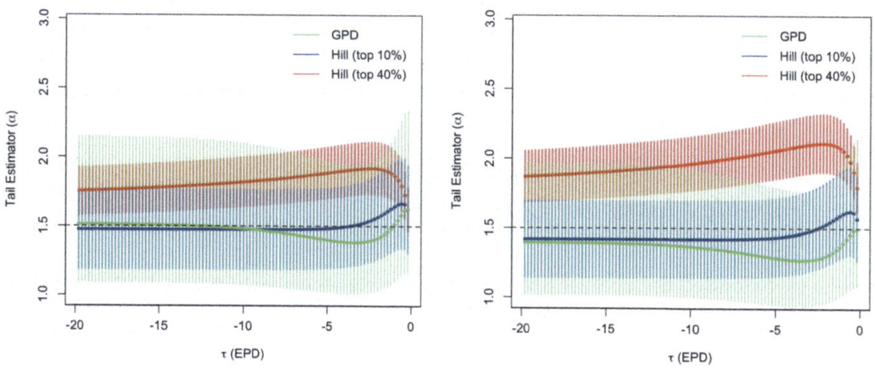

Fig. 4 Simulation of EPD samples (with the same Monte Carlo seed on the left, and another seed on the right), with $\alpha = 1.5$ and τ varying from -20 to 0. The red curve (—) is Hill's estimator $\widehat{\alpha}_u$ obtained when u is the 60% quantile, while the blue curve (—) is obtained when u is the 90% quantile. The green curve (—) is the profile likelihood estimator of a GPD distribution

one (on the right), as on Fig. 3, and we simply change the value of τ. The red curve (—) is Hill's estimator $\widehat{\alpha}_u$ obtained when u is the 60% quantile, while the blue curve (—) is obtained when u is the 90% quantile. The green curve (—) is the profile likelihood estimator of a GPD distribution. Horizontal segments are confidence intervals. For Hill's estimator, observe that if the threshold u is not large enough, estimator can be substantially biased. For the GPD, the fit is better, but the confidence interval is very large. Observe further that when $\tau \in (-2, 0)$, Hill's estimator in the higher tail can actually be overestimated.

5 Modeling Large Events

Given n i.i.d. random losses $X_1, \ldots, X_n$, let S_n and M_n, respectively, denote the sum and the maximum

$$S_n = X_1 + \cdots + X_n \quad \text{and} \quad M_n = \max\{X_1, \ldots, X_n\}. \tag{44}$$

A classical concept is the *probable maximum loss* (Sect. 5.1) which is the distribution of M_n, or an affine transformation (since the distribution depends on n, the number of claims in insurance, or the time horizon in finance). It is possible, as in Sect. 3 of Beirlant et al. (2004), to look "*close to the maximum*," by studying the limiting distribution of the kth largest value[8] $X_{k:n}$, when n goes to infinity, as well as k. If $k/n = O(1)$, we study the asymptotic distribution of some high quantile (Sect. 5.2). It is also possible to focus on large losses with respect to the entire portfolio, i.e., S_n. This

[8] Given a sample $\{x_1, \ldots, x_n\}$, let $\{x_{1:n}, \ldots, x_{n:n}\}$ denote the ordered version, with $x_{1:n} = \min\{x_1, \ldots, x_n\}$, $x_{n:n} = \max\{x_1, \ldots, x_n\}$ and $x_{1:n} \leq \ldots x_{n-1:n} \leq x_{n:n}$.

will yield the concepts of *subexponential* distribution (Sect. 5.3) when comparing M_n and S_n, and comparing a high quantile, or the sum above a high quantile will be related to *top share* (Sect. 5.4), also called *large claim index* in insurance applications.

5.1 Probable Maximum Loss

Paul Levy extended the central limit theorem to variables with a non-finite variance by considering non-degenerate distributions G such that

$$\lim_{n\to\infty} \mathbb{P}\left(\frac{\overline{X}_n - b_n}{a_n} \leq x\right) = G(x) \tag{45}$$

where $\overline{X}_n = n^{-1}S_n$. It is an extension of the central limit theorem in the sense that, if X_i's have a finite variance and if $b_n = \mathbb{E}[X]$ while $a_n = \sqrt{n^{-1}\text{Var}[X]}$, then G is the $\mathcal{N}(0, 1)$ distribution, as well as any sequences such that $b_n \to \mathbb{E}[X]$ and $\sqrt{n}\, a_n \to \sqrt{\text{Var}[X]}$, as $n \to \infty$. In the case of variables with infinite variance, then G is called a stable distribution.

At the same period, Fisher–Tippett investigated a rather similar problem, searching for possible limiting distribution for a standardized version of the maximum

$$\lim_{n\to\infty} \mathbb{P}\left(\frac{M_n - b_n}{a_n} \leq x\right) = G(x). \tag{46}$$

It was shown—in Gnedenko (1943)—that the only possible limit was the so-called extreme value distribution

$$G_\alpha(x) = \exp(-x^{-\alpha}) \tag{47}$$

including the limiting case $G_0(x) = \exp(-e^{-x})$, on some appropriate interval. For instance, assume that X is $\mathcal{P}_1(1, \alpha)$ distributed, let $b_n = 0$ and $a_n = U(n) = n^{1/\alpha}$, then

$$\lim_{n\to\infty} \mathbb{P}\left(\frac{M_n - b_n}{a_n} \leq x\right) = \lim_{n\to\infty} [F(a_n x + b_n)]^n = \lim_{n\to\infty} \left[F(n^{1/\alpha}x)\right]^n \tag{48}$$

$$= \lim_{n\to\infty} \left(1 - \frac{x^{-\alpha}}{n}\right)^n = \exp(-x^{-\alpha}) \tag{49}$$

More generally, assume that X is Pareto-type with tail index α, and consider $b_n = 0$ and $a_n \sim U(n)$ as $n \to \infty$. From Proposition 3.3.2 in Embrechts et al. (1997),

$$\lim_{n\to\infty} [F(a_n x + b_n)]^n = \exp\left(\lim_{n\to\infty} -n\overline{F}(a_n x + b_n)\right) \tag{50}$$

and since

$$n\overline{F}(a_n x) \sim \frac{\overline{F}(a_n x)}{\overline{F}(a_n)} \to x^{-\alpha} \text{ as } n \to \infty \tag{51}$$

therefore, we can obtain

$$\lim_{n\to\infty} \mathbb{P}\left(\frac{M_n - b_n}{a_n} \leq x\right) = \exp(-x^{-\alpha}) \tag{52}$$

as previously. Such a distribution is known as the Fréchet distribution with index α.

5.2 *High Quantiles and Expected Shortfall*

Given a distribution F, the quantile of level p of the distribution is

$$Q(p) = \inf\left\{y \in \mathbb{R} : F(y) \geq p\right\}. \tag{53}$$

Since we focus on high quantiles, i.e., $p \to 1$, a natural-related function is

$$U(x) = Q\left(1 - \frac{1}{x}\right), \tag{54}$$

and to study properties of that function as $x \to \infty$.

In the previous section, we studied limiting behavior of $a_n^{-1}(M_n - b_n)$ as $n \to \infty$. Let $\widehat{F}_n$ denote the empirical cumulative distribution of $\{X_1, \cdots, X_n\}$,

$$\widehat{F}_n(x) = \frac{1}{n}\sum_{i=1}^{n} \mathbf{1}(X_i \leq x). \tag{55}$$

Observe that $M_n = \widehat{F}_n^{-1}(1 - 1/n) = \widehat{U}_n(n)$, so M_n should be "close" to $U(n)$. So it could make sense to study

$$\lim_{n\to\infty} \frac{U(nx) - U(n)}{a_n} \tag{56}$$

Beirlant et al. (2009, Sect. 3.3) show that the only possible limit is

$$h(x) = \frac{x^\gamma - 1}{\gamma} \tag{57}$$

If U is the quantile function of $\mathcal{P}(1, \alpha)$, $U(x) = x^{1/\alpha}$, with auxiliary function[9] $a(t) = \alpha^{-1}t^{1/\alpha}$, then

$$\frac{U(tx) - U(t)}{a(t)} = \alpha(x^{1/\alpha} - 1) \tag{58}$$

for all x, i.e., $\gamma = \alpha^{-1}$.

Assume that X is $\mathcal{P}(u, \alpha)$, then

$$Q(p) = u \cdot (1 - p)^{-1/\alpha} \tag{59}$$

and if X is $\mathcal{GPD}(u, \sigma, \alpha)$, then

$$Q(p) = u + \sigma\big[(1 - p)^{-1/\alpha} - 1\big] \tag{60}$$

For extended Pareto distributions, quantiles are computed numerically.[10]

Another important quantity is the average of the top p%, called expected shortfall,

$$\text{ES}(p) = \mathbb{E}(X|X > Q(1 - p)) = \frac{1}{p}\int_{1-p}^{1} Q(u)\text{d}u. \tag{61}$$

This quantity is closely related to the mean excess function, since $\text{ES}(p) = Q(1 - p) + e(Q(1 - p))$, where e is the mean excess function. Assume that X is $\mathcal{P}_1(u, \alpha)$, then

$$\text{ES}(p) = \frac{\alpha}{\alpha - 1}u(1 - p)^{-1/\alpha} = \frac{\alpha}{\alpha - 1}Q(p), \tag{62}$$

and if X is $\mathcal{GPD}(u, \sigma, \alpha)$, then $X|X > Q(1 - p)$ is $\mathcal{GPD}(Q(1 - p), \sigma + Q(1 - p) + u, \alpha)$ from Eq. (15), and therefore

$$\text{ES}(p) = \frac{\alpha Q(1 - p) + \sigma - u}{\alpha - 1}. \tag{63}$$

For the extended Pareto model, only numerical approximations can be obtained.

On Fig. 5, we have the evolution of $Q(1 - p)$ for various values of p, from $1/10$ down to $1/1000$. A strict Pareto (—) with tail index $\alpha = 1.5$ is considered. On the left, we plot the quantile function of a GPD distribution (- - -), for small and large σ's, and on the right, we consider an EPD distribution (- - -) with different values of τ.

[9]The study of the limiting distribution of the maximum of a sample of size n made us introduce a normalizing sequence a_n. Here, a continuous version is considered—with $U(t)$ instead of $U(n)$—and the sequence a_n becomes the auxiliary function $a(t)$.

[10]See the R packages `ReIns` or `TopIncomes`.

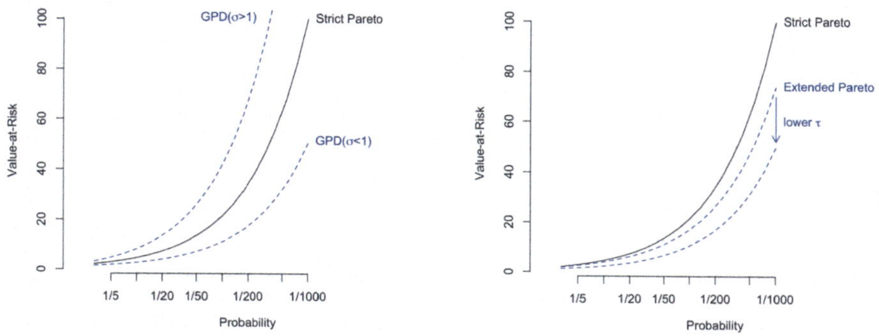

Fig. 5 Quantile $Q(1-p)$ as a function of the probability p (on a log scale), with a strict Pareto (—) with tail index $\alpha = 1.5$, and a GPD distribution (- - -) on the left, and an EPD distribution (- - -) on the right

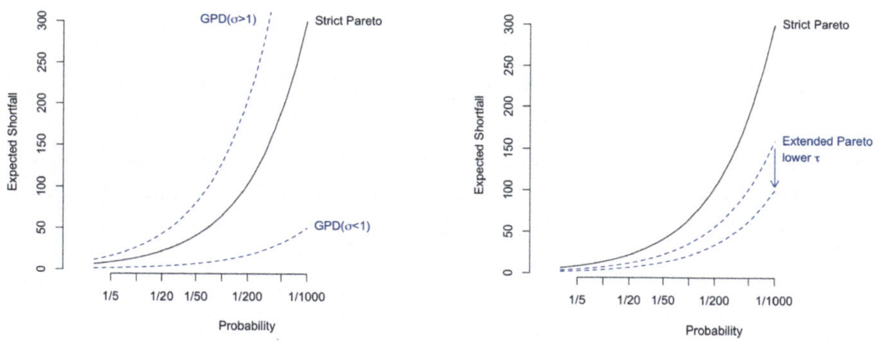

Fig. 6 Expected shortfall ES(p) as a function of the probability (on a log scale), with a strict Pareto (—) with tail index $\alpha = 1.5$, and a GPD distribution (- - -) on the left, and an EPD distribution (- - -) on the right

On Fig. 6, we have the evolution of ES(p) for various values of p, from $1/10$ down to $1/1000$. The strict Pareto distribution (—) still has tail index $\alpha = 1.5$. On the left, we plot the expected shortfall of a GPD distribution (- - -), for small and large σ's, and on the right, we consider an EPD distribution (- - -) with different values of τ.

Finally, on Fig. 7, we compare $Q(1-p)$ and ES(p) for various values of p, and plot the line $(Q(1-p), \text{ES}(p))$. As expected, it is a straight line in the strict Pareto case (—), above the first diagonal (- - -), meaning that there can be substantial losses above any quantile. Two GPD distributions with the same tail index α are considered on the left (- - -), and an EPD case is on the right (- - -). The GPD case is not linear, but tends to be linear when p is large enough (in the upper tail, we have a strict Pareto behavior). The EPD case is linear, with an expected shortfall smaller than the strict Pareto case.

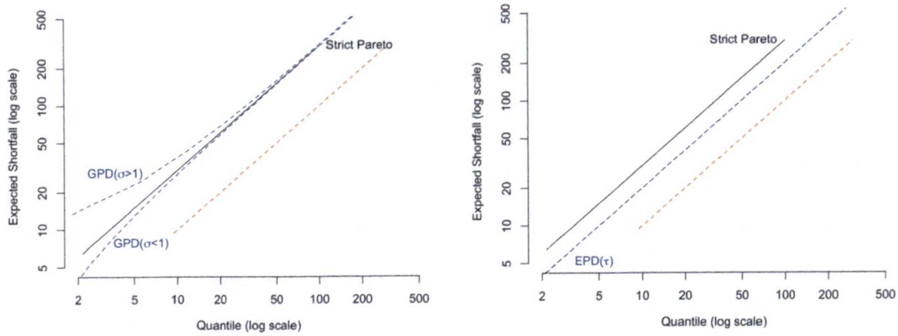

Fig. 7 Expected shortfall against quantile for different values of the probability (i.e., $(Q(1-p)), \text{ES}(p))$, with a strict Pareto (——) with tail index $\alpha = 1.5$, and a GPD distribution (- - -) on the left, and an EPD distribution (- - -) on the right. The straight line below (- - -) is the identity curve, where the expected shortfall would be equal to the associated quantile

5.3 Maximum, Sum, and Subexponential Distributions

In insurance, one possible way to define *large* losses is to say that a single claim is large when the total amount (over a given period of time) is predominantly determined by that single claim. The dominance of the largest value can be written

$$\mathbb{P}[S_n > x] \sim \mathbb{P}[M_n > x] \text{ as } x \to \infty. \tag{64}$$

Heuristically, the interpretation is that the sum will exceed a large value x only when (at least) one single loss X_i also exceeds x. If losses X_i are i.i.d. with CDF F we will say that F belongs to the class of subexponential distributions, $F \in \mathcal{S}$. As proved in Albrecher et al. (2017, Sect. 3.2) strict Pareto and Pareto-type distributions belong to class $\mathcal{S}$. Hence, this class can be seen as an extension of the Pareto models. See Goldie and Klüppelberg (1998) for applications of such distributions in a risk management context.

In Sects. 6.2.6 and 8.2.4 of Embrechts et al. (1997), another quantity of interest is introduced, the ratio of the maximum and the sum, M_s/S_n. As proved in O'Brien (1980), $\mathbb{E}(X)$ is finite if and only if the ratio converges (almost surely) toward 0. A non-degenerated limit is obtained if and only if X is Pareto-type, with index $\alpha \in (0, 1)$. An alternative would be to consider the ratio of the sum of the largest claims (and not only the largest one) and the total sum.

5.4 Top Share and Large Claim Index

In Sect. 5.2, we discussed how to move from M_n to the average of the top $p\%$ of the distribution. It could be possible to normalize by the total sum. The top p-% loss share can be defined as

$$\text{TS}(p) = \frac{p\,\mathbb{E}(X|X > Q(1-p))}{\mathbb{E}(X)} = \frac{p\text{ES}(p)}{\mathbb{E}(X)} = 1 - L(1-p) \tag{65}$$

where L denotes the Lorenz curve, i.e.,

$$L(u) = \frac{1}{\mathbb{E}(X)} \int_0^{Q(u)} \overline{F}(y)\text{d}y = \frac{1}{\mathbb{E}(X)} \int_0^{u} Q(y)\text{d}y. \tag{66}$$

The empirical version of TS_p is

$$\widehat{\text{TS}}(p) = \frac{x_{n-\lfloor np\rfloor:n} + x_{n:n}}{S_n} \tag{67}$$

where $(x_{i:n})$ denotes the order version of sample (x_i), with $x_{1:n} \leq x_{2:n} \leq \cdots \leq x_{n:n}$. See Sect. 8.2.2 of Embrechts et al. (1997) for convergence properties of that estimator. In the case where X is a strict Pareto $\mathcal{P}(u, \alpha)$, then

$$\text{TS}(p) = p^{(\alpha-1)/\alpha}, \tag{68}$$

while if X is $\mathcal{GPD}(u, \sigma, \alpha)$,

$$\text{TS}(p) = 1 - \frac{u(1-p) + \alpha \cdot \big(I_1(\alpha-2, 2) - I_{p^{1/\alpha}(\alpha-1,2)}\big)}{u + \sigma\alpha I_1(\alpha-1, 2)} \tag{69}$$

as in Sect. 6 of Arnold (2008), where $I_z(a, b)$ is the incomplete Beta function,

$$I_z(a, b) = \int_0^{z} t^{a-1}t^{(}1-t)^{b-1}\text{d}t. \tag{70}$$

No simple expression can be derived for the extended Pareto case, but numerical computations are possible, as on Fig. 8, where strict Pareto distribution with tail index $\alpha = 1.5$ is plotted against Pareto-type distributions, GPD and EPD, with the same tail index.

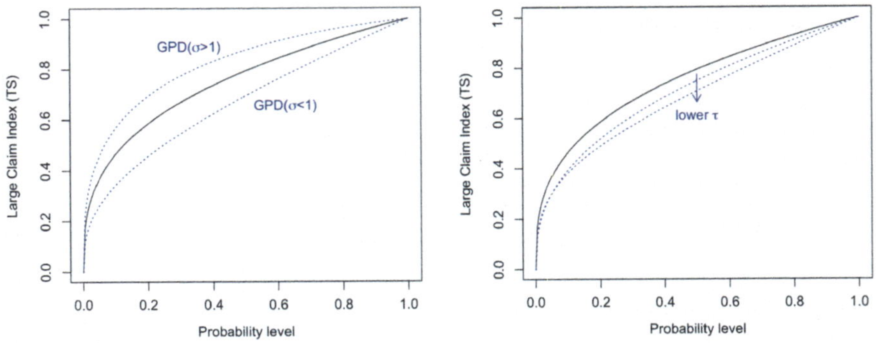

Fig. 8 Large claim index TS(p) as a function of the probability p, with a strict Pareto (——) with tail index $\alpha = 1.5$, and a GPD distribution (- - -) on the left, and an EPD distribution (- - -) on the right

5.5 *From Pareto to Pareto-Type Models*

The concepts introduced earlier were based on the assumption that the distribution F of the observations was known. But actually, for small enough p, we might only have a Pareto distribution in the upper tail, above some threshold u, with $\mathbb{P}[X > u] \geq p$, as suggested in Smith (1987). And let q_u denote the probability to exceed that threshold q_u.

Note that if $x > u$, $\mathbb{P}[X > x] = \mathbb{P}[X > x|X > u] \cdot \mathbb{P}[X > u] = q_u \mathbb{P}[X > x|X > u]$. Thus, if a $\mathcal{P}_1(u, \alpha)$ distribution is considered in the tails,

$$\mathbb{P}[X > x] = q_u \left(\frac{x}{u}\right)^{-\alpha} \tag{71}$$

and therefore,

$$Q_u(1-p) = u \cdot \left(\frac{p}{q_u}\right)^{-1/\alpha}, \quad \text{where } q_u = \mathbb{P}[X > u], \tag{72}$$

and

$$\text{ES}_u(p) = \frac{\alpha}{\alpha - 1} Q_u(1-p). \tag{73}$$

If a $\mathcal{GPD}(u, \sigma, \alpha)$ distribution is considered in the tails,

$$Q_u(1-p) = u \cdot \sigma \left[\left(\frac{p}{q_u}\right)^{-1/\alpha} - 1\right]^{-1}, \text{ where } q_u = \mathbb{P}[X > u], \tag{74}$$

and

$$\text{ES}_u(p) = \frac{\alpha Q_u(1-p) + \sigma - u}{\alpha - 1}. \tag{75}$$

Nevertheless, if a simple update of formulas was necessary for downside risk measures (such as a quantile and an expected shortfall), expression can be more complicated when they rely on the entire distribution, such as the large claim index TS. From Eq. (65), if p is small enough, write

$$\text{TS}_u(p) = \frac{p\text{ES}_u(p)}{\mathbb{E}(X)} \tag{76}$$

where $\mathbb{E}(X)$ is either approximated using $\overline{x}$, or a more robust version would be $(1-q_u)\overline{x}_u + q_u\text{ES}_u(q_u)$ where $\overline{x}_u$ is the empirical average of values below threshold u, while $\text{ES}_u(q_u)$ is the parametric mean from a Pareto-type model, with threshold u.

6 Insurance and Reinsurance

Classical quantities in a reinsurance context are the return period (that can be related to quantiles, and the financial value-at-risk) and a stop-loss premium (related to the expected shortfall).

6.1 Return Period and Return Level

Consider $Y_1, \ldots, Y_n$, a collection of i.i.d. random variables, with distribution F. Consider the sequence of i.i.d. Bernoulli variables $X_i = \mathbf{1}_{Y_i > z(t)}$ for some threshold $z(t)$, such that $\overline{F}(z(t)) = 1/t$. The first time threshold $z(t)$ reached is the random variable N_t defined as

$$N_t = \min\{i \in \mathbb{N}_* : Y_i > z(t)\}. \tag{77}$$

Then, N_t has a geometric distribution,

$$\mathbb{P}[N_t = k] = \frac{1}{(1-t)^{k-1}}\frac{1}{t}. \tag{78}$$

The return period of the events $\{Y_i > z(t)\}$ is then $\mathbb{E}[N_t] = t$, or conversely, the threshold that is reached—on average—t events is $z(t) = Q(1 - 1/t) = U(t)$, called also return level.

Hence, with strict Pareto losses $\mathcal{P}_1(u, \alpha)$

$$z(t) = ut^{1/\alpha}, \tag{79}$$

while with $\mathcal{GPD}(u, \sigma, \alpha)$ losses,

$$z(t) = u + \sigma \left[t^{1/\alpha} - 1 \right]. \quad (80)$$

In the case of non-strict Pareto distribution, where only the top $q\%$ is $\mathcal{P}_1(u, \alpha)$ distributed,

$$z_u(t) = u(q_u t)^{1/\alpha}, \text{ where } q_u = \mathbb{P}[X > u]. \quad (81)$$

Thus, if we assume that the top 10% is Pareto distributed, the return level of a centennial event ($t = 100$) corresponds to the return level of a decennial event ($q_u t = 10$) for the top 10% observations. If the top $q\%$ is $\mathcal{GPD}(u, \sigma, \alpha)$ distributed,

$$z_u(t) = u + \sigma\big((q_u t)^{1/\alpha} - 1\big), \text{ where } q_u = \mathbb{P}[X > u]. \quad (82)$$

From an empirical perspective, consider a dataset with n observations, i.i.d., $y_1, \ldots, y_n$. Assume that above threshold u, the (conditional) distribution is $\mathcal{GPD}(u, \sigma, \alpha)$. Denote n_u the number of observations above u, so that $\widehat{F}_n(u) = 1 - n_u/n$. Then

$$\widehat{z}_u(t) = u + \widehat{\sigma} \left[\left(\frac{n_u}{n} t \right)^{1/\widehat{\alpha}} - 1 \right]. \quad (83)$$

6.2 Reinsurance Pricing

Let Y denote the loss amount of an accident. Consider a contract with deductible d, so that the indemnity paid by the insurance company is $(Y - d)_+$, where $\cdot_+$ denotes the positive part, i.e. $(y - d)_+ = \max\{0, y - d\}$. The pure premium of such a contract, with deductible d, is

$$\pi(d) = \mathbb{E}\big[(Y - d)_+\big] \quad (84)$$

and therefore, if $e(d)$ denotes the mean excess function,

$$\pi(d) = \mathbb{E}\big[Y - d | Y > d\big] \cdot \mathbb{P}(Y > d) = e(d) \cdot \mathbb{P}(Y > d). \quad (85)$$

If we assume that losses above some threshold u have a $\mathcal{GPD}(u, \sigma, \alpha)$ distribution, for any $d > u$,

$$e_u(d) = \left(\frac{\sigma - u}{\alpha - 1} + \frac{\alpha}{\alpha - 1} d \right) \quad (86)$$

and therefore

$$\pi_u(d) = \frac{n_u}{n} \left[1 + \left(\frac{d - u}{\sigma} \right) \right]^{-\alpha} \cdot \left(\frac{\sigma - u}{\alpha - 1} + \frac{\alpha}{\alpha - 1} d \right). \quad (87)$$

If we plug in estimators of σ and α, we derive the estimator of the pure premium $\pi_u(d)$. Note that in Sect. 4.6 in Albrecher et al. (2017), approximations are given for extended Pareto distributions.

As previously, consider a dataset with n observations, i.i.d., $y_1, \dots, y_n$, assume that above threshold u, the (conditional) distribution is $\mathcal{GPD}(u, \sigma, \alpha)$ and let n_u denote the number of observations above u. Then, if we fit a GPD distribution, the estimated premium would be

$$\widehat{\pi}_u(d) = \frac{n_u}{n}\left[1 + \left(\frac{d-u}{\widehat{\sigma}}\right)\right]^{-\widehat{\alpha}} \cdot \left(\frac{\widehat{\sigma}-u}{\widehat{\alpha}-1} + \frac{\alpha}{\widehat{\alpha}-1}d\right), \tag{88}$$

for some deductible d, higher than u, our predefined threshold.

6.3 Application on Real Data

In order to illustrate, two datasets are considered, with a large fire loss dataset (the "Danish dataset" studied in Beirlant and Teugels 1992; McNeil 1997)[11] and large medical losses (from the SOA dataset studied in Cebrián et al. 2003). On Fig. 9, Pareto plots are considered for losses above some threshold u, with $u = 10$ for fire losses, and $u = 1$ for medical losses (here in \$'00,000), i.e.,

$$\left(x_{i:n}, 1 - \frac{i-(n-n_u)}{n_u}\right)_{i=n-n_u+1,\dots,n} \tag{89}$$

on a log–log scatterplot. In Pareto models, points should be on a straight line, with slope $-\alpha$. The plain line (—) is the Pareto distribution (fitted using maximum likelihood techniques) and the dotted line (- - -) corresponds to a generalized Pareto distribution.

On Fig. 10, we can visualize estimates of α, as a function of the number of tail events considered. The strong line (**—**) is the Pareto distribution, the thin line (—) is the GPD distribution while the dotted line (- - -) corresponds to a extended Pareto distribution. Here, the three distributions were fitted using maximum likelihood techniques, and only $\widehat{\alpha}$ is plotted.

On Fig. 11, we can visualize return levels for various return periods (on a log scale), when generalized Pareto distributions are fitted above level u ($u = 10$ for fires and $u = 2$ for medical claims). Those values are quantiles, then return periods are interpreted as probabilities (e.g., a return period of 200 is a quantile of level 99.5%).

On Fig. 12, we can visualize the mean excess function of individual claims, for fire losses on the left, and medical claims on the right (in \$ '00,000). The dark line

[11] It is the `danishuni` dataset in the `CASdatasets` package, available from http://cas.uqam.ca/.

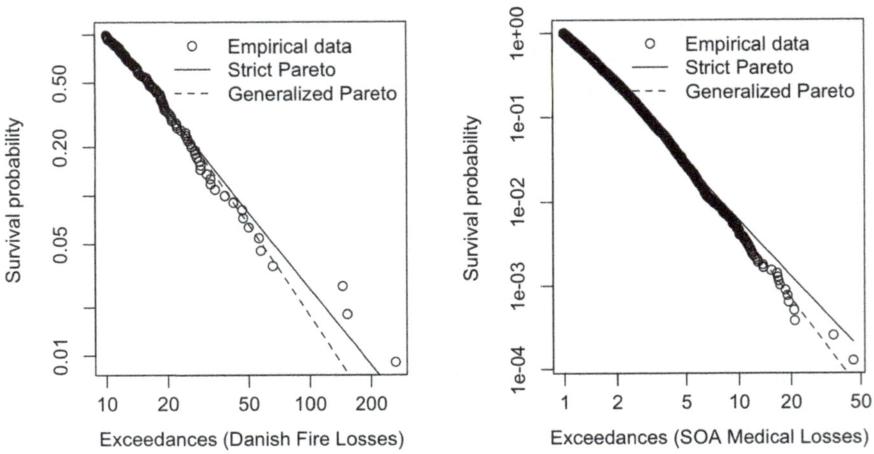

Fig. 9 Pareto plot, with the Danish fire data on the left, and the SOA medical claims data on the right

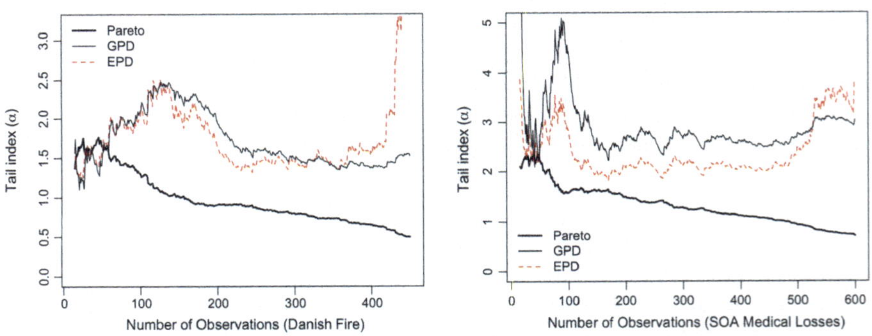

Fig. 10 Estimation of α, with a Pareto model (—), a generalized Pareto model (—) and an extended Pareto model (- - -), as a function of n_u, the number of tail events considered

is the empirical version (average above some threshold) and the line (—) is the GPD fit

$$\widehat{e}_u(d) = \left(\frac{\widehat{\sigma} - u}{\widehat{\alpha} - 1} + \frac{\widehat{\alpha}}{\widehat{\alpha} - 1} d\right) \tag{90}$$

respectively, with $u = 10$ on the left and $u = 2$ on the right, for various values of the threshold d ($\geq u$).

On Fig. 13, we can visualize the stability of estimators with respect to the choice of the threshold u, with the estimation of $e(d)$, with $d = 20$ for Danish fires and $d = 12$ for medical losses. The empirical average is the horizontal dashed line (- - -), the estimator derived from a Pareto distribution above threshold u is the plain strong line (—), the one from a generalized Pareto model above threshold u is the plain line (—), and the extended Pareto model above threshold u is the dashed line (- - -), as a

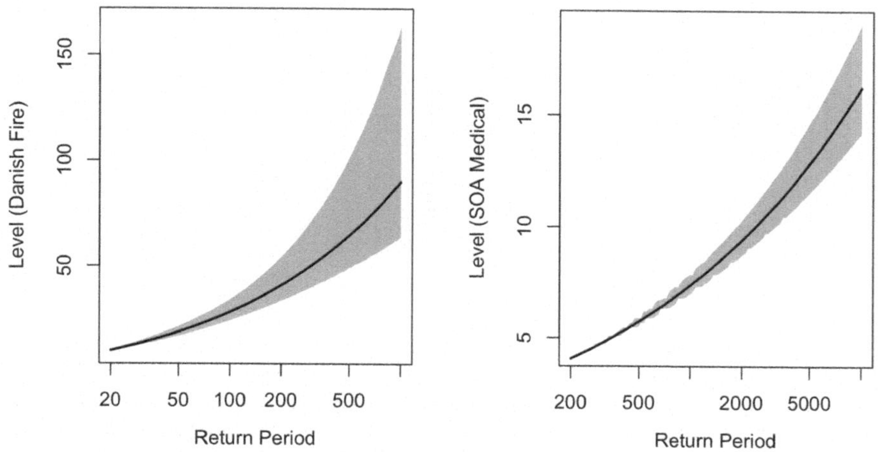

Fig. 11 Estimation return levels for various return periods, for with the Danish fire data on the left, and the SOA medical claims data on the right. The gray area is the confidence region

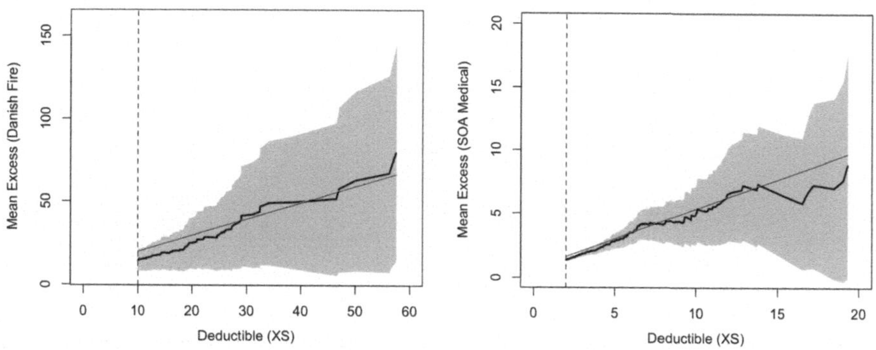

Fig. 12 Mean excess functions Danish fire data on the left, and the SOA medical claims data on the right, with confidence intervals on the empirical version. The line (——) is the GPD fit, for a given threshold (respectively, 10 and 2)

function of u. The estimator obtained from the extended Pareto model is quite stable and close to the empirical estimate.

Finally, on Fig. 14, we have the evolution of the (annual) pure premium, which is simply the mean excess function, multiplied by the (annual) probability to have a loss that exceed the deductible. For the Danish fire losses, since we have 10 years of data, the premium takes into account this 10-factor (which is consistent with a Poisson's process assumption for claims occurrence): on average there are 11 claims per year above 10 and 2.5 above 25. For the SOA medical losses, there are 213 claims above \$ 500,000 ($u = 5$) for a mean excess function slightly below 3, so the pure premium is close to 600, for a deductible (per claim) of 5.

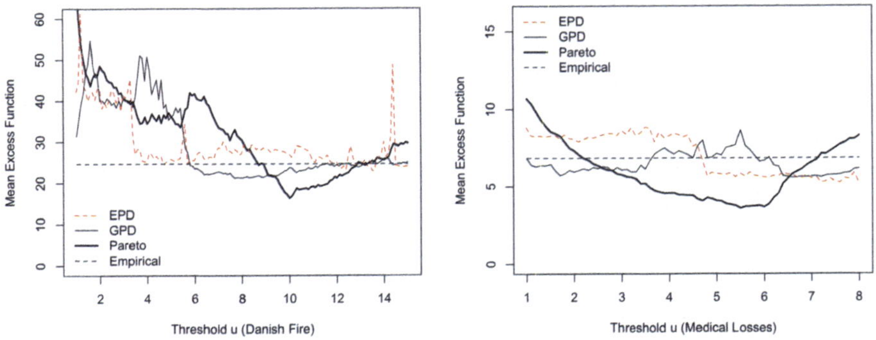

Fig. 13 Estimation of $e(d)$, with $d = 20$ for Danish fires and $d = 12$ for medical losses, with the empirical average (- - -), the estimator derived from a Pareto distribution above threshold u (——), from a generalized Pareto model above threshold u (——) and from an extended Pareto model above threshold u (- - -), as a function of u

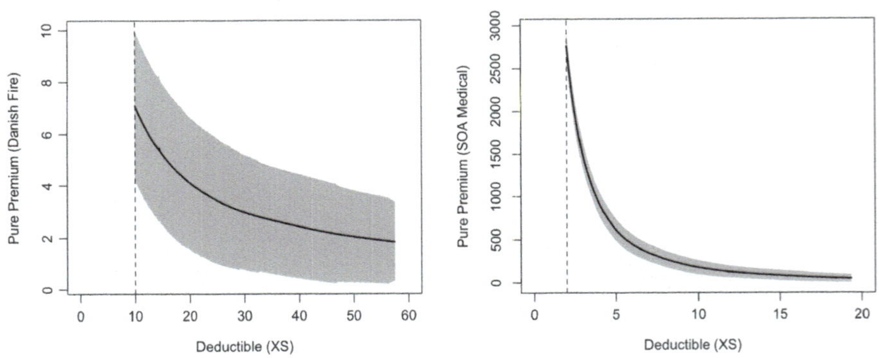

Fig. 14 Annual pure premium for Danish fire data on the left and the SOA Medical claims data on the right

7 Finance and Risk Measures

7.1 Downside Risk Measures

Classical problems in risk management in financial applications involve (extreme) quantile estimation, such as the value-at-risk and the expected shortfall. Let Y denote the negative returns of some financial instrument, and then the value-at-risk VaR(p) is defined as the quantile of level $1 - p$ of the distribution,

$$\text{VaR}(p) = Q(1-p) = \inf\{y \in \mathbb{R}: F(y) \geq 1-p\}, \tag{91}$$

while the expected shortfall (or tail conditional expectation) is the potential size of the loss exceeding the value-at-risk,

$$\mathrm{ES}(p) = \mathbb{E}\big[X \,\big|X > \mathrm{VaR}(p)\big] \tag{92}$$

If X_t is the (random) loss variable of a financial position from time t to time $t + h$ (with a given time horizon h), classical distributions have been intensively consider in financial literature, starting from the Gaussian case, $X_t \sim \mathcal{N}(\mu_t, \Sigma_t)$. Then

$$\mathrm{VaR}(p) = \mu_t + \Phi^{-1}(1-p)\Sigma_t \quad \text{and} \quad \mathrm{ES}_p = \mu_t + \frac{\phi(\Phi^{-1}(1-p))}{p}\Sigma_t, \tag{93}$$

where ϕ and Φ denote, respectively, the density and the c.d.f. of the centered standard Gaussian distribution.

Since Pareto models have a simple expression to model losses above a threshold, with a GPD above a threshold u, for $y > u$,

$$\begin{aligned}\widehat{F}_u(y) &= \mathbb{P}[Y \leq y] \\ &= \mathbb{P}[Y \leq y|Y > u] \cdot \mathbb{P}[Y > u] + \mathbb{P}[Y \leq u] \\ &= G_{\widehat{\sigma},\widehat{\alpha}}(y-u) \cdot \frac{n_u}{n} + \frac{n-n_u}{n}\end{aligned}$$

where n_u denotes the number of observations above threshold u, and

$$G_{\sigma,\alpha}(x) = 1 - \left[1 + \left(\frac{x}{\sigma}\right)\right]^{-\alpha} \qquad \text{for } x \geq u, \tag{94}$$

is the $\mathcal{GPD}(0, \sigma, \alpha)$ CDF. Then for an event rare enough—i.e., a probability small enough ($p < n_u/n$)—the value-at-risk VaR_p can be approximated by

$$\widehat{\mathrm{VaR}}_u(p) = u \cdot \widehat{\sigma}\left[\left(\frac{p}{q_u}\right)^{-1/\widehat{\alpha}} - 1\right]^{-1}, \quad \text{where } q_u = \mathbb{P}[Y > u], \tag{95}$$

as well as the expected shortfall,

$$\widehat{\mathrm{ES}}_u(p) = \frac{\widehat{\alpha}\widehat{\mathrm{VaR}}_u(p) + \widehat{\sigma} - u}{\widehat{\alpha} - 1}. \tag{96}$$

7.2 *Application on Real Data*

In finance, log-returns are rarely independent, and it is necessary to take into account the dynamics of the volatility. Consider some GARCH-type process for log-returns, $y_t = \mu_t + \Sigma_t x_t$, where (x_t) are i.i.d. variables, and where (Σ_t) is the stochastic volatility. Then, as mentioned in McNeil and Frey (2000)

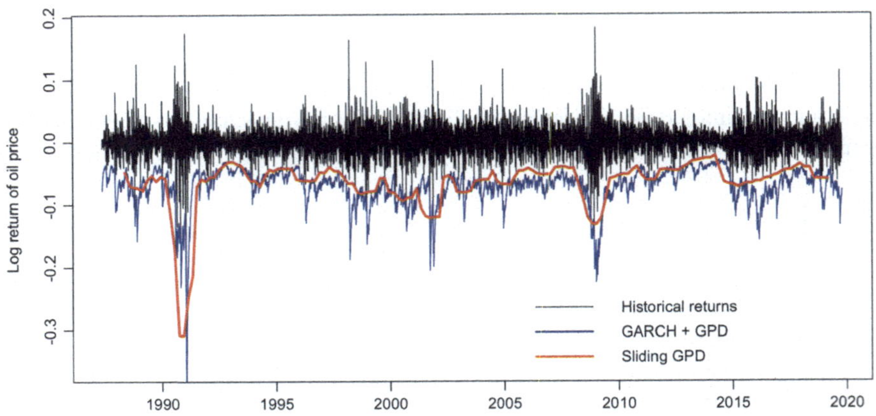

Fig. 15 Value-at-risk at level 0.5% for the daily log-return of oil prices (——), with a GARCH+GPD model (——) and a sliding window estimator GPD model (——)

$$\widehat{\text{VaR}}_{Y_t|\Sigma_t}(p) = \mu_t + \Sigma_t \cdot \widehat{\text{VaR}}_{X_t}(p) \tag{97}$$

$$\widehat{\text{ES}}_{Y_t|\Sigma_t}(p) = \mu_t + \Sigma_t \cdot \widehat{\text{ES}}_{X_t}(p) \tag{98}$$

To estimate (Σ_t), several approaches can be considered, from the exponential weighted moving average (EWMA), with

$$\widehat{\Sigma}_{t+1}^2 = \beta\widehat{\Sigma}_t + (1-\beta)y_t^2, \quad \text{for some } \beta \in (0,1), \tag{99}$$

to the GARCH(1,1) process,

$$\widehat{\Sigma}_{t+1}^2 = \alpha_0 + \alpha_1 y_t^2 + \beta_1 \widehat{\Sigma}_t \text{ for some } \alpha_1, \beta_1 \in (0,1), \text{ with } \alpha_1 + \beta_1 < 1. \tag{100}$$

To illustrate, we use returns of the Brent crude price of oil,[12] extracted from the North Sea prices and comprises Brent Blend, Forties Blend, Oseberg and Ekofisk crudes.

On Fig. 15, for the plain line (GARCH+GPD, ——) we use

$$\widehat{\text{VaR}}_{Y_t|\Sigma_t}(p) = \widehat{\Sigma}_t \cdot \widehat{\text{VaR}}_{X_t}(p) \tag{101}$$

where

$$\widehat{\text{VaR}}_{X_t}(p) = u + \frac{\widehat{\sigma}}{\widehat{\alpha}}\left[\left(\frac{n}{n_u}(1-q)\right)^{-\widehat{\alpha}} - 1\right] \tag{102}$$

[12] Available from https://www.nasdaq.com/market-activity/commodities/bz%3Anmx.

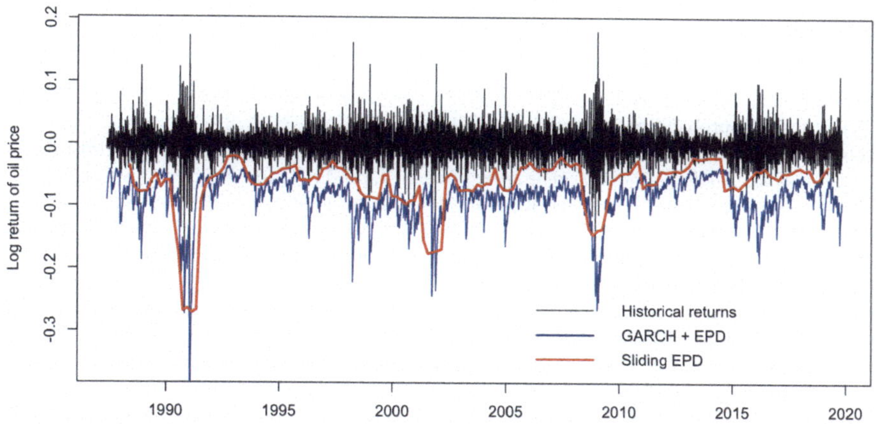

Fig. 16 Value-at-risk at level 0.5% for the daily log-return of oil prices (——), with a GARCH+EPD model (——) and a sliding window estimator EPD model (——)

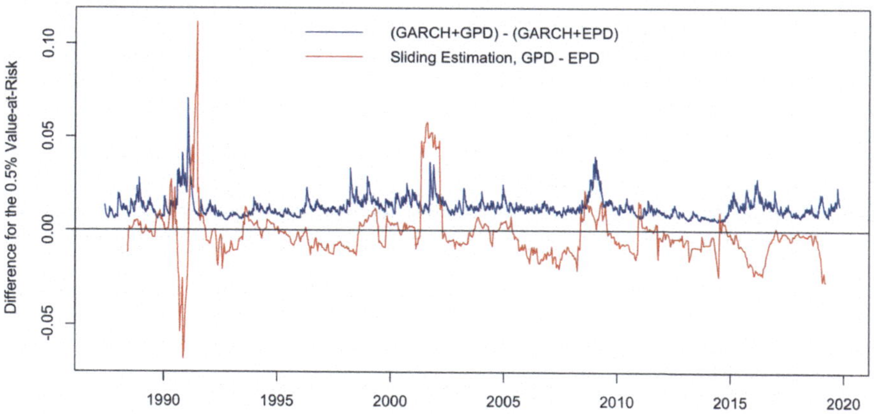

Fig. 17 Difference of the value-at-risk at level 0.5% for the daily log-return of oil prices between GARCH + Pareto models (——) and a sliding window estimator of Pareto models (——)

with $u = 2\%$ and $\widehat{\Sigma}_t$ is the fitted volatility with a GARCH(1,1) process. The red line is obtained using a generalized Pareto fit on a sliding window $[t \pm 150]$.

On Fig. 16, an extended Pareto model is considered for $\widehat{\text{VaR}}_{X_t}(p)$.

Finally, on Fig. 17, we compare the GARCH + Pareto models, with either a generalized Pareto model (as in McNeil and Frey 2000) or an extended Pareto model. Observe that the 0.05% value-at-risk is always over estimated with the generalized Pareto model. Nevertheless, with Pareto models estimated on sliding windows (on $[t \pm 150]$) the difference can be negative, sometimes. But overall, it is more likely that the generalized Pareto overestimates the value-at-risk.

References

Albrecher, H., Beirlant, J., & Teugels, J. L. (2017). *Reinsurance: Actuarial and statistical aspects*. Wiley series in probability and statistics.

Arnold, B. C. (2008). Pareto and generalized Pareto distributions. In D. Chotikapanich (Ed.), *Modeling income distributions and Lorenz curves* (Chap. 7, pp. 119–146). New York: Springer.

Balkema, A., & de Haan, L. (1974). Residual life time at great age. *Annals of Probability, 2*, 792–804.

Beirlant, J., Goegebeur, Y., Segers, J., & Teugels, J. (2004). *Statistics of extremes: Theory and applications*. Wiley series in probability and statistics.

Beirlant, J., Joossens, E., & Segers, J. (2009). Second-order refined peaks-over-threshold modelling for heavy-tailed distributions. *Journal of Statistical Planning and Inference, 139*, 2800–2815.

Beirlant, J., & Teugels, J. L. (1992). Modeling large claims in non-life insurance. *Insurance: Mathematics and Economics, 11*(1), 17–29.

Bingham, N. H., Goldie, C. M., & Teugels, J. L. (1987). *Regular variation*. Encyclopedia of mathematics and its applications. Cambridge: Cambridge University Press.

Cebrián, A. C., Denuit, M., & Lambert, P. (2003). Generalized Pareto fit to the society of actuaries' large claims database. *North American Actuarial Journal, 7*(3), 18–36.

Charpentier, A., & Flachaire, E. (2019). Pareto models for top incomes. hal id: hal-02145024.

Davison, A. (2003). *Statistical models*. Cambridge: Cambridge University Press.

de Haan, L., & Ferreira, A. (2006). *Extreme value theory: An introduction*. Springer series in operations research and financial engineering.

de Haan, L., & Stadtmüller, U. (1996). Generalized regular variation of second order. *Journal of the Australian Mathematical Society, 61*, 381–395.

Embrechts, P., Klüppelberg, C., & Mikosch, T. (1997). *Modelling extremal events for insurance and finance*. Berlin, Heidelberg: Springer.

Fisher, R. A., & Tippett, L. H. C. (1928). Limiting forms of the frequency distribution of the largest or smallest member of a sample. *Proceedings of the Cambridge Philosophical Society, 24*, 180–290.

Gabaix, X. (2009). Power laws in economics and finance. *Annual Review of Economics, 1*(1), 255–294.

Ghosh, S., & Resnick, S. (2010). A discussion on mean excess plots. *Stochastic Processes and Their Applications, 120*(8), 1492–1517.

Gnedenko, B. (1943). Sur la distribution limite du terme maximum d'une serie aleatoire. *Annals of Mathematics, 44*(3), 423–453.

Goldie, C. M., & Klüppelberg, C. (1998). Subexponential distributions. In R. J. Adler, R. E. Feldman, & M. S. Taqqu (Eds.), *A practical guide to heavy tails* (pp. 436–459). Basel: Birkhäuser.

Guess, F., & Proschan, F. (1988). 12 mean residual life: Theory and applications. In *Quality control and reliability. Handbook of statistics* (Vol. 7, pp. 215–224). Amsterdam: Elsevier.

Hagstroem, K. G. (1925). La loi de pareto et la reassurance. *Skandinavisk Aktuarietidskrift, 25*.

Hagstroem, K. G. (1960). Remarks on Pareto distributions. *Scandinavian Actuarial Journal, 60*(1–2), 59–71.

Hall, P. (1982). On some simple estimate of an exponent of regular variation. *Journal of the Royal Statistical Society: Series B, 44*, 37–42.

Jessen, A. H., & Mikosch, T. (2006). Regularly varying functions. *Publications de l'Institut Mathématique, 19*, 171–192.

Klüppelberg, C. (2004). Risk management with extreme value theory. In B. Finkenstädt, & H. Rootzén (Eds.), *Extreme values in finance, telecommunications, and the environment* (Chap. 3, pp. 101–168). Oxford: Chapman & Hall/CRC.

Kremer, E. (1984). *Rating of non proportional reinsurance treaties based on ordered claims* (pp. 285–314). Dordrecht: Springer.

Lomax, K. S. (1954). Business failures: Another example of the analysis of failure data. *Journal of the American Statistical Association, 49*(268), 847–852.

McNeil, A. (1997). Estimating the tails of loss severity distributions using extreme value theory. *ASTIN Bulletin*, *27*(27), 117–137.
McNeil, A. J., & Frey, R. (2000). Estimation of tail-related risk measures for heteroscedastic financial time series: An extreme value approach. *Journal of Empirical Finance, 7*(3), 271–300. Special issue on Risk Management.
O'Brien, G. L. (1980). A limit theorem for sample maxima and heavy branches in Galton-Watson trees. *Journal of Applied Probability, 17*(2), 539–545.
Pareto, V. (1895). La legge della domanda. In Pareto (Ed.), *Ecrits d'économie politique pure* (Chap. 11, pp. 295–304). Genève: Librairie Droz.
Peng, L., & Qi, Y. (2004). Estimating the first- and second-order parameters of a heavy-tailed distribution. *Australian & New Zealand Journal of Statistics*, *46*, 305–312.
Pickands, J. (1975). Statistical inference using extreme order statistics. *Annals of Statistics*, *23*, 119–131.
Resnick, S. (2007). *Heavy-tail phenomena: Probabilistic and statistical modeling* (Vol. 10). New York: Springer.
Resnick, S. I. (1997). Discussion of the Danish data on large fire insurance losses. *ASTIN Bulletin*, *27*(1), 139–151.
Reynkens, T. (2018). ReIns: Functions from "Reinsurance: Actuarial and statistical aspects". R package version 1.0.8.
Rigby, R. A., & Stasinopoulos, D. M. (2005). Generalized additive models for location, scale and shape (with discussion). *Applied Statistics*, *54*, 507–554.
Roy, A. D. (1952). Safety first and the holding of assets. *Econometrica*, *20*(3), 431–449.
Schumpeter, J. A. (1949). Vilfredo Pareto (1848–1923). *The Quarterly Journal of Economics*, *63*(2), 147–173.
Scollnik, D. P. M. (2007). On composite lognormal-Pareto models. *Scandinavian Actuarial Journal*, *2007*(1), 20–33.
Smith, R. L. (1987). Estimating tails of probability distributions. *Annals of Statistics, 15*(3), 1174–1207.
Vajda, S. (1951). Analytical studies in stop-loss reinsurance. *Scandinavian Actuarial Journal*, *1951*(1–2), 158–175.

The manufacturer's authorised representative in the EU is Springer Nature Customer Service Centre GmbH, Europaplatz 3, 69115 Heidelberg, Germany. If you have any concerns regarding our products, please contact ProductSafety@springernature.com

Printed and bound by CPI Group (UK) Ltd, Croydon, CR0 4YY
30/09/2026
02225508-0016